Trends in Logic

Volume 42

TRENDS IN LOGIC
Studia Logica Library

VOLUME 42

SCOPE OF THE SERIES

The book series Trends in Logic covers essentially the same areas as the journal Studia Logica, that is, contemporary formal logic and its applications and relations to other disciplines. The series aims at publishing monographs and thematically coherent volumes dealing with important developments in logic and presenting significant contributions to logical research.

The series is open to contributions devoted to topics ranging from algebraic logic, model theory, proof theory, philosophical logic, non-classical logic, and logic in computer science to mathematical linguistics and formal epistemology. However, this list is not exhaustive, moreover, the range of applications, comparisons and sources of inspiration is open and evolves over time.

More information about this series at http://www.springer.com/series/6645

Janusz Czelakowski

Freedom and Enforcement in Action

A Study in Formal Action Theory

Springer

Janusz Czelakowski
Department of Mathematics
and Informatics
University of Opole
Opole
Poland

ISSN 1572-6126 ISSN 2212-7313 (electronic)
Trends in Logic
ISBN 978-94-017-9854-9 ISBN 978-94-017-9855-6 (eBook)
DOI 10.1007/978-94-017-9855-6

Library of Congress Control Number: 2015938093

Springer Dordrecht Heidelberg New York London

Printed on acid-free paper

Springer Science+Business Media B.V. Dordrecht is part of Springer Science+Business Media
(www.springer.com)

Preface

The aim of this book is to present a formal theory of action and to show the relations of this theory with logic and other disciplines. The book concerns the semantic, mathematical, and logical aspects of action.

In contemporary logic, reflections on actions and their performers (agents) have assumed a growing and expanding importance. The theme of action, particularly that of effective and rational action, is heavily rooted in the praxeological tradition. From the viewpoint of logic, the problem of action goes beyond traditional branches of logic such as syntactics and semantics. The center of gravity of the issues the problem of action raises is situated on the borderline between logical pragmatics and praxeology.

The book focuses on the following tasks:

A. Description and Formalization of the Language of Action

In the contemporary literature, one can distinguish at least seven approaches to action. Each of them outlines a certain perspective of action theory by bringing out some specific aspects of human actions.

1. *The linguistic framing*, initiated by Maria Nowakowska. Atomic (in other words: elementary) actions (procedures), as well as compound actions are distinguished. Atomic actions are primitive and non-reducible to others. Compound actions are sets of finite sequences of atomic actions. If one identifies the set of atomic actions with an alphabet, in the sense of formal linguistics, each compound action becomes a language over this alphabet, that is, it becomes a set of words. For a description of compound actions one can then apply the methods of mathematical linguistics. Compound actions can be—in particular cases—regular languages, context-free languages, etc. This formulation goes deeply into the theory of algorithms and is appropriate for describing routine, algorithmizable actions, such as the manufacture of cars or the baking of bread.

2. *The dynamic logic approach.* The view that human action is modeled on their resemblance of computer programs can be found in the works of many researchers (Boden, Segerberg, Suppes to mention a few). In this formulation, an action is identified with a binary relation defined on a set of states. This relation, called the resultant relation, assigns to each state a set of possible outcomes of the action, when the action is being performed in this state. Each pair of states belonging to this relation is called a possible performance of the action. The formulation, in a natural way, links action theory with (fragments of) set theory, whose main components are graph theory and the theory of relations.
3. *Stit semantics* gives an account of action from a perspective in which actions are seen *not* as operations performed in an action system and which yield new states of affairs, but rather as selections of pre-existent histories (or trajectories) of the system in time. Stit semantics is therefore time oriented, and time, being a situational component of action, plays a special role in it.
4. A special framing of the subject area of action is offered by *deontology* and *deontic logic*. It is from the deontological perspective that a typology of actions is determined; here, forbidden, permitted, and obligatory actions are distinguished. This formulation binds action theory with jurisprudence and the theory of norms.
5. The fifth perspective of action originates from *Dynamic Epistemic Logic* (DEL), the logic of knowledge change. DEL is concerned with actions which change states of agents' knowledge and beliefs. DEL builds models of the dynamics of inquiry and accompanying flows of information. It provides insight into the properties of individual or group agents (knowers) and analyzes consequences of epistemic or verbal actions. Public announcements may serve as an example [see van Benthem (2011); van Benthem et al. (2013)].
6. A pragmatic approach to action is developed by *decision theory*. From the perspective of this theory, 'decision making is a cognitive process resulting in the selection of a course of actions usually among several alternative scenarios'—see Wikipedia. Decision theory therefore differentiates between problem analysis, which is a part of the cognitive process, and the selection of an appropriate course of actions by the agent(s). The information gathered in problem analysis at each stage of decision making is then used toward making further steps. Decision theory is not concerned with the performability of actions but rather with their costs—some actions turn out to be less or more profitable than others. In other words, decision theory views actions as decisions and assesses the latter in terms of costs or losses.
7. *Game theory* is a study of strategies and decision making. There is no strict division line between game theory and decision theory. It is said that game theory may be viewed as *interactive* decision theory because it builds mathematical models of conflict and cooperation between rational decision-makers.

B. Models of the Action Theory

In view of the difficulty in determining the adequate language of actions, one should not expect a theory to be defined in an axiomatic way. The natural compromise consists in defining some intended models of action. In this book, two classes of models are discussed:

- the class of elementary action systems,
- the class of situational action systems.

The second of the classes includes the first one as a limit case. The models allow for unification of most of the formulations of action theory mentioned in (A). On the ground of the above-mentioned models, one can define compound actions (as it is done in the models considered by Nowakowska); likewise, one can reconstruct models for dynamic logic. In terms of action systems it is possible to determine notions of permitted, obligatory, and forbidden actions that are fundamental to deontology.

The book also outlines the relations obtaining between situational action systems and situational semantics.

C. Performability of Actions

The central problem that action theory poses for itself to solve is to provide an adequate concept of the performability of action. The performability of actions depends on the parameter which is the state of an action system (see point B). What is more, the very notion of performability itself in not of an absolute character but is relativized to a possible manner (aspect) of performing an action: for example, an action can be physically performable (e.g., driving a car along a one-way road in the opposite direction), when one takes into account technical limitations while being legally non-performable—if one takes into account (as in the example given) the limitations arising from the regulations contained in the highway code (actions *in fraudem legis*). (In this example, the non-performability of action in the legal sense means that the action is forbidden.) One of the aims of the book is to present the definition of performability (atomic and compound ones) formulated in terms of elementary and situational action systems.

D. Actions and Their Agents

Actions are performed by single persons (individuals), teams of people (collective groups), robots, and groups, these being combinations of collective bodies, robots, and machines. Performers of actions are referred to with the collective term of *agent of action*. The literature on the subject is quite extensive and is focused on providing truth-conditions for sentences of the form: *a is the agent of the action A*.

The above problem has been analyzed by Brown, Chellas, Horthy, Kanger, Pacuit, Segerberg, van Benthem, von Wright, and many others. An initial discussion of the issue requires accepting certain ontological assumptions first. Actions (and acts) are correlated with changes in states of affairs. Besides the states, categories of actions and their agents are distinguished. Moreover, in this book, there is introduced the notion of the situational envelope of an action that takes into account such parameters accompanying an action as: time, location, order of actions, etc.

A difficult part of the theory is the question of the intentionality of an action, e.g., when the intention of the agent is to perform an action, yet—for a variety of reasons—the agent desists from performing it. The notional apparatus permits the introduction of clear-cut criteria for differentiating between single actions (when the agent is an individual) and collective actions (when the agent is a collective body), as well as between one-time actions (such as the stabbing of Julius Caesar) and actions understood as a type (e.g., stabbing as a type of criminal action).

The problem area of verbal actions and models of information flow which accompanies actions requires special treatment. This question is not studied in the book; nevertheless, the models of action to be developed allow for their extension over verbal actions. In this context, one can modify the existing models of belief systems deriving from Alchourrón, Gärdenfors, and Makinson.

E. Probabilistic Models of Action

The notion of performability mentioned in C is not probabilistic but binary: a given action A is either performable or not in a definite state u. This notion does not encompass some aspects of the performability of actions as, e.g., quality grading (poor, medium, good performance, etc.). One framework that brings theories of action closer to probability calculus and decision theory introduces a quantitative measure of the degree of performability of an action. It is the probability of performability of an action in a given state u. Also introduced are other measures of performabilty such as the probability of the transition of the system from one state to another on the condition that an action is performed. The measure is, to use the simplest example, the probability of hitting—in the determined initial conditions u—the 'bull's-eye' with an arrow shot from a bow (the intended state v). The performed action is here shooting an arrow at a target.

Two types of probabilistic models of action are distinguished. The emphasis is put on their practical applications. The relevant models are constructed from elementary systems (point B) by introducing (conditional) probabilities of transition from one state to another, under the assumption that the given action is performed. (The notion of the performability of an action is distinguished from a possible performance as well as from a performance of an action—the latter being a one-time act, belonging to the situational surrounding of the system of action.)

F. Relationship with Deontic Logic

A significant feature of action theory is its firm rooting in the theory of law and theory of legal and moral norms. This part of action theory is called deontology. The central place in it is occupied by deontic logic. This is still an area which is characterized by the existence of disparities concerning fundamental matters, the proliferation of formal logical systems, as well as a lack of mathematical and logical results of generally recognized significance and depth. In the formulations of deontic logic known from the literature, deontic operators are considered as unary sentence-generating functors defined on sentences. In semantic stylizations, these formulations distinguish permitted, forbidden, and obligatory *states of affairs*.

In this book, a formulation of deontic logic is presented, according to which the deontic operators belong to quite a different category: they are defined on actions, and not on states of affairs. Thus, in this formulation, these are actions that are permitted, forbidden, or obligatory entities, and not sets of states. It leads to two simple formalized systems of deontic logic, whose semantics is founded on elementary action systems. The difference between the systems consists in the fact that the first one validates the so-called closure principle, while the other rejects it. The closure principle says that every action which is not forbidden is permitted. These systems are free from deontic paradoxes. Completeness theorems for these systems are proved. Deontic models of compound actions are also considered.

The book presents a new approach to norms. Norms in the broad sense are viewed as certain rules of action. In the simplest case they are instructions which, under given circumstances, permit, forbid, or order the performance of an action.

G. Relations with the Theory of Algorithms and Programming

The theory of algorithmizable actions is a vital part of action theory. Here, algorithmizable actions are set against actions that are creative, single, and unique in their nature.

There is no satisfactory definition of algorithmizable actions. According to an informal definition, an algorithm is a set of rules that precisely defines sequence of actions. Instances of algorithmizable actions are regular or context-free compound actions. According to the above linguistic approach, regular (respectively, context-free) compound actions are defined as regular (context-free) languages over the alphabet consisting of atomic actions.

A part of the book establishes certain results on algorithmizable actions referring to the notion of an action program. The prototype here is the meaning of the term "program", with which it is invested by computer science. The above-mentioned problem area displays relations with algorithmic logic in the sense of Salwicki and the theory of algorithms; yet, it is not identical with them.

H. Non-monotonic Reasonings and Action

An approach to non-monotonic reasonings which links them with the theory of action is outlined. A general *semantic* scheme of defining non-monotonic reasonings in terms of frames is presented. Each frame $\boldsymbol{F}$ is a set of states W endowed with a family $\boldsymbol{R}$ of relations of a certain kind. If S is a propositional language, then each frame determines in a natural way a consequence-like operation on S. The latter does not generally exhibit all properties of consequence operations as, e.g., monotonicity. Such operations exemplify non-monotonic patterns of reasoning. The class of resulting structures encompasses preferential model structures and supra-classical reasonings.

I. The Existing State of Knowledge in the Scope of the Study Area

Reflection on the rational and irrational actions of human beings is not alien to the Polish logical tradition. Books by Tadeusz Kotarbiński took the lead. The works of Kazimierz Ajdukiewicz, directed toward practical application of logic, and the activity of Kotarbiński's disciples (Nowakowska, Stonert, Konieczny, Gasparski) testify to this only too well. It seems research in this area needs new impulses so as to permit it ultimately to penetrate to a broader extent and more profoundly into the scientific and technical achievements of recent decades, especially as regards dynamic logic, theory of automata, and programming.

The present book puts the emphasis on the mathematical and formal-logical aspects of an action. Despite being firmly grounded in the tradition of Lvov-Warsaw School, to a broad extent it takes into account the output of many schools, including the Scandinavian, Dutch or Pittsbourgh ones, to mention a few. It also takes into consideration the author's own modest contributions. In the years 2001–2003, the author was in charge of the research project "Logika i działanie (Logic and Action)", signature 1 H01A 011 18, financed within a generous grant obtained from the then *Komitet Badań Naukowych* (*State Committee for Scientific Research*). The support from the Committee resulted in the first drafts of this book.

Fragments of the book have been presented by the author over many years at different conferences, both at home and abroad. The first presentation took place at the University of Konstanz in October 1992, and the next during a meeting at Umeå University (Sweden) in September 1993. Among other conferences the following must be noted: *Logic and its Applications in Philosophy and the Foundations of Mathematics*, organized annually since 1996 in Karpacz or Szklarska Poreba (The Giant Mountains, Poland), where the author presented his papers relating to various aspects of action a number of times. Throughout all this time the author has received valuable advice and criticism from many people whose individual names shall not be mentioned, but who are all cordially thanked.

It remains to hope that the book will (modestly) contribute to the formulation of a generally accepted paradigm of action theory.

Acknowledgments

As a non-native speaker of English, the author is indebted to Dr. Matthew Carmody for proofreading the whole text of the book and for making appropriate linguistic recommendations and amendments.

Opole, Poland — Janusz Czelakowski
September 2014

Contents

Part I
Elements of Formal Action Theory

Chapter 1
Elementary Action Systems

Abstract This chapter expounds basic notions. An elementary action system is a triple consisting of the set of *states*, the *transition relation* between states, and a family of binary relations defined on the set of states. The elements of this family are called *atomic actions*. Each pair of states belonging to an atomic action is a *possible performance* of this action. This purely extensional understanding of atomic actions is close to dynamic logic. Compound actions are defined as sets of finite sequences of atomic actions. Thus compound actions are regarded as languages over the alphabet whose elements are atomic actions. This chapter is concerned with the problem of performability of actions and the algebraic structure of the set of compound actions of the system. The theory of probabilistic action systems is also outlined.

1.1 Introductory Remarks

Any practical activity requires a strict distinction between (material or mental) the objects and states ascribed to them. A purposeful activity is always undertaken by an agent who has some capacity to predict possible interactions between objects—particularly himself and material objects—and is an activity aimed at bringing about certain desirable states of those objects.

A man facing a problem situation first tries to produce a clear image of all the factors defining the situation. In other words, he wants to realize in what state of things he finds himself at the moment. This subjective world image then undergoes a certain intellectual process—utilizing the available tools, he determines possible variants of a further course of events. This is a preliminary phase of solving the problem, consisting in finding all the factors determining the current state of affairs, and then defining a set of actions to be performed to reach the intended state of things. The second, practical phase aims to find the techniques to affect the object and bring about a desirable change in it. This is achieved by working out an action plan which is to lead to the goal. And, finally, the last stage—verification of whether the actual effects of the performed actions correspond to the intended ones. We may thus speak about the three-phase structure of an action: a clear recognition of the situation, the

J. Czelakowski, *Freedom and Enforcement in Action*, Trends in Logic 42,
DOI 10.1007/978-94-017-9855-6_1

defining of an action plan, i.e., a set of actions which are to be performed in a definite order, and lastly the checking of the effects of the action.

In the contemporary literature one may distinguish at least seven approaches to action. They are briefly discussed in the preface. Each of them is centered around a specific aspect of action. We shall present here in greater detail the three that are pertinent to the content of this book. One, having its roots in dynamic logic, treats human action as a computer program. The conviction that the notion of a computer program may be a useful analogy for human action is developed in the papers by Boden (1977), Suppes (1984), Segerberg (1989), and others. As Segerberg writes (1989, p. 327): "(...) in understanding what happens when, for example John performs the action of eating an apple it might be useful to view John as running a certain 'routine' (...)." ('Routine' is the term suggested by Segerberg for the informal counterpart of the notion of a program when speaking of human action.)

In dynamic logic, a universe of possible states is considered. Corresponding to each program α, there is a set $|\alpha|$ of paths (i.e., sequences of states), each path representing a possible computation according to that program. The set $|\alpha|$ is called the intension of the program α. In propositional dynamic logic (PDL), the intension of a program is usually thought to be a set of pairs of states than a set of paths—in the perspective of PDL it is only the starting point and the endpoint of a path that matter. Thus to each program α a certain binary relation on the set of states is assigned—one may call it the resultant relation of a program. The operation that identifies a program with its resultant relation leads, in consequence, to distinguishing a family of binary relations on the set of states. In the context of human action theory modeled on computer science, the relations are just called actions (see Segerberg 1989).

The approach presented in Nowakowska's book (1979) describes human action as a complex system of irreducible simple actions ('elementary' or atomic actions). Compound actions are viewed as sets of sequences of simple actions but the latter generally cannot be performed in random order. *Simple action* is here a primitive notion. It is an approach with formal-linguistic origin: simple actions can be performed only in a certain order just as letters are used only in certain orders to make up words. This idea allows us to use the concepts of formal linguistics for the analysis of the 'grammar' of action. The starting point in Nowakowska's action theory is an ordered pair

$$(D, L)$$

where D is a finite set and L is a subset of D^*, the set of all finite sequences of elements of D. The members of D are interpreted as simple actions; L in turn is the set of physically-feasible finite strings of simple actions. The set L, in analogy with linguistics, is called the language of actions.

Many of the notions presented in Nowakowska's book such as distributive classes and parasite sequences, are drawn from the arsenal of formal linguistics. They are a helpful tool in the analysis of actions. Her theory also contains specific notions which have no linguistic counterparts, e.g., the complete ability, key actions, etc.

The third approach to action originates from deontic logic. It dates back to the seminal work of von Wright (1963). This approach deals with an analysis of the notion of a norm and the typology of norms and actions. In the literature devoted to deontic logic one distinguishes the following three categories of actions:

1. obligatory actions
2. permissive actions
3. prohibitive actions.

Norms define the ways and circumstances in which actions are performed, e.g., by specifying the place, time, and order in which they should be executed. Some norms may oblige the agent to perform an action while others may prohibit performing it in the given situation.

This book comes in between the above-mentioned approaches to action. As in Nowakowska's book, the notion of an atomic action is the starting point in our investigation. Compound actions are defined as sets of finite strings of atomic actions. The latter are binary relations on so-called discrete systems, i.e., every atomic action A is identified with a set of ordered pairs (u, w), where u and w are possible states of the system. Every pair (u, w) belonging to A is called a *possible performance* of the atomic action A; w is a possible effect of the action A in the state u. The above purely extensional understanding of atomic actions is close to that of dynamic logic. In turn, compound actions are regarded as languages over the alphabet whose elements are atomic actions.

Throughout the book we use the standard notation adopted in Zermelo-Fraenkel set theory (ZF). $\mathbb{N}$ is the set of natural numbers with zero. It is often marked as ω. A binary *relation* is any set of ordered pairs. A binary *relation on a set* W is any subset of the Cartesian product $W \times W$. $\wp(W)$ is the power set of W (= the set of all subsets of W.)

Let P and Q be binary relations on a set W. We define:

Set-theoretic join:	$P \cup Q := \{(u, w) : P(u, w) \text{ or } Q(u, w)\}$
Set-theoretic intersection:	$P \cap Q := \{(u, w) : P(u, w) \text{ and } Q(u, w)\}$
Composition:	$P \circ Q := \{(u, w) : \text{for some } v \in W,\ P(u, v) \text{ and } Q(v, w)\}$
Iteration:	$P^0 := E_W, \qquad P^{n+1} := P^n \circ P$ for $n \geqslant 0$, where
	$E_W := \{(u, u) : u \in W\}$ is the diagonal relation on W
Iterative closure:	$P^* := \bigcup\{P^n : n \geqslant 0\}$
Transitive closure:	$P^+ := \bigcup\{P^n : n \geqslant 1\}$
Inversion:	$P^{-1} := \{(u, w) : (w, u) \in P\}$.

The set

$$Dom(P) := \{x : (\exists y)(x, y) \in P\}$$

is the *domain* of P while

$$CDom(P) := \{y : (\exists x)(x, y) \in P\}$$

is the *co-domain* of P.

1.2 Elementary Action Systems

Before introducing the formal apparatus of action theory, let us consider the following simple (hypothetical) example. It is midnight. Mrs. Smith is sitting in the living room and listening to the radio. She wants to go to bed. But first she has to perform two actions:

A	Turning off the radio
B	Switching off the light

The current state u_0 is described by the sentence $\alpha \wedge \beta$, where

α :	The radio is turned on
β :	The light in the living room is switched on

She wants to achieve the state u_f in which the radio and the light are turned off. It is assumed that the sentence "The radio is turned off" is equivalent to "The radio is not turned on." The same applies to the second sentence concerning light switch. The final state u_f is therefore described by the sentence $\neg\alpha \wedge \neg\beta$.

u_f may be attained from u_0 in two ways—by turning off the radio, and then switching off the light, or vice versa. There are two intermediate states between u_0 and u_f:

$$u_1 : \quad \alpha \wedge \neg\beta$$

in which the radio is on and the light is off, and

$$u_2 : \quad \neg\alpha \wedge \beta$$

in which the radio is off but the light is still on. We therefore distinguish the following four-element set W of states:

$$W := \{u_0, u_1, u_2, u_f\}.$$

According to the formalism adopted in this book, each atomic action is represented by a set of ordered pairs of states—the first element of the pair representing the state of the system just before the given action is being executed, and the second element representing the state of the system immediately after performing the action. In this example the actions A and B are identified with the following sets of ordered pairs of states:

$$A = \{(u_0, u_2), (u_1, u_f)\},$$
$$B = \{(u_0, u_1), (u_2, u_f)\},$$

respectively.

When it is completely dark in the room, Mrs. Smith is so scared that she cannot localize the radioset in the room. It is not possible for her to reach the state u_f from the state u_1 (she cannot turn off the radio after switching the light off). To capture such a situation, there is another key ingredient introduced in the action formalism—the relation R of direct transition from one state to another. The relation R points out which direct transitions between states are physically possible. Here "physical possibility" refers to any transition which is effected by performing one of the above actions and taking into account Mrs. Smith's abilities to do them. Accordingly:

$$R := \{(u_0, u_2), (u_2, u_f), (u_0, u_1)\}.$$

(There is no direct transition from u_1 to u_f.)

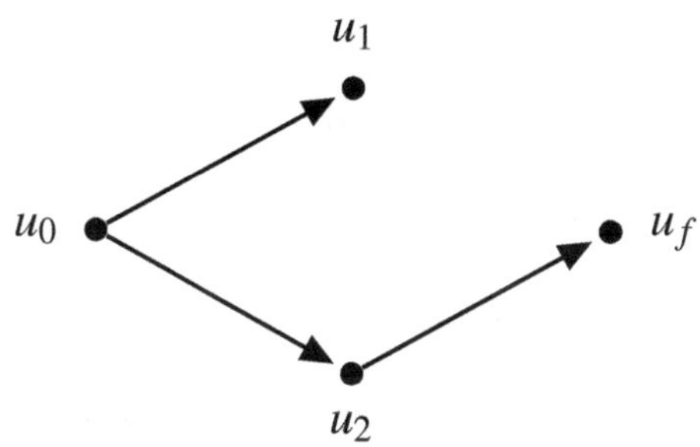

Fig. 1.1

Collecting together the above ingredients one arrives at the system

$$M = (W, R, \{A, B\}), \tag{1.2.1}$$

which encodes Mrs. Smith's current states and the action she has to perform so that she could go to the bedroom.

A second hypothetical model is one in which Mrs. Smith sees perfectly in darkness. In this model there is a direct transition from u_1 to u_f—the state u_f is achieved by turning off the radio in the darkness of the living room. Here

$$R = \{(u_0, u_2), (u_1, u_f), (u_2, u_f), (u_0, u_1)\}.$$

(1.2.1) is an instantiation of the general concept of an elementary action system—the notion which we shall discuss in this chapter.

Yet another model is taken from everyday life. You want to get dressed in the morning and then go to the office. You are to perform two actions: A—putting on the trousers, and B—putting on the shoes. It is clear that you first put the trousers and then the shoes. (It is rather unusual to perform these actions in the reverse order, especially when the trousers are skintight.) The corresponding action system is similar to the above one. There are four states distinguished: u_0—you are undressed, u_1—the trousers are on, u_2—the shoes are on, u_f—the trousers and the shoes are on. We may assume that

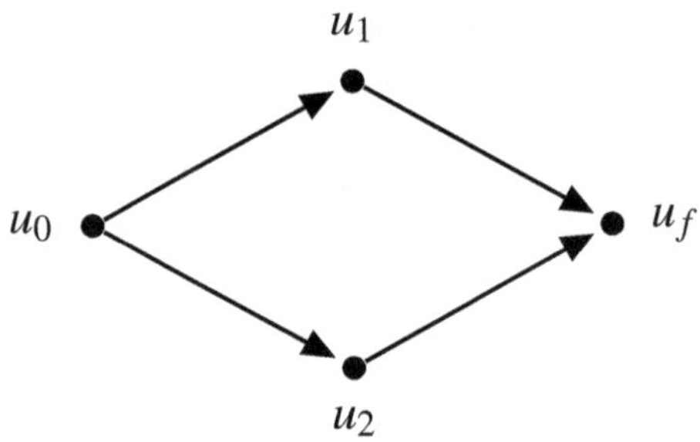

Fig. 1.2

$$A = \{(u_0, u_1), (u_2, u_f)\} \quad \text{and} \quad B = \{(u_0, u_2), (u_1, u_f)\}.$$

The relation R reflects the above limitations. Thus

$$R = \{(u_0, u_1), (u_1, u_f), (u_0, u_2)\}.$$

Intuitively, A is performable in u_0 but unperformable in u_2. B is performable in u_0 and in u_1. Thus, if you first perform B in u_0, thus arriving at u_2, there is no possibility of performing A in u_2, because R excludes it (there are no R-transition from u_2 to the final state u_f).

A *discrete system* is a pair

$$\boldsymbol{U} = (W, R) \tag{1.2.2}$$

where W is a nonempty set called the set of *states*, and R is a binary nonempty relation on W, $R \subseteq W \times W$, called the *direct transition relation* between states. If $u, w \in W$ then the fact that (u, w) belongs to R is also marked by '$u \to_R w$' or '$R(u, w)$', or, more succinctly, by '$u \; R \; w$', and read: 'The system $\boldsymbol{U}$ directly passes from the state u to the state w'.

An *atomic action* (on the system $\boldsymbol{U}$) is an arbitrary nonempty subset $A \subseteq W \times W$, i.e., it is a nonempty binary relation on W.

Definition 1.2.1 An *elementary discrete action system* is a triple

$$\boldsymbol{M} = (W, R, \boldsymbol{\mathcal{A}}) \tag{1.2.3}$$

where $\boldsymbol{U} = (W, R)$ is a discrete system and $\boldsymbol{\mathcal{A}}$ is a nonempty family of atomic actions on $\boldsymbol{U}$. The members of the family $\boldsymbol{\mathcal{A}}$ are called *atomic actions of the action system* (1.2.3). □

Instead of '$A(u, w)$' we often write '$u \to^A w$' or simply '$u \; A \; w$' and read: 'The action A carries the system $\boldsymbol{U}$ from the state u to the state w.' Any pair (u, w) such that $A(u, w)$ holds, is called a *possible performance* of the action A (in the state u). Thus, any atomic action is identified with the set of its possible performances.[1]

Definition 1.2.1 does not assume that the atomic actions in $\mathcal{A}$ are mutually disjoint relations. Hence, if $(u, w) \in A \cap B$ for some $A, B \in \mathcal{A}$, the pair (u, w) is a possible performance of A and a possible performance of B as well. (We shall return to this issue later.)

Every elementary action system is a Kripke frame $\boldsymbol{U} = (W, R)$ enriched with a bunch $\mathcal{A}$ of binary relations on it. Thus action systems are multi-relational Kripke frames in which one relation is distinguished, viz., R. The relation R encodes a possible evolution of the frame, i.e., it imposes limitations on the possibility of direct transitions of the system $\boldsymbol{U}$ from some states to others. A variety of possible interpretations of R are obtained by choosing (in a proper way) interesting classes of discrete systems. To mention only the most important of these interpretations: the relation R can be interpreted as the physical possibility of a transition from one state to another, or as compatibility with a social role, or as compatibility with labor regulations of a given institution. Apart from physical limitations, it is often necessary in some action systems to take into account restrictions that are imposed by law. These systems may be called *deontic* action systems—some actions in such a system are legally forbidden, e.g., on the strength of traffic regulations, though they are physically feasible (actions *in fraudem legis*). Finally, R can be viewed as compatibility with the principles or empirical procedures of one or any other theory. Each of the mentioned interpretations is bound with a certain set of discrete systems.

The relation R also determines when certain transitions $u \; R \; w$ are irreversible. Hence, R, in general, need not be symmetric. The relation R^*, the transitive and reflexive closure of R, is called the *relation of indirect transitions* between states of the system (W, R), or in short, the *transition relation*.

A direct transition $u \; R \; w$ is said to be *accomplished* by an atomic action $A \in \mathcal{A}$ if (u, w) is a possible performance of A, i.e., $u \; A \; w$ holds.

A possible performance $u \; A \; w$ of A is *realized* in the system $\boldsymbol{M} = (W, R, \mathcal{A})$ if (u, w) is also an element of R, i.e., $u \; R \; w$ holds. The fact that a performance $u \; A \; w$ is realized thus depends on the selection of the direct transition relation R.

Action theory presupposes a definite ontology of the world. We accept that all individuals are divided into *actual*, *potential*, and *virtual* ones (see Scott 1970). The above division may vary within a flow of time—a given individual, treated initially as a virtual or potential entity, may become an actual individual, i.e., an element of physical reality. In the context of this work, where the notion of a system and its states are understood as widely as possible, we assume that the system can exist in actual, potential, or virtual states. The distinction between particular parts of this division will not be clear-cut but it can be useful. When, for example, one intends to build a house, all actions (whether they are atomic or compound) such as preparing a project, finding a site to build on, the purchase of materials, can be regarded as

[1]Strictly speaking, the adopted set-theoretic formalism *represents* everyday situations in which actions and states of affairs are involved in much the same way as probabilistic spaces model random events.

transformations between virtual or potential states of the system, which is the house to be built; in these states the house is not a physical reality. One can assume that the actual states of the house are designated by a number of actions starting with laying the foundations and closing with the completion of the house. A similar division of states appears while designing and sewing a dress, making a new model of a car, and so on.

However, when dealing with elementary action systems, we do not distinguish explicitly a separate category of actual states of the system as a subset of W. We shall uniformly speak of *states* of the action system.

Ota Weinberger (1985, p. 314) writes: "An action is a transformation of states within the flow of time (including the possibility of an identity transformation—a standstill) which fulfills the following conditions:

(i) the transformation of states (or behaviour) is accounted to a subject (to an agent)
(ii) the subject has at his disposal a range for action, i.e., a class of at least two states of affairs which are possible continuations of a given trajectory in the system of states
(iii) an appropriate information process underlying the subjects' decision-making causes the future development of the system to fall into one of the alternative possibilities within the range of action."

The concept of action presented here is convergent with that of Weinberger; it is however more comprehensive—the notions of a situation and the situational envelope of action are extensively highlighted in it. The issue is discussed at greater length in Chap. 2.

Atomic actions need not be partial functions from the set W of states into W. But each binary relation A on W defines a total function from W into the power set $\wp(W)$, namely to A the mapping $\delta_A : W \to \wp(W)$ is assigned, where

$$\delta_A(u) := \{w \in W : A(u, w)\}.$$

The set $\delta_A(u)$ may be empty. $\delta_A(u)$ is called the set of *possible effects* (on the system) of the action A in the state u.

The correspondence $A \to \delta_A$ is bijective: every function $\delta : W \to \wp(W)$ is of the form δ_A for some unique binary relation A on W. (A is defined as follows: $A(u, w)$ if and only if $w \in \delta(u)$, for all $u, w \in W$.)

Since every binary relation on W can be uniformly replaced by a function δ from W to $\wp(W)$, one can move from the 'relational' image to the 'functional' image of action theory, i.e., instead of considering elementary action systems $\boldsymbol{M} = (W, R, \boldsymbol{\mathcal{A}})$ given as in Definition 1.2.1, one may examine systems of the form

$$(W, \delta_R, \{\delta_A : A \in \boldsymbol{\mathcal{A}}\})$$

as well with δ_R and δ_A defined as above.

The pair (W, δ_A) is customarily called the *directed graph* of the relation A (the *digraph* of A, for short). If W is finite, the graph (W, δ_A) is often represented by means of its diagram. The elements of W are depicted by means of points on the plane called *vertices* of the graph. If $w \in W$ and $\delta_A(w)$ is nonempty, the vertex w is linked with each element $u \in \delta_A(w)$ by means of an arrow. Each such arrow is called an *oriented edge*; w is its beginning. For example, if $W = \{w_1, w_2, w_3\}$ and $A = \{(w_1, w_1), (w_1, w_2), (w_2, w_3)\}$, the graph (W, δ_A) is represented by the following diagram.

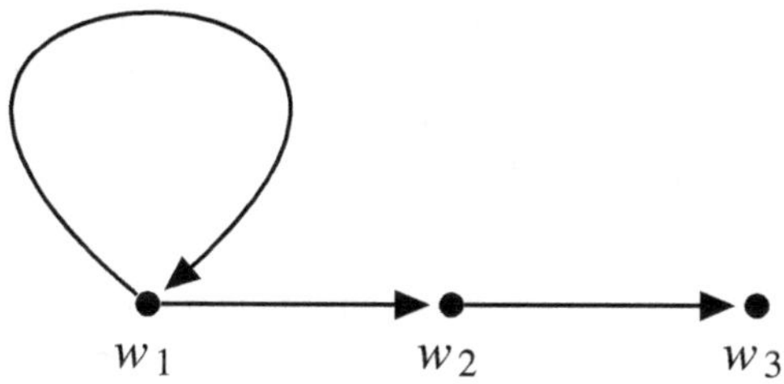

Fig. 1.3

The directed graph (W, δ_A) has no multiple edges. This means that it does not contain—as a subgraph—a graph of the form.

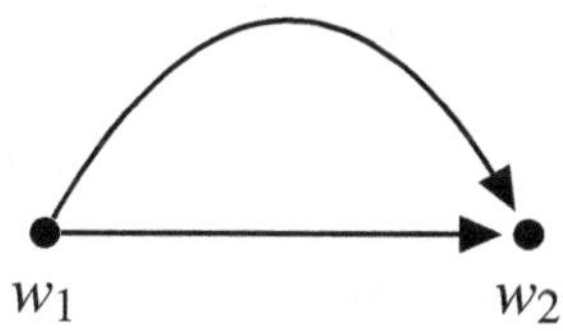

Fig. 1.4

More formally, for any two distinct vertices w_1, w_2 there exists at most one arrow from w_1 to w_2 and at most one arrow leading from w_2 to w_1.

The directed graph (W, δ_A) may contain loops, i.e., it may contain subgraphs of the form.

Fig. 1.5

This may happen if A contains pairs (w, w), for some $w \in W$.

The fact that atomic actions and the relation R of direct transitions between states are represented in the form of digraphs (on the set W of vertices) has two consequences. On the one hand, the possibility of graphic representation of the action system as a set of points and arrows on the plane makes the structure of the system more lucid and simpler to grasp intuitively. (This remark concerns only the action systems which can be represented by means of families of planar graphs.) On the other hand, defining an action system as a set of digraphs (without multiple edges) enables us to apply methods and results that belong to graph theory.

We shall now give a few key definitions in the theory of elementary action systems.

Definition 1.2.2 The (finite) *reach* of an elementary action system $\boldsymbol{M} = (W, R, \boldsymbol{\mathcal{A}})$ is the binary relation $\boldsymbol{Re_M}$ on W which is defined as follows: $\boldsymbol{Re_M}(u, w)$ if and only if *either* $u = w$ and $R(u, u)$ *or* there exists a finite nonempty sequence of states $u_0, \dots, u_n$ of W satisfying the following conditions:

(i) $u_0 = u, \quad u_n = w$;
(ii) for each i $(0 \leqslant i \leqslant n-1)$ there exists an atomic action A_i such that $A_i(u_i, u_{i+1})$ and $R(u_i, u_{i+1})$. □

Condition (ii) is more succinctly written as

(iii) $u_0\ A_0, R\ u_1\ A_1, R\ u_2 \dots A_{n-2}, R\ u_{n-1}\ A_{n-1}, R\ u_n$.

The reach $\boldsymbol{Re_M}$ of the system $\boldsymbol{M} = (W, R, \boldsymbol{\mathcal{A}})$ is always a subrelation of R^*. (In fact, $\boldsymbol{Re_M} \subseteq R^+ \cup (E_W \cap R)$. $\boldsymbol{Re_M}$ is reflexive if and only if R is reflexive. The reach of $\boldsymbol{M}$ is the set of indirect transitions between states which are accomplished by means of finite strings of atomic actions of $\boldsymbol{\mathcal{A}}$ (including the empty string).

The fact that $A_0, \dots, A_{n-1}$ is a finite sequence of atomic actions of $\boldsymbol{\mathcal{A}}$, $u_0, \dots, u_n$ is a nonempty sequence of states states of W such that $A_i(u_i, u_{i+1})$ for all i $(0 \leqslant i \leqslant n-1)$ is depicted as a figure

$$u_0\ A_0\ u_1\ A_1\ u_2 \dots u_{n-2}\ A_{n-2}\ u_{n-1}\ A_{n-1}\ u_n. \tag{1.2.4}$$

Any such figure (1.2.4) is called a (possible) *operation* of the action system $\boldsymbol{M} = (W, R, \boldsymbol{\mathcal{A}})$. If the sequence $A_0, \dots, A_{n-1}$ is empty, then (1.2.4) reduces to the sequence u_0 with $n = 0$.

The operation (1.2.4) is *realizable* in the elementary action system $\boldsymbol{M}$ if and only if $u_i\ R\ u_{i+1}$ holds for $i = 0, \dots, n-1$. If $A_0, \dots, A_{n-1}$ is empty, then (1.2.4) is realizable if and only if $u_0\ R\ u_0$. The realizable operation (1.2.4) is written as in (iii), that is, as

$$u_0\ A_0, R\ u_1\ A_1, R\ u_2 \dots u_{n-2}\ A_{n-2}, R\ u_{n-1}\ A_{n-1}, R\ u_n. \tag{1.2.5}$$

Thus, the reach of $\boldsymbol{M}$ is the resultant relation of the set of all (finite) realizable operations of $\boldsymbol{M}$: $u\ \boldsymbol{Re_M}\ w$ holds if and only if there exists a realizable operation of $\boldsymbol{M}$, which leads from u to w.

The following definition isolates some classes of elementary action systems.

Definition 1.2.3 Let $\boldsymbol{M} = (W, R, \boldsymbol{\mathcal{A}})$ be an elementary action system. The system $\boldsymbol{M}$ is

(a) *separative* if $A \cap B = \emptyset$ for any two distinct actions $A, B \in \boldsymbol{\mathcal{A}}$;
(b) *normal* if $A \subseteq R$, for all $A \in \boldsymbol{\mathcal{A}}$;
(c) *strictly normal* if $\bigcup \boldsymbol{\mathcal{A}} = R$;
(d) *equivalential* if it is separative and strictly normal; that is, the actions of $\boldsymbol{\mathcal{A}}$ form a partition of R;
(e) *strictly equivalential* if the function assigning $Dom(A)$ to each $A \in \boldsymbol{\mathcal{A}}$ is one-to-one and the family $\{Dom(A) : A \in \boldsymbol{\mathcal{A}}\}$ forms a partition of $Dom(R)$;
(f) *complete* if the reach $\boldsymbol{Re_M}$ of the system is equal to $R^+ \cup (E_W \cap R)$;
(g) *reversible* if for any atomic action $A \in \boldsymbol{\mathcal{A}}$ and every pair (u, w) such that uA, Rw, the pair (w, u) belongs to the reach of the system.
(h) *deterministic* if the actions of the system are functional, that is, for each $A \in \boldsymbol{\mathcal{A}}$, if $(u, v) \in A$ and $(u, w) \in A$, then $v = w$. □

The following observations are immediate:

Proposition 1.2.4 *Let $\boldsymbol{M} = (W, R, \boldsymbol{\mathcal{A}})$ be an elementary action system.*

*(1) If **M** is strictly normal, then it is normal.*
*(2) If **M** is strictly equivalential, then it is equivalential.*
*(3) **M** is reversible if and only if the reach of **M** is a symmetric relation.*
*(4) If **M** is complete and the relation R of direct transition is symmetric, the system **M** is reversible.* □

Definition 1.2.5 Let $\boldsymbol{M} = (W, R, \boldsymbol{\mathcal{A}})$ be an elementary action system. A *task for* $\boldsymbol{M}$ is any pair (Φ, Ψ) such that $\Phi \subseteq W$ and $\Psi \subseteq W$. The set Φ is called the *initial condition* of the task while Ψ is the *terminal condition* of the task. □

Elementary action systems are designed as set-theoretic or diagrammatic constructs, whose role is to assure the implementation of a given task, that is, to provide a bunch of strings of atomic actions which would lead from Φ-states to Ψ-states. This aspect makes the theory of atomic action systems similar to automata theory. But the focus of action theory is rather on the performability of actions than the collecting together of accepted strings of atomic actions leading from initial to terminal states.

1.3 Examples

To warm up we give an analysis of the following well-known logical puzzle.

Example 1.3.1 (Crossing a river) Three missionaries and three cannibals are gathered together on the left bank of a river. There is a boat large enough to carry one or two of them. They all wish to cross to the right bank. However, if cannibals outnumber missionaries on either bank, the cannibals indulge in their natural propensities and eat the missionaries. Is it possible to cross the river without a missionary being eaten?

The problem is modeled as a simple action system with a task. Possible states are labeled by figures of the form

$$(+)\mathrm{CCMM} \mid \mathrm{CM},$$

where the symbol (+) indicates that the boat is on the left bank, two cannibals and two missionaries are on the left bank, and one cannibal and one missionary are on the other bank.

The symbol $(+)\mathrm{CCCMMM} \mid \varepsilon$ marks the initial state, and $\varepsilon \mid (+)\mathrm{CCCMMM}$— the required final state.

There are altogether 18 states that meet the above requirements. Apart from the initial state

$$(+)\mathrm{CCCMMM} \mid \varepsilon,$$

there are 8 states more in which the boat is on the left bank:

$$\begin{array}{ll} (+)\mathrm{CCMMM} \mid \mathrm{C}, & (+)\mathrm{C} \mid \mathrm{CCMMM}, \\ (+)\mathrm{CCMM} \mid \mathrm{CM}, & (+)\mathrm{CM} \mid \mathrm{CCMM}, \\ (+)\mathrm{CCC} \mid \mathrm{MMM}, & (+)\mathrm{MMM} \mid \mathrm{CCC}, \\ (+)\mathrm{CC} \mid \mathrm{CMMM}, & (+)\mathrm{CMMM} \mid \mathrm{CC} \end{array}$$

and 8 dual states, where the boat is on the right bank:

$$\begin{array}{ll} \mathrm{CCMMM} \mid (+)\mathrm{C}, & \mathrm{C} \mid (+)\mathrm{CCMMM}, \\ \mathrm{CCMM} \mid (+)\mathrm{CM}, & \mathrm{CM} \mid (+)\mathrm{CCMM}, \\ \mathrm{CCC} \mid (+)\mathrm{MMM}, & \mathrm{MMM} \mid (+)\mathrm{CCC}, \\ \mathrm{CC} \mid (+)\mathrm{CMMM}, & \mathrm{CMMM} \mid (+)\mathrm{CC} \end{array}$$

together with the final state

$$\varepsilon \mid (+)\mathrm{CCCMMM},$$

which is the dual of the initial state.

Some other states as, e.g., $(+)\mathrm{CCM} \mid \mathrm{CMM}$, are fatal and therefore excluded. In this state two cannibals consume the missionary who has remained on the left bank. After this action the state $(+)\mathrm{CCM} \mid \mathrm{CMM}$ would turn into $(+)\mathrm{CC} \mid \mathrm{CMM}$. The states $(+)\varepsilon \mid \mathrm{CCCMMM}$ and $\mathrm{CCCMMM} \mid (+)\varepsilon$ are also excluded.

In the above description of states cannibals and missionaries are not individuated, that is, cannibals and missionaries are treated as two three-element collectives, whose members are not discriminated.

There are 5 atomic actions labeled, respectively, as M, C, CM, MM, CC. These labels inform who is sailing in the boat from one bank to the other. For, e.g., CM

marks the action according to which one cannibal and one missionary travel across the river.

According to the arrow depiction of atomic actions, M consists of the following pairs of states:

$$\begin{aligned}&(+)\mathrm{CCMMM} \mid \mathrm{C} \to^{M} \mathrm{CCMM} \mid (+)\mathrm{CM},\\&(+)\mathrm{CM} \mid \mathrm{CCMM} \to^{M} \mathrm{C} \mid (+)\mathrm{CCMMM}\end{aligned}$$

as well as their inverses

$$\begin{aligned}&\mathrm{CCMM} \mid (+)\mathrm{CM} \to^{M} (+)\mathrm{CCMMM} \mid \mathrm{C},\\&\mathrm{C} \mid (+)\mathrm{CCMMM} \to^{M} (+)\mathrm{CM} \mid \mathrm{CCMM}.\end{aligned}$$

The action C consists of the pairs

$$\begin{aligned}&(+)\mathrm{CCCMMM} \mid \varepsilon \to^{C} \mathrm{CCMMM} \mid (+)\mathrm{C},\\&(+)\mathrm{CCMMM} \mid \mathrm{C} \to^{C} \mathrm{CMMM} \mid (+)\mathrm{CC},\\&(+)\mathrm{C} \mid \mathrm{CCMMM} \to^{C} \varepsilon \mid (+)\mathrm{CCCMMM}\\&(+)\mathrm{CCC} \mid \mathrm{MMM} \to^{C} \mathrm{CC} \mid (+)\mathrm{CMMM},\\&(+)\mathrm{CC} \mid \mathrm{CMMM} \to^{C} \mathrm{C} \mid (+)\mathrm{CCMMM},\end{aligned}$$

and their inverses

$$\begin{aligned}&\mathrm{CCMMM} \mid (+)\mathrm{C} \to^{C} (+)\mathrm{CCCMMM} \mid \varepsilon,\\&\mathrm{CMMM} \mid (+)\mathrm{CC} \to^{C} (+)\mathrm{CCMMM} \mid \mathrm{C}.\\&\varepsilon \mid (+)\mathrm{CCCMMM} \to^{C} (+)\mathrm{C} \mid \mathrm{CCMMM}.\\&\mathrm{CC} \mid (+)\mathrm{CMMM} \to^{C} (+)\mathrm{CCC} \mid \mathrm{MMM}\\&\mathrm{C} \mid (+)\mathrm{CCMMM} \to^{C} (+)\mathrm{CC} \mid \mathrm{CMMM}.\end{aligned}$$

The action CM consists of the following pairs:

$$\begin{aligned}&(+)\mathrm{CCCMMM} \mid \varepsilon \to^{CM} \mathrm{CCMM} \mid (+)\mathrm{CM},\\&(+)\mathrm{CCMM} \mid \mathrm{CM} \to^{CM} \mathrm{CM} \mid (+)\mathrm{CCMM},\\&(+)\mathrm{CM} \mid \mathrm{CCMM} \to^{CM} \varepsilon \mid (+)\mathrm{CCCMMM}\end{aligned}$$

and their inverses

$$\begin{aligned}&\mathrm{CCMM} \mid (+)\mathrm{CM}\varepsilon \to^{CM} (+)\mathrm{CCCMMM} \mid \varepsilon,\\&\mathrm{CM} \mid (+)\mathrm{CCMM} \to^{CM} (+)\mathrm{CCMM} \mid \mathrm{CM},\\&\varepsilon \mid (+)\mathrm{CCCMMM} \to^{CM} (+)\mathrm{CM} \mid \mathrm{CCMM}\end{aligned}$$

In turn, the action MM consists of the pairs

$$\text{(+)CCMM} \mid \text{CM} \to^{MM} \text{CC} \mid \text{(+)CMMM},$$
$$\text{(+)CMMM} \mid \text{CC} \to^{MM} \text{CM} \mid \text{(+)CCMM},$$

and their inverses

$$\text{CC} \mid \text{(+)CMMM} \to^{MM} \text{(+)CCMM} \mid \text{CM},$$
$$\text{CM} \mid \text{(+)CCMM} \to^{MM} \text{(+)CMMM} \mid \text{CC}$$

The action CC consists of the pairs

$$\text{(+)CCCMMM} \mid \varepsilon \to^{CC} \text{CMMM} \mid \text{(+)CC},$$
$$\text{(+)CCC} \mid \text{MMM} \to^{CC} \text{C} \mid \text{(+)CCMMM},$$
$$\text{(+)CC} \mid \text{CMMM} \to^{CC} \varepsilon \mid \text{(+)CCCMMM}$$
$$\text{(+)CCMMM} \mid \text{C} \to^{CC} \text{MMM} \mid \text{(+)CCC}$$

and their inverses

$$\text{CMMM} \mid \text{(+)CC} \to^{CC} \text{(+)CCCMMM} \mid \varepsilon$$
$$\text{C} \mid \text{(+)CCMMM} \to^{CC} \text{(+)CCC} \mid \text{MMM},$$
$$\varepsilon \mid \text{(+)CCCMMM} \to^{CC} \text{(+)CC} \mid \text{CMMM},$$
$$\text{MMM} \mid \text{(+)CCC} \to^{CC} \text{(+)CCMMM} \mid \text{C}.$$

The relation R of direct transition between states respects the above rules for safely crossing the river. When two states u and w are connected by R, i.e., $u \to_R w$ holds, then w is achieved from u by performing one of the above actions. There are no direct transitions from the above states to fatal states. (Since fatal states have been deleted, each of the above actions A obeys the principle that missionaries are not outnumbered by cannibals in the states entering possible performances of A. If we decided to adjoin fatal states to the above list of possible states, we would also be enforced to add a new atomic action, that of eating missionaries by cannibals.) This fact guarantees that the above system is normal. We therefore have that R is simply the set-theoretic union of the above actions,

$$R := C \cup M \cup CC \cup CM \cup MM.$$

R thus contains 32 pairs. R is a symmetric relation because the relations C, M, CC, CM, MM are all symmetric.

It immediately follows from the above definitions that the resulting action system

$$\boldsymbol{M} = (W, R, \boldsymbol{\mathcal{A}})$$

with $\mathcal{A} := \{C, M, CC, CM, MM\}$ is separative, normal, complete, and reversible. In fact, for every pair (u, w) such that $u\ A, R\ w$, it is the case that $w\ A, R\ u$, for any action $A \in \mathcal{A}$.

There are dead states: (+)MMM | CCC and CCC | (+)MMM are examples. If u_0 represents any of these two states, there is no state w such that $u_0 \rightarrow_R w$. But these two conceivable dead states have another characteristic feature: they are isolated. This means that there is no state v by means of which one may enter u_0, that is, there in no state v such that $v \rightarrow_R u_0$. □

Example 1.3.2 This is a simplified description of washing in an automatic washing machine.

Let us consider three sentences:

α:	The bed linen is not washed
β:	The linen is in the machine
γ:	The machine is turned on.

In the (unavoidably incomplete) automatic linen washing scheme we wish to present, one can assume that each state is described by one sentence of the form $c_1\alpha\ \&\ c_2\beta\ \&\ c_3\gamma$, where $c\delta = \delta$ if $c = 1$ and $c\delta = \neg\delta$ if $c = 0$. ($\neg\delta$ is negation of the sentence δ.) Thus, the set W of states consists of eight elements. They are:

w_1: The linen is not washed & the linen is not in the machine & the machine is not turned on (i.e., it is turned off)

w_2: The linen is not washed & the linen is in the machine & the machine is turned on

w_3: The linen is not washed & the linen is in the machine & the machine is turned off

w_4: The linen is not washed & the linen is not in the machine & the machine is turned on

w_5: The linen is washed & the linen is not in the machine & the machine is turned off

w_6: The linen is washed & the linen is not in the machine & the machine is turned on

w_7: The linen is washed & the linen is in the machine & the machine is turned off

w_8: The linen is washed & the linen is in the machine & the machine is turned on.

The set $\mathcal{A}$ of atomic actions consists of five elements:

A:	Putting the linen into the machine
B:	Taking the linen out of the machine
C:	Turning on the machine
D:	Turning off the machine
E:	Washing the linen in the machine

The action E is exclusively performed by the machine, i.e., the linen is washed by having put it first dry into the machine and then turning on the machine. The remaining actions are performed by the person operating the machine. Such actions as programming adjustment, providing a quantity of detergent, starching, spinning, etc., are omitted.

Thus the process of washing can be defined as the following operation:

$$w_1 \; A \; w_3 \; C \; w_2 \; E \; w_8 \; D \; w_7 \; B \; w_5.$$

The aim of washing is the implementation of the task $(\{w_1\}, \{w_5\})$, i.e., moving the system from the w_1-state to the w_5-state.

The states w_4 and w_6, though attainable by means of the above actions, are irrelevant to the process of washing. The w_6-state is attained from the w_8-state by turning off the machine, taking out the linen, and turning it on again, which can be depicted as follows:

$$w_8 \; D \; w_7 \; B \; w_5 \; C \; w_6.$$

Similarly, the state w_4 is attained from w_1 by performing the action $C : w_1 \; C \; w_4$.

All the actions A, B, C, D, and E are binary relations on W. It can be assumed that the relation E consists only of one pair:

$$E := \{(w_2, w_8)\}.$$

(The action E is treated here as atomic; in a more accurate description of washing, E is a complex system of actions.) The action A is reduced to putting the dry linen into the turned off machine:

$$A := \{(w_1, w_3)\}.$$

The machine can be turned on or turned off when the linen is in the machine or outside as well; at the same time the linen can be washed or dry. Thus

$$C := \{(w_1, w_4), (w_3, w_2), (w_5, w_6), (w_7, w_8)\},$$
$$D := \{(w_2, w_3), (w_4, w_1), (w_6, w_5), (w_8, w_7)\}.$$

The washed linen (if the machine has been on already) or the dry linen can be taken out from the machine (if the agent operating the machine decides to take it out again as soon as it had been put in, giving up washing). Thus

$$B := \{(w_7, w_5), (w_3, w_1)\}.$$

All the actions A, B, C, D, and E are partial functions. Also, $C = D^{-1}$.

What is the relation R of direct transition on the set W? It cuts out from the set $W \times W$ those pairs (u, w) not physically realized. This means that the direct transition from u into the state w is impossible because of technical or physical limitations. First of all, the relation R excludes in the above model any transitions from "wet" to "dry" states—we assume that the dry linen can be washed out but its drying in the machine after washing is impossible. (In the above example the washing machine is not equipped with a dryer.) Secondly, the relation R reflects the fact that it is not possible (or at least undesirable) to put in or take out directly the linen from the working, turned on machine. Consequently, taking into account the above limitations we have that

$$R := \{(w_1, w_3), (w_1, w_4), (w_2, w_3), (w_3, w_1), (w_3, w_2), (w_4, w_1), (w_5, w_6), (w_5, w_7), (w_6, w_5), (w_7, w_5), (w_7, w_6), (w_7, w_8), (w_8, w_7), (w_2, w_8)\}.$$

This relation is graphically depicted below:

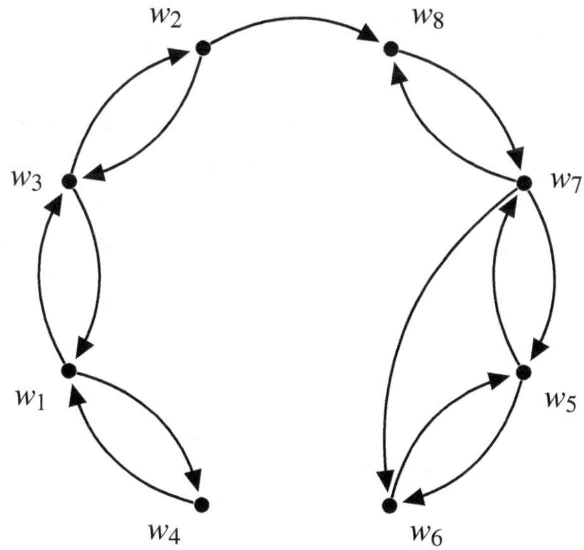

Fig. 1.6

The action system thus defined

$$(W, R, \{A, B, C, D, E\}) \tag{1.3.2}$$

is deterministic and separative. The system is also normal. However, the system is not complete—the transition $w_5 \, R \, w_7$ does not belong to the reach of the system. This is due to the fact that only unwashed linen can be put into the machine. (If we extend action A with the pair (w_5, w_7), the system would become complete.) □

Example 1.3.3 The third example we present is more sophisticated and it refers directly to proof theory.

Let S be a nonempty set, whose elements are called *sentential formulas*. The nature and the internal structure of the propositions of S are inessential here. The set S itself will be called a *sentential language*.

A finitary Hilbert-type *rule of inference* in S is any set of pairs of the form (X, ϕ), where X is a finite set of formulas of S and ϕ is an individual formula of S. (X, ϕ) is read as "*Infer* ϕ *from* X."

Now let T be a fixed set of sentential formulas of S which will be called a *theory* in S, and let Θ be a fixed set of inference rules. A *proof* from T carried out by means of the rules of Θ is any nonempty finite sequence of formulas $(\phi_0, \phi_1, \ldots, \phi_n)$ such that for each i $(0 \leqslant i \leqslant n)$ either $\phi_i \in T$ or there exists a subset J of $\{0, \ldots, i-1\}$ and a rule $r \in \Theta$ such that $(\{\phi_j : j \in J\}, \phi_i) \in r$. The notion of a proof is strictly relativized to the set of rules Θ.

Let W_T be the set of all proofs from T. We define the relation R of direct transition on W_T as follows:

> $R(u, w)$ if and only if the proof w is obtained from the proof u by adjoining a single formula at the end of u, i.e., $w = u^\wedge(\phi)$ for some ϕ.

($u_1{}^\wedge u_2$ is the concatenation of the sequences u_1 and u_2.) R is therefore irreflexive.

The following atomic action A_r is assigned to each rule r in Θ:

> $A_r(u, w)$ if and only if there exists a sentential formula ϕ and a subsequence $(\phi_1, \ldots, \phi_k)$ of u such that $w = u^\wedge(\phi)$ and $(\{\phi_1, \ldots, \phi_k\}, \phi) \in r$.

In other words, the 'action' of A_r on a proof u consists in adjoining a new formula at the end of u according to the rule r (provided that such a formula can be found). A_r thus suffixes proofs by adding one new element. Each action A_r is therefore an irreflexive relation.

Let $\boldsymbol{\mathcal{A}} := \{A_r : r \in \Theta\}$. The triple $\boldsymbol{M} := (W, R, \boldsymbol{\mathcal{A}})$ is an elementary action system. The system is normal but it need not be deterministic. The reach of the system is equal to R^+; this follows from the fact that $\bigcup\{A : A \in \boldsymbol{\mathcal{A}}\} = R$ and R is irreflexive. Therefore the system $\boldsymbol{M}$ is complete.

The system need not be separative—it may happen that a pair (u, w) belongs to $A_r \cap A_s$ for some different rules r and s.

For every formula ϕ belonging to T, the sequence (ϕ) of length 1 is a proof. The proofs of length 1 are called *initial*. Let Φ_T be the totality of all initial proofs. (Without loss of generality the members of Φ_T may be identified with the formulas of T.)

For an arbitrary formula ϕ of T, we define Ψ_ϕ to be the set of all proofs w such that ϕ occurs as the last element of w. The pair (Φ_T, Ψ_ϕ) is a task for the system $\boldsymbol{M}$. In order to achieve a state of the system which is in Ψ_ϕ one has to find a proof of ϕ from T. Thus, the system $\boldsymbol{M}$ can be moved from a state of Φ_T to a state of Ψ_ϕ if and only if Ψ_ϕ is nonempty. □

The Paradox of Two Agents This paradox concerns rather the notion of agency than performability of actions. Let $\boldsymbol{M} := (W, R, \boldsymbol{\mathcal{A}})$ be an elementary action system operated by two different agents a and b. Suppose $A, B \in \mathcal{A}$ are two atomic actions which are not disjoint, i.e., $A \cap B \neq \emptyset$ (*vide* Example 1.3.3). Let us assume that a is the agent of the action A and b is the agent of the action B. Suppose there are states $u, w \in W$ such that u A, R w and u B, R w, i.e., (u, w) is a realizable performance of both A and B. This fact may seem to be paradoxical. The paradox is in the fact that performing the action A in the state u by the agent a, resulting in w, is at the same time performing the action B (by the agent b) although the agents a and b are entirely different and while a is busy with A, b is at rest.

One may argue that it resembles the situation in which competences criss-cross, i.e., the situation where the obligations (and actions) of people working in an office overlap—in such a situation one person does what is some other person's duty and vice versa.

The paradox is merely apparent. In the framework of elementary action system discussions concerning agency are premature. The notion of an agent belongs to a richer theory, where situational aspects of performability of actions, in particular agency, are taken into account. The situational notion of the performability of an action (which will be discussed later) takes into account the situational environment in which the system is—in particular it introduces the notion of an agent and describes the ways the work of the agents is organized and how they operate the system. The notion of an agent is thus incorporated into the situational envelope of an elementary action system. Within the framework of elementary systems, talk of agents operating the system is devoid of meaning. On the other hand, it may happen that one agent may be assigned to *many* atomic actions in a given action system and he may perform them simultaneously.

Actions are not the same objects as acts of performing actions. The phrase *Driving a car*, being a name of a class of compound actions, bears a meaning that is comprehensible by any car driver. *Driving a car by Adam* is a name of a subclass of actions belonging to the former class. In turn, the utterance *Adam is driving a car* refers to an act of performing an action from this subclass.

1.4 Performability of Atomic Actions in Elementary Action Systems

Let $\boldsymbol{M} := (W, R, \mathcal{A})$ be an elementary action system and let A be a fixed action of $\mathcal{A}$. We wish to define a precise mathematical notion of performability of the action A. What does it mean to say that A is performable? Suppose we are given states u, w such that $A(u, w)$ holds. A quick answer to the question is this: the agent of A is able to move the system from the u-state to the w-state. But this answer does not fully satisfy us. First, the meaning of the phrase 'is able' is unclear. We want the meaning of the phrase to be related directly to the intrinsic properties of the system. Second, the term 'agent' is not incorporated into the language of elementary action systems. Third, the above explication does not take into account the fact that performability depends on the *type* of limitations pertaining to the system. For example, the action may be performable if only physical limitations are taken into account and at the same time it may be unperformable with respect to legal regulations. According to the discussion of definition of an action system (Definition 1.2.1), the relation R specifies a definite type of limitation imposed on the functioning of the system. Therefore, a reliable definition of performability should take on a form relativized to the fixed relation R of direct transition. Third, we want performability to be the property of an atomic action dependent on states of the system. This means that performability is treated here as a 'parameterized' property defined with respect to a given state of the system. It is clear that the states the system operates may vary with the passage of time. In fact we distinguish two forms of performability—the strong and the weak ones. We call them $\forall$-performability and $\exists$-performability, respectively. Here are the formal definitions.

Definition 1.4.1 Let $\boldsymbol{M} := (W, R, \boldsymbol{\mathcal{A}})$ be an elementary action system. Let A be an atomic action in $\boldsymbol{\mathcal{A}}$, and $u \in W$.

(i) The action A is *performable* in the state u if and only if there exists a state $w \in W$ such that $u\ A, R\ w$ (i.e., the transition $u\ R\ w$ is accomplished by A);
(ii) A is *totally performable* in the state u if A is performable in u and for every state $w \in W$, if $u\ A\ w$, then $u\ R\ w$. □

The performability of A is thus tantamount to the existence of a direct R-transition from u to a state w such that (u, w) is a possible performance of A. We shall use the terms "A is performable in u" and "A is $\exists$-performable in u" interchangeably. The second conjunct of the definition of total performability of A in u states that every possible performance of A in u is admissible by R. We shall henceforth also use the term "A is $\forall$-performable in u" when we speak of the total performability of A in u.

The notion of the performability of an action (in a given state) is relativized to the relation R of direct transition and so one should rather speak of R-performability. Thus, different interpretations of R determine different meanings of the performability of an action.

The above definition offers a concept of zero-one performability. This definition does not differentiate, say, poor performance from good performance or medium performance of an action. A probabilistic conception of performability is outlined in Sect. 1.5.

One may argue that Definition 1.4.1 does not take into account the epistemic abilities of the agents operating the system. The agent of an action A may know how A is defined, i.e., it is conceivable for him which pairs (u, w) belong to A and which do not. But at the same time he may not know the limitations imposed by R, i.e., he may not know if the transition from a state u to another state w is admissible in the system by R or not. (We shall return to this issue in the last chapter devoted to the epistemic status of the agent of an action.) In situational action systems agents may be 'components' of the situational envelope. In these systems the relation of direct transition may take into account the agents' praxeological and epistemic abilities.

Observation 1.4.2

(1) An atomic action A is performable in u if and only if $\delta_A(u) \cap \delta_R(u) \neq \emptyset$.
(2) A is totally performable in u if and only if $\emptyset \neq \delta_A(u) \subseteq \delta_R(u)$. □

Definition 1.4.1 is extended by defining the performability of an atomic action in the system.

Definition 1.4.3 Let A be an atomic action of the elementary action system $\boldsymbol{M} := (W, R, \boldsymbol{\mathcal{A}})$.

(i) A is *performable in the system* $\boldsymbol{M}$ if and only if a is performable in every state $u \in W$ such that $\delta_A(u) \neq \emptyset$.
(ii) A is *totally performable in* $\boldsymbol{M}$ if and only if A is totally performable in every state u such that $\delta_A(u) \neq \emptyset$. □

Just as in Definition 1.4.1 instead of the phrase "A is performable in $\boldsymbol{M}$" we shall sometimes use the term "A is $\exists$-performable in $\boldsymbol{M}$", and analogously, we shall say that A is $\forall$-performable in $\boldsymbol{M}$ whenever A is totally performable in $\boldsymbol{M}$.

Observation 1.4.4 *The atomic action A is totally performable in* $\boldsymbol{M}$ *if and only if* $A \subseteq R$. □

The dual property to performability is that of the unperformability of an action. The atomic action A is said to be *unperformable in a state* $u \in W$ if and only if $\delta_R(u) \cap \delta_A(u) = \emptyset$. In particular, if the set $\delta_A(u)$ is empty, A is unperformable in u.

A is said to be *totally unperformable* in the system $\boldsymbol{M}$ if and only if A is unperformable in every state of W.

Observation 1.4.5 *The atomic action A is totally unperformable in* $\boldsymbol{M}$ *if and only if* $A \cap R = \emptyset$. □

A weaker variant of the notion of unperformability is the negation of total performability. It does not seem, however, that the latter notion has any noteworthy properties or applications.

Total performability always implies performability. The converse holds for deterministic actions, i.e., actions that are partial functions on W. The distinction between performability and total performability is essential in the case of indeterministic atomic actions.

Definition 1.4.1.(i) of performability of an atomic action A in a state u does not specify a state w for which $u\ A, R\ w$ holds; w may be an arbitrary element of $\delta_A(u) \cap \delta_R(u)$. We define another property that can provide more detailed information about a given action.

Let (u, w) be a pair of states such that $R(u, w)$ holds. We recall that the transition $u\ R\ w$ is *accomplished by an action* A if and only if $A(u, w)$ holds, i.e., the transition $u\ R\ w$ is a possible performance of A. If $u\ R\ w$ is accomplished by A, then A is $\exists$-performable in u. This property of transitions (let us call it *accomplishment*) has the following feature.

Coherence Principle *If a transition $u\ R\ w$ is accomplished by A and (u, w) is a possible performance of another action B, then $u\ R\ w$ is accomplished by B.*

The Coherence Principle yields the following conclusion which may be regarded as paradoxical:

> *If $u\ R\ w$ is accomplished by A (and hence A is performable in u), then for every atomic action $B \in \mathcal{A}$ for which (u, w) is a possible performance is also performable in u.*

The 'paradox' is caused by the fact that if a transition $u\ R\ w$ is accomplished by A then *at the same time* $u\ R\ w$ is accomplished by any action B for which (u, w) is a possible performance. This conclusion seems bizarre for one would expect that the accomplishment of the transition $u\ R\ w$ by one action A should not affect the accomplishment of this transition by any other action of the system. (This is essentially the same problem which we discussed at the end of Sect. 1.3.)

The 'paradox' arises again from confusing the notion of performability with the act of performing an action. In the language of elementary action system a discourse on actions involving time parameters is excluded. In particular, the notion of the simultaneity of accomplishment of performability of actions is beyond this language. We shall look at the problem in the next chapter, where issues centering around the situational envelope of an elementary action system are discussed. If the system is separative, i.e., $A \cap B = \emptyset$ for any two distinct atomic actions $A, B \in \mathcal{A}$, then the above 'problem' disappears. On the other hand, from the viewpoint of elementary action theory, it makes sense to speak of a succession of elementary actions in the context of compound actions—this notion will be presented in Sect. 1.7.

We close this section with some observations on equivalent elementary action systems.

Definition 1.4.6 Let (W, R) be a discrete system. Two elementary action systems $\boldsymbol{M} := (W, R, \mathcal{A})$ and $\boldsymbol{N} := (W, R, \mathcal{B})$ (on (W, R)) are *equivalent* if and only if they have the same reach, i.e., $\boldsymbol{Re_M} = \boldsymbol{Re_N}$. □

Examples 1.3.1–1.3.3 provide action systems, in which all atomic actions are subsets of the relation R of direct transition, i.e., the atomic actions are totally performable. This fact is not unexpected—every action system is equivalent to a system in which the atomic actions are totally performable.

Proposition 1.4.7 *Every elementary action system $\boldsymbol{M} := (W, R, \mathcal{A})$ is equivalent to a normal action system $\boldsymbol{N} := (W, R, \mathcal{B})$.*

Proof Given an action system $\boldsymbol{M}$, define $\mathcal{B} := \{A \cap R : A \in \mathcal{A}\} \setminus \{\emptyset\}$. Then $\boldsymbol{M}$ and $(W, R, \mathcal{B})$ are equivalent. □

The above proposition gives rise to the question of whether every atomic action can be identified with the set of its *realizable* performances in the system. (By analogy, in logic any proposition is identified with the set of all possible states of affairs ('worlds') in which the proposition is true.) Every action would be thus totally performable in the system and we would not bother with the situations in which an action cannot be performed. It seems that such a solution would lead us too far. In the case of action systems the problem is more involved—the set W of states may admit various possible relations R of direct transition defining different aspects (e.g., legal, physical, etc.) of the performability of the same action. The identification of an action A with its realizable performances (relative to a fixed relation R) will make it impossible to analyze the properties of the action with respect to other admissible interpretations of the relation of direct transition.

As each atomic action A is a binary relation on the set of states W, the domain $Dom(A) = \{u \in W : (\exists w \in W)\ u\ A\ w\}$ may be thought of as the set of states in which the agent(s) intend to perform A. Some states u in $Dom(A)$ may be impossible from the perspective of performability of A—the agent may declare the intention of performing A in u but his abilities are insufficient; the action may simply be unperformable. In turn, the set $CDom(A) = \{w \in W : (\exists u \in W)\ u\ A\ w\}$ encapsulates the states which are the possible results of performing A. A state w in $CDom(A)$ is therefore reachable by means of the action A if it is the case that $u\,A\,w$ and $u\,R\,w$ hold for some $u \in Dom(A)$. From the praxeological point of view one may safely assume in Definition 1.4.1 that all atomic actions are total relations, i.e., $Dom(A) = W$ for all A. This assumption may be interpreted to mean that the agents may attempt to perform A in whichever state of the system. But the actual performability of A takes place in only few states u, viz., those that are linked with some state in $CDom(A)$ by the relation R.

Proposition 1.4.8 *For every elementary action system* $\boldsymbol{M} := (W, R, \boldsymbol{\mathcal{A}})$ *there exists an equivalent deterministic action system* $\boldsymbol{N} := (W, R, \boldsymbol{\mathcal{B}})$.

Proof Assuming the Axiom of Choice, every binary relation is the set-theoretic union of partial unary functions. Hence, for every action $A \in \boldsymbol{\mathcal{A}}$, there exists a family $\boldsymbol{\mathcal{B}}(A) := \{B_i : i \in I(A)\}$ of partial functions on W such that $A = \bigcup\{B_i : i \in I(A)\}(= \bigcup \boldsymbol{\mathcal{B}}(A))$. Let $\boldsymbol{\mathcal{B}} := \bigcup\{\boldsymbol{\mathcal{B}}(A) : A \in \boldsymbol{\mathcal{A}}\}$. The action system $\boldsymbol{N} := (W, R, \boldsymbol{\mathcal{B}})$ is deterministic and equivalent to $\boldsymbol{M}$. □

Note The Axiom of Choice, abbreviated as AC, states that:

> *For every nonempty family* $\boldsymbol{X}$ *of nonempty sets there exists a function* f *defined on* $\boldsymbol{X}$ *such that* $f(A) \in A$ *for all* $A \in \boldsymbol{X}$.

In set theory functions are uniformly defined as certain sets of ordered pairs. □

Yet another procedure, contrasting with the above 'normalizing' of an action system $\boldsymbol{M} = (W, R, \boldsymbol{\mathcal{A}})$, consists in extending the atomic actions of $\boldsymbol{\mathcal{A}}$ to total relations on W. (A relation $B \subseteq W \times W$ is *total* (or *serial*) on W if $Dom(B) = W$.) To present it, it is assumed that the relation R of direct transition satisfies the condition: for every state $u \in W$, $\delta_R(u)$ is a proper subset of W. (This condition is not, however, excessively restrictive.)

Let A be a fixed relation in $\boldsymbol{\mathcal{A}}$ and let $u \in W$. We define the relation A_u on W as follows. If $\delta_A(u) \neq \emptyset$, we put: $A_u := A$. If $\delta_A(u) = \emptyset$ then we select an arbitrary element $w \in W$ such that $w \notin \delta_R(u)$ and put: $A_u := A \cup \{(u, w)\}$. Assuming the Axiom of Choice, we see that the iteration of this procedure enables us to extend A to a total relation $\breve{A}'$ on W. Moreover, if A is a partial function, then the extension $\breve{A}'$ is a total function on W. The resulting action system $\boldsymbol{N} := (W, R, \{\breve{A}' : A \in \boldsymbol{\mathcal{A}}\})$ is not normal. It has, however, another property: for every $A \in \boldsymbol{\mathcal{A}}$, the action $\breve{A}'$ is performable (totally performable) in the sense of $\boldsymbol{N}$ in the same states in which A is performable (totally performable) in the sense of $\boldsymbol{M}$. This is an immediate consequence of the fact that A and $\breve{A}'$ have the same realizable performances. Thus, from the viewpoint of action theory, the abilities of the agents of A and $\breve{A}'$ are the same.

Without loss of generality, the scope of the action theory presented so far can be restricted to elementary action systems in which all atomic actions are total relations on the sets of states. In particular, in the case of deterministic action systems, it can be assumed that all members of $\boldsymbol{\mathcal{A}}$ are unary total functions on W. (Deterministic action systems would be then represented by monadic algebras $(W, R, f_1, f_2, \ldots)$ together with a distinguished binary relation R.) This assumption however is not explicitly adopted in this book.

1.5 Performability and Probability

We remarked in Sect. 1.4 that the distinction between performability and total performability is relevant in the case of atomic actions which are not partial functions. A simple illustration of such a situation is the shooting of an arrow at a board. We present here a simplified model of shooting which omits a number of aspects of this action. It is assumed that the shooting takes place in one fixed state u_0, which is determined by the following factors: the distance between the archer and the board is 70 m, there is no wind and the equipment of the archer is in good working order. The space W of states consists of the state u_0 and the states which represent the possible effects of shooting. It is also assumed that there are only two possible effects (outcomes) of the shooting—missing the board (state w_0) and hitting the board (state w_1). Thus, $W = \{u_0, w_0, w_1\}$. (We neglect the eventuality of the breaking of the bowstring, etc. One may also consider wider and more natural spaces of states with a richer gradation of hits. The hit value is measured by the number of points scored in a hit. Assuming that for a hit on the board one may score up to 10 points, the set of possible effects would contain 10 elements, say $w'_1, \ldots, w'_{10}$, where w'_k denotes the hit with k points.) Shooting an arrow from the bow is treated here as an atomic action A with two possible performances (u_0, w_0) and (u_0, w_1) only, i.e.,

$$A := \{(u_0, w_0), (u_0, w_1)\}. \tag{1.5.1}$$

In the first case the shot is a miss, in the second a hit. In other words, a possible performance of A is simply the shooting of an arrow, regardless of whether the arrow hits the target or not.

Disregarding the regulations of the game and taking into account only the physical limitations determined by the environment of the archer, the relation R will take the following form, within the same space W of states:

$$R := \{(u_0, w_0), (u_0, w_1)\}. \tag{1.5.2}$$

Then $A = R$ and so A would be an atomic action totally performable in the action system

$$(W, R, \{A\}). \tag{1.5.3}$$

Under the interpretation of R provided by (1.5.2), the shot from the bow is performed irrespective of hitting or missing the board.

Generally, the fact that an arbitrary action A in a system $\boldsymbol{M}$ is stochastic manifests itself in two ways: either as the tendency to a course other than the intended course of performing A in u or as the tendency to perform, by the agent in a state u, an action other than A. These propensities of the system are beyond the agent's control and independent of his will. In the former case the planned action A is actually performed but the course of action is different from the intended one (the arrow may miss the target). In the latter case the agent actually performs an action B which is different from the intended action A. In model (1.5.3), with the relation R defined by means of formula (1.5.2), the first tendency is dominant—the archer is always able to perform the action A, but does not always hit the target. In other words, in model (1.5.3), the action A is fully performable at u_0, that is, irrespective of the result of shooting an arrow, the agent indeed performs A.

System (1.5.3) is an exemplification of the first option in the probabilistic action theory: the agent knows what action he is carrying out, but the results (outcomes) of this particular action are stochastic and remain out of his control.

But the rules of the game recognize only those shots that hit the board. If the relation R represents the obedience to the rules of the game, then R admits only one transition—from u_0 to w_1, i.e.,

$$R := \{(u_0, w_1)\}. \tag{1.5.4}$$

According to Definition 1.4.1.(i) and the formulas (1.5.1) and (1.5.2), the shot from the bow is performed if and only if the arrow actually hits the board. Therefore, shooting an arrow is a performable action but not totally performable in the elementary action system

$$(W, R, \{A\}), \tag{1.5.5}$$

with R defined as in (1.5.4).

While the definition of the relation R given in formula (1.5.2) and the definition of performability based on it do not seem to cause interpretative difficulties, formula (1.5.4) and the understanding of performability of A as hitting the target give rise to questions and doubts. According to formulas (1.5.1) and (1.5.4), the action A is performable in u_0 but it is not totally performable in this state. How is it, however, guaranteed that the shot by the archer will hit the board? In other words, is the fact of the performability of A vacuously satisfied, i.e., the state of things consisting in hitting the board is conceivable but, in fact, the state is achieved after no actual shot? Similar questions and doubts also arise in the case of such familiar actions as tossing a coin or throwing a dice.

What does action performability really mean? What is the process that brings the system to a state that is the effect of a given action? Two options arise here. One looks for the possibility of control over the course of events brought about by the agent, no matter who he is. Such a notion of performability is close, though not tantamount, to the notion of controlling the system. The conception of performability, provided by Definition 1.4.1 is devoid of probabilistic or stochastic connotations. Performing an action thus resembles a chess player's deliberate move across the chessboard—it is a conscious choice of one of many possibilities of the direct continuation of the game. If such possibilities do not exist, e.g., when the nearest squares where the white knight could move are taken by other pieces, performing the action—white chessman's move—is impossible in this case. The eventuality of unintended and accidental disturbances of the undertaken actions can be entirely neglected. Nature is not involved in these actions so long as it exerts any uncontrolled influence on the course of planned actions.

The second option—let us call it *probabilistic*—assesses the performability of an action in terms of the *chances* of bringing about the intended effect of the action in the given state or situation. The agent has no possibility of having full control over his actions; the performability of an action is the resultant of the agent's abilities and the influence of external forces on the course of events. It is a kind of game between the agent and nature. One may say that the agent always shares, though unintentionally, his power with nature. In fact, it is a kind of diarchy—the agent has no absolute control over the situation. For example, when shooting an arrow, the archer takes into account gravitational forces, air resistance, and other factors that guide the arrow's flight; however, the role of the archer after the moment the arrow is shot is of no significance.

The probabilistic character of the action A is revealed in action system (1.5.5); this situation is well expressed by the proverb: "Man proposes, God disposes." The performability of action A, i.e., hitting the board with an arrow, should be described not only in terms of the one-zero relation R but in terms of probability that assesses the chances of hitting the target in the state u_0. The performability of the action A should just be identified with this probability, i.e., a number from the closed interval $[0, 1]$. At the same time one should not speak of the performability or unperformability of the action A in the state u_0, but rather of the probability of performability (or unperformability) of this action in this state.

Instead of model (1.5.5), let us consider the following model

$$(W, R, \{B, C\}), \tag{1.5.6}$$

where $B := \{(u_0, w_0)\}$, $C := \{(u_0, w_1)\}$ and $R := \{(u_0, w_0), (u_0, w_1)\}$, i.e., R is defined as in (1.5.4). The actions B and C are totally performable in u_0 in the sense of Definition 1.4.1. B is the action of shooting an arrow *and* missing the target; C is shooting an arrow *and* hitting the board. The problem is, however, that the archer is unable to foresee which action is actually being performed at u_0: is it B or C? The archer committing himself to perform, for example, the action C may not be able to do it; instead, a performance of the action B may take place. Model (1.5.6) then illustrates the second of the mentioned tendencies, i.e., the tendency to perform, at a given state, an action other than the intended one.

If the relation R is equal to $\{(u_0, w_1)\}$, then, in model (1.5.5), the action B is unperformable in the state u_0; this means that missing the target after each shot is excluded. This may seem paradoxical as we are convinced that even an experienced archer can miss the target, i.e., B is *sometimes* performable in this state. The solution is simple—the relation $R = \{(u_0, w_1)\}$ does not mirror the physical aspects of shooting an arrow but only defines a 'legal' qualification of the action B—the shot is successful (i.e., it is qualified as a realizable performance of B) only if the arrow hits the target. This condition is thus vacuously satisfied.

The division line between the probabilistic and the 'non-probabilistic' options in the action theory is not sharp. Speaking of the probabilistic and non-probabilistic concepts of action performability we mean extreme, 'pure' situations which—in practice—are not fully realized. Natural phenomena such as gravitational or electromagnetic forces cannot be completely eliminated as factors affecting the performance of a large number of actions. Nature then is omnipresent. The assumption that the tool the agent uses (be it a bow or an engine) is *completely* subordinated to his will is not strictly speaking even true. However, the assumption can be safely made when the agent is experienced in the use of a tool that works perfectly.

The deterministic description of actions and their performability is an ideal one and as such particularly desirable. It is visible in situations associated with a mass production, for example, where the complete repeatability of the manufacture cycle and the elimination of disturbances whatsoever are factors treated with the greatest consideration by the producer. In these cases the relation between the states in which the actions involved in certain production phases are performed and the results of these actions is treated as a one-to-one map. Nevertheless, none of the action theories should in advance exclude the concept of randomness from its domain. This would lead to conceptual systems which would differ from the ones we have discussed. We mean here *probabilistic* action systems.

From the mathematical point of view probabilistic action systems are built on 'ordinary' elementary action systems $\boldsymbol{M} = (W, R, \mathcal{A})$ by enriching them with families of probabilities providing a quantitative measure of the performability of atomic actions. It is assumed that the system $\boldsymbol{M}$ is nontrivial, which means that R is nonempty, and, for each $A \in \mathcal{A}$, the relation $A \cap R$ is nonempty as well. To facilitate the discussion we assume that R is the union of the set of atomic actions $\mathcal{A}$, i.e., $R = \bigcup_{A \in \mathcal{A}} A$. (This assumption can be dropped but it simplifies the discussion.) It follows that the system $\boldsymbol{M}$ is normal. We additionally assume that the set W of states is countable.

To each atomic action $A \in \boldsymbol{\mathcal{A}}$ and each state $u \in W$ a measure p_u^A with the supporting set $\delta_A(u)$ is assigned, i.e., p_u^A is a function from the power set $\wp(W)$ to the closed interval [0, 1] which satisfies the following conditions:

(1.5.7) p_u^A is σ-additive, i.e., $p_u^A(\bigcup_{n\in\omega} X_n) = \sum_{n\in\omega} p_u^A(X_n)$, for every denumerable family $\{X_n : n \in \omega\}$ of disjoint subsets of W;
(1.5.8) $p_u^A(\{w\}) \neq 0$ for all $w \in \delta_A(u)$ and $p_u^A(\{w\}) = 0$ whenever $w \notin \delta_A(u)$;
(1.5.9) If $\delta_A(u) \neq \emptyset$, then $p_u^A(\delta_A(u)) \leqslant 1$.

If the set $\delta_A(u)$ is empty, we assume that p_u^A is identically equal to 0. It follows from the above clauses that if $\delta_A(u) \neq \emptyset$, then $p_u^A(W) = p_u^A(\delta_A(u))$ and $p_u^A(W)$ is a positive number in the interval [0, 1]. Thus, for each u for which $\delta_A(u)$ is nonempty, the function p_u^A is a nontrivial σ-measure in the sense of measure theory.

For $X \subseteq W$, the number $p_u^A(X)$ is interpreted as the probability that performing the action A in u will lead the system to a state belonging to X. The positive number $p_u^A(W)$ is therefore called the *probability of the performability of the action* A at u. This number may be less than 1.

In a given state u various actions of $\boldsymbol{\mathcal{A}}$ are available and each of them is usually performed with different probabilities.

The above formulation of the probability of the performability of an atomic action A results in distinguishing a family of measures $\{p_u^A : u \in W\}$, defined on the σ-algebra $\wp(W)$ of subsets of W. It is difficult to detect any regularities between the above measures.

Supplementing (1.5.7)–(1.5.9) with the following, rather technical condition:

(1.5.10) For every atomic action A the series $\sum_{u\in W} p_u^A(W)$ is convergent to a positive number

enables, however, a neat formalization of our probabilistic action theory. Condition (1.5.10) is trivially satisfied if the space W is finite. The sum $\sum_{u\in W} p_u^A(W)$, which is equal to $\sum_{u\in W} p_u^A(\delta_A(u))$, does not bear probabilistic connotations—it is simply the sum of all probabilities of performing A in particular states of W.

Conditions (1.5.7)–(1.5.10) are jointly equivalent to the notion of a *bi-distribution*. The latter is understood to be any function $m^A(\cdot|\cdot)$ from $\wp(W)\times\wp(W)$ into a bounded interval $[0, a]$ of nonnegative real numbers satisfying the following conditions:

(1.5.7)* $0 \leqslant m^A(\{u\}|W) \leqslant 1$ for all $u \in W$ and the series $\sum_{u\in W} m^A(\{u\}|W)$ is convergent,
(1.5.8)* m^A is a σ-additive in both arguments, i.e., for any two families $\{X_n : n \in \omega\}$, $\{Y_n : n \in \omega\}$ of disjoint subsets of W, and for all $X, Y \subseteq W$,

$$m^A(\bigcup_{n\in\omega} X_n|Y) = \sum_{n\in\omega} m^A(X_n|Y)$$

and

$$m^A(X|\bigcup_{n\in\omega} Y_n) = \sum_{n\in\omega} m^A(X|Y_n);$$

(1.5.9)* for any $u, w \in W$,

$$m^A(\{u\}|\{w\}) \neq 0 \quad \text{if} \quad w \in \delta_A(u)$$

and

$$m^A(\{u\}|\{w\}) = 0 \quad \text{otherwise.}$$

If $m^A(\cdot\,|\,\cdot)$ is a bi-distribution on $\wp(W) \times \wp(W)$, then the function

$$p_u^A(\cdot) := m^A(\{u\}|\,\cdot),$$

for any fixed state $u \in W$, is a measure on $\wp(W)$ satisfying conditions (1.5.7)–(1.5.9) above. Moreover, the family $\{p_u^A : u \in W\}$ satisfies condition (1.5.10), because $\sum_{u\in W} p_u^A(W) = \sum_{u\in W} m^A(\{u\}|\,\cdot\,)$ and the last series is convergent by (1.5.7)*.

Conversely, if a family $\{p_u^A : u \in W\}$ of σ-measures on $\wp(W)$ is singled out so that (1.5.7)–(1.5.10) hold, a bi-distribution m^A on $\wp(W) \times \wp(W)$ is defined by putting:

$$m^A(X|Y) := \sum_{u\in X} p_u^A(Y). \tag{1.5.11}$$

(If X is empty, it is assumed that $m^A(X|Y) := 0$.)

The easy proof that m^A is well-defined and that it indeed satisfies conditions (1.5.7)*–(1.5.9)* is omitted. (The positive-term series in the formula (1.5.11) is convergent since it is majorized by the convergent series $\sum_{u\in X} p_u^A(W)$. The latter series is convergent by (1.5.10). We thus see that the notion of a bi-distribution is co-extensive with the notion of a family of measures $\{p_u^A : u \in W\}$ satisfying postulates (1.5.7)–(1.5.10).

Every bi-distribution $m(\cdot\,|\,\cdot)$ on $\wp(W) \times \wp(W)$ with W countable is uniquely determined by its values on the pairs $\{u\} \times \{w\}$, or, which amounts to the same—on the ordered pairs (u, w), where u and w range over W.

If $m^A(\cdot\,|\,\cdot)$ is a bi-distribution and $u \in W$, $Y \subseteq W$, then the number $m^A(\{u\}|Y)$ has the following interpretation—it is the *probability that performing the action A will move the system from u to an Y-state*. $m^A(\{u\}|Y)$ should not be confused with a conditional probability. It seems that there does not exist a strict relationship between these two notions.

The notion of a bi-distribution enables us to generalize the above theory to the case of uncountable sets of states W. It requires, however, prior modification of the condition (1.5.7)* and also the dropping of the assumption that the function m^A is defined on the entire product $\wp(W) \times \wp(W)$. It should be assumed that the domain of the function m^A is the Cartesian product $\boldsymbol{F_0} \times \boldsymbol{F_0}$, where $\boldsymbol{F_0}$ is a suitably selected σ-field of subsets of W. We shall not, however, discuss this issue here.

Let $\boldsymbol{F}$ denote the σ-field of subsets of $W \times W$ generated by the sets of the form $X \times Y$, where $X \subseteq W$, $Y \subseteq W$. (W is assumed to be a countable set of states.) It follows from a well-known result of measure theory (see Halmos 1950) that if m_1 and m_2 are two (σ-additive) finite measures on $\wp(W)$, then there exists a unique finite measure m defined on the σ-field $\boldsymbol{F}$ such that $m(X \times Y) = m_1(X) \cdot m_2(Y)$, for all $X, Y \subseteq W$.

If $m(\cdot\,|\,\cdot)$ is a bi-distribution on $\wp(W) \times \wp(W)$, then the maps $\mu_1(\cdot)$ and $\mu_2(\cdot)$ defined by the formulas

$$\mu_1(\cdot) := m(\cdot\,|W) \quad \text{and} \quad \mu_2(\cdot) := m(W|\,\cdot),$$

are finite measures on the σ-field $\wp(W)$. According to the above result, there exists a σ-additive measure μ on the σ-field $\boldsymbol{F}$ such that

$$\mu(X \times Y) = \mu_1(X) \cdot \mu_2(Y), \tag{1.5.12}$$

for all $X, Y \subseteq W$.

If the bi-distribution $m(\cdot\,|\,\cdot)$ satisfies the equation

$$m(X|Y) = m(X|W) \cdot m(W|Y), \tag{1.5.13}$$

for all $X, Y \subseteq W$, then trivially (1.5.12) and (1.5.13) imply that

$$\mu(X \times Y) = m(X|Y), \tag{1.5.14}$$

for all $X, Y \subseteq W$. In other words, if $m(\cdot\,|\,\cdot)$ satisfies (1.5.13), then it uniquely extends to an 'ordinary' measure on the σ-field $\boldsymbol{F}$. This observation would reduce the theory of bi-distributions to classical measure theory.

It is an open question if, given a bi-distribution m there always exists a measure μ on $\boldsymbol{F}$ so that (1.5.14) holds, for all $X, Y \subseteq W$. We also do not know whether (1.5.13) is well substantiated. In case W is countable, (1.5.13) is equivalent to another identity. Writing $m(u|w)$ instead of $m(\{u\}|\{w\})$, it is not difficult to check that (1.5.13) holds if and only if

$$m(u|w) = \sum_{v,v' \in W} m(u|v') \cdot m(v|w),$$

for all $u, w \in W$.

We thus arrive at the following definition:

Definition 1.5.1 Let $\boldsymbol{M} = (W, R, \boldsymbol{\mathcal{A}})$ be a nontrivial action system in which R is the union of the actions of $\boldsymbol{\mathcal{A}}$. A *probabilistic elementary action system* over $\boldsymbol{M}$ is a pair

$$\big(W, \{m^A(\cdot\,|\,\cdot) : A \in \boldsymbol{\mathcal{A}}\}\big) \tag{1.5.15}$$

where $\{m^A(\cdot|\cdot) : A \in \mathcal{A}\}$ is a family of bi-distributions on $\wp(W) \times \wp(W)$ assigned to the atomic actions of $\mathcal{A}$. □

Note In the above definition the "action" component $\mathcal{A}$ is in fact redundant. The symbol "$\mathcal{A}$" plays merely the role of an index set. This is due to the fact that, given the family $\{m^A(\cdot \mid \cdot) : A \in \mathcal{A}\}$, one can reconstruct each action A of $\mathcal{A}$ by means of the formula: $u\ A\ w$ if and only if $m^A(u|w) > 0$—see (1.5.9)*. (We write here $m^A(u|w)$ instead $m^A(\{u\}|\{w\})$.) Consequently, $u\ R\ w$ if and only if $m^A(u|w) > 0$ for some $A \in \mathcal{A}$.

The following definition is formulated entirely in probabilistic terms (it is formulated in terms of probabilistic transitions p_u^A, cf. conditions (1.5.7)–(1.5.9)) and it abstracts from the notion of an atomic action viewed as a binary relation on a set of states.

Definition 1.5.2 A countable *probabilistic atomic action system* is a triple

$$\boldsymbol{M}^{pr} = \big(W, \mathcal{A}, \{p_u^A : u \in W, A \in \mathcal{A}\}\big),$$

where W is a countable set (the set of *states* of the system), $\mathcal{A}$ is a set (the set of *atomic actions* of the system), and for any $u \in W$, $A \in \mathcal{A}$, p_u^A is a finite σ*-measure* on the power set $\wp(W)$ such that $p_u^A(W) \leqslant 1$. □

The situation where p_u^A is the zero function for some $u \in W$, $A \in \mathcal{A}$ is allowed. It may also happen that p_u^A is a normalized measure, i.e.,$p_u^A(W) = 1$. For any $X \subseteq W$, the number $p_u^A(X)$ is the *probability of the performability of the action A in the state u resulting in an X -state*. $p_u^A(W)$ is interpreted as the probability of the *performability* of A in u.

If one additionally assumes that for any $A \in \mathcal{A}$, the series $\sum_{u \in W} p_u^A(W)$ is convergent to a positive number, then the above definition can be also expressed in terms of bi-distributions $m^A(\cdot \mid \cdot)$.

Once a probabilistic system $\boldsymbol{M}^{pr}$ is defined, the crucial problem probabilistic action theory faces is that of assigning probabilities to sequences $A_0, A_1, \ldots, A_n$ of atomic actions of the system, and more generally to compound actions, relative to a given state u. (Compound actions are investigated in Sect. 1.7.) □

As mentioned earlier, the bi-distribution $m^A(\cdot \mid \cdot)$ need not be a normalized measure, i.e., the number $m^A(u|W)$ which measures the probability of the performability of the action A in the system at u need not be equal to 1. This is caused by the fact that the agent, in a given state u, usually has a few stochastic actions to perform at choice, say $A_1, \ldots, A_n$. The actions are not under his full control. Assuming that with probability 1 the agent always performs one of the actions $A_1, \ldots, A_n$ in the state u, it is not possible to guarantee that the agent will perform with certainty in u the intended and initiated action, say A_1. (The archer with probability 1 hits or misses the target in u_0; at the moment of shooting it is not, however, possible to decide which of the possible two actions is actually performed.) We assume that the bi-distribution $m^A(\cdot \mid \cdot)$, assigned to the action A, reflects the above phenomenon. The number $m^A(u|Y)$ expresses the probability that performing A in u leads the system to a Y-state. The probability takes into account two facts: first, that in u the agent of A may actually perform, irrespective of his will and his declared intentions, an action other than A, and, second, the fact that though the system is in u and the action A is actually being performed, the system will not always be transferred to a Y-state. The latter fact is also independent of his will. (We have to assume here that Y is a proper subset of W.) The bi-distribution $m^A(\cdot \mid \cdot)$ is then the measure of the resultant of these two factors. In the formalism accepted here the above two factors are not separated, i.e., we do not single out two distinct probabilistic distributions which separately measure the probability of the performability of A and the probability that performing A leads from u to Y-states (provided that A is actually performed).

The following simple illustration sheds some light on that issue. An agent has an unloaded dice at his disposal. Moreover, there are two urns containing balls. Urn I contains five balls (3 white and 2 red), and urn II also contains five balls (1 white and 4 red). If the outcome of throwing the dice is 1, 2, 3 or 4, the agent selects a ball from Urn I; if the outcome is 5 or 6, the agent picks out a ball from Urn II. The above situation can be expressed in terms of elementary action systems. The space of states W contains four elements. u_0 is the initial state, before the dice is tossed. u_1 and u_2 are two possible states just after rolling the dice; u_1 represents the state in which a $1, 2, 3$ or 4 is rolled, and u_2 marks the state showing 5 or 6. In turn, w and r mark two possible states after picking out a ball from one of the urns, that is, picking out a white ball or the red ball, respectively.

There are two actions being performed; A—tossing a dice, and B—selecting a ball from one of the two urns. A contains two pairs, $A = \{(u_0, u_1), (u_0, u_2)\}$ and $B = \{(u_1, w), (u_2, w), (u_1, r), (u_2, r)\}$. Intuitively, A is performable with probability 1 at u_0 because tossing the coin is performable and no other option is available at u_0. Analogously, B is performable with probability 1 at each of the states u_1, u_2, because drawing a ball from whichever urn is performable and no other possibility occurs. The direct transition relation merely ensures that the actions A and B are to be successively performed in the right order. Accordingly, R consists of pairs

$$(u_0, u_1), (u_0, u_2), (u_1, w), (u_2, w), (u_1, r), (u_2, r).$$

The functioning of this normal action system $(W, R, \{A, B\})$ is represented by the following picture.

The system satisfies the condition that $R = A \cup B$. Moreover, the actions A and B are disjoint. The probability 4/6 is assigned to the transition (u_0, u_1) and the weight of the transition (u_0, u_2) is 2/6. In other words, the probability $m^A(u_0|u_1)$ that performing the action A in u_0 will lead the system to the state u_1 is 4/6 and the probability $m^A(u_0|u_2)$ that performing the action A in u_0 will lead the system to the state u_2 is 2/6. (We write here $m^A(u|w)$ instead $m^A(\{u\}|\{w\})$.)

One assigns the probabilities 3/5, 1/5, 2/5, and 4/5 to the transitions (u_1, w), (u_2, w), (u_1, r), and (u_2, r), respectively. This means that the probability $m^B(u_1|w)$ that performing the action B in u_1 will lead the system to the state w is 3/5, the probability $m^B(u_2|w)$ that performing the action B in u_2 will lead the system to the state w is 1/5, and so on.

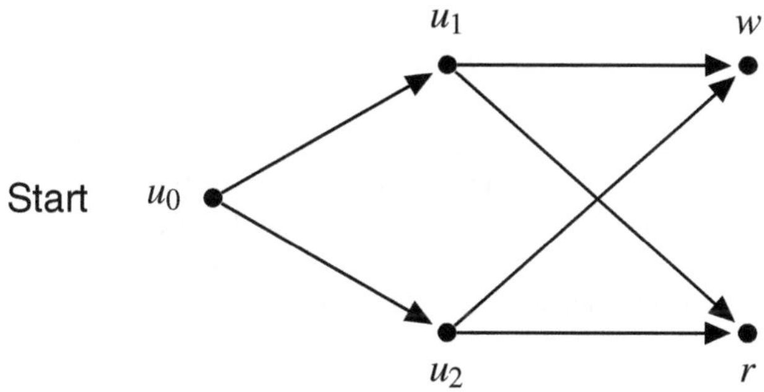

Fig. 1.7

As the above action system should meet conditions (1.5.7)*, (1.5.8)*, and (1.5.9)*, we assume that $m^A(u|w) = 0$ whenever $w \notin \delta_A(u)$, and $m^B(u|w) = 0$ whenever $w \notin \delta_B(u)$, for any states u, w.

The probability of performability of the action A in u_0 equals 1. This is due to the fact that the sum of the probabilities assigned to the transitions (u_0, u_1) and (u_0, u_2) gives 1, $m^A(u_0|u_1) + m^A(u_0|u_1) = 4/6 + 2/6 = 1$. Similarly, the probability of performability of B in each of the states u_1 and u_2 equals 1. Indeed, $m^B(u_1|w) + m^B(u_1|r) = 3/5 + 2/5 = 1$ and $m^B(u_2|w) + m^B(u_2|r) = 1/5 + 4/5 = 1$. We may therefore say that the agent has the power to perform A in u_0 (but the results of performing A are random and out of his/her control, i.e., he/she does not know at u_0 the urn that will be drawn). Likewise, the agent is able with certainty to perform B in each of the states u_1 and u_2. (But again the possible outcomes of B are random in the sense that the agent does not know whether the ball he/she will draw is red or white. But the color of the drawn balls is irrelevant from the perspective of performability of B!) The above system satisfies the condition that for any state u, $m^A(u|w)$ equals 0 or 1 and similarly, $m^B(u|w)$ equals 0 or 1.

Despite the probabilistic wording, the above action system $(W, R, \{A, B\})$ has the property that the probability of the performability of A in u_0 is 1 and the performability of B in either state u_1, u_2 is also 1.

In a more accurate model, instead of one action B there are two actions B_r and B_w of picking out a red ball and a white ball, respectively. Thus $B_r = \{(u_1, r), (u_2, r)\}$ and $B_w = \{(u_1, w), (u_2, w)\}$. In the resulting normal model $(W, R, \{A, B_r, B_w\})$ the probabilities assigned to performability of B_r in the states u_1 and u_2 are equal to 2/5 and 4/5, respectively, and the probabilities assigned to performability of B_w in the states u_1 and u_2 are equal to 3/5 and 1/5, respectively. Consequently, the agent is no longer able to perform with certainty the action B_r in each of the states u_1 and u_2 (i.e., in each state he/she is unable to foresee the color of the ball that will be drawn). A similar remark applies to the action B_w. In this model the actions B_r, B_w are genuinely probabilistic. We may also split the action A into two probabilistic actions: A_{I} of rolling a 1, 2, 3 or 4 and the action of rolling a 5 or 6. Thus $A_{\mathrm{I}} = \{(u_0, u_1)\}$ and $A_{\mathrm{II}} = \{(u_0, u_2)\}$.

As mentioned earlier, from the perspective of probabilistic action theory the crucial problem concerns the assignment of probabilities to compound actions; in particular to strings of atomic actions once all probabilities pertinent to atomic actions are known. Let

$$(W, R, \{A_{\mathrm{I}}, A_{\mathrm{II}}, B_r, B_w\})$$

be the system defined as above. What is the probability of the performability of the string of actions A_{I}, B_r in u_0? (The string A_{I}, B_r is identified with the compound action $\{A_{\mathrm{I}} B_r\}$ containing only one word of length 2—see Sect. 1.7.) There is only one string of performances of $\{A_{\mathrm{I}} B_r\}$, viz. $u_0\ A_{\mathrm{I}}\ u_1\ B_r\ r$. We may treat the actions A_{I} and B_r as stochastically independent, which means that the probability of performing the string $A_{\mathrm{I}} B_r$ in u_0 is equal to

$$m^{A_{\mathrm{I}}}(u_0|u_1) \cdot m^{B_r}(u_1|r) = 4/6 \cdot 2/5 = 8/30.$$

Following this pattern, we determine the probabilities of the performability of other strings $A_{\mathrm{II}} B_r$, A_{I}, B_w, and A_{II}, B_w in u_0:

$$\begin{aligned} m^{A_{\mathrm{II}}}(u_0|u_2) \cdot m^{B_r}(u_2|r) &= 2/6 \cdot 4/5 = 8/30, \\ m^{A_{\mathrm{I}}}(u_0|u_1) \cdot m^{B_w}(u_1|w) &= 4/6 \cdot 3/5 = 12/30, \\ m^{A_{\mathrm{II}}}(u_0|u_2) \cdot m^{B_w}(u_2|w) &= 2/6 \cdot 1/5 = 2/30. \end{aligned}$$

They all add up to 1.

One may of course define more complicated models built on directed graphs with urns being edges of a digraph. The urns may contain balls of different sorts.

The above simple examples provide mere samples of the problems probabilistic action theory poses. They are not tackled here from the mathematical perspective.

The above definition of a probabilistic system does not fully resolve the issue of the relationship holding between the bi-distributions m^A and m^B for different actions $A, B \in \mathcal{A}$. In some cases they may be treated as stochastically independent actions. Another option that comes to light is to assume that for all $u, w \in W$,

$$m^A(u|w) = m^B(u|w) \quad \text{whenever} \quad w \in \delta_A(u) \cap \delta_B(u). \tag{1.5.16}$$

It follows from (1.5.16) that for all $u \in W$ and $Y \subseteq W$, $m^A(u|Y) \leqslant m^B(u|Y)$ whenever the action A is a subset of B. In particular, $m^A(u|W) \leqslant m^B(u|W)$ whenever $A \subseteq B$. This agrees with the intuition that the larger an atomic action, the higher the probability of its performability.

Let (W, R) be a countable discrete system. Yet another probabilistic approach to action assigns probabilities to direct transitions between states of (W, R). Accordingly, to each ordered pair $(u, w) \in W \times W$ a number $p^R(u, w) \in [0, 1]$ is assigned so that the following conditions are met:

$$p^R(u, w) = 0 \quad \text{if} \quad (u, w) \notin R \qquad (1.5.17)$$
$$p^R(u, w) \neq 0 \quad \text{otherwise,}$$

and, for every $u \in W$ such that $\delta_R(u) \neq \emptyset$,

$$\sum \{p^R(u, w) : w \in \delta_R(u)\} = 1. \qquad (1.5.18)$$

The number $p^R(u, w)$ is interpreted as the probability of the direct transition of the discrete system (W, R) from the state u to the state w.

Condition (1.5.17) says that the probability of the direct transition from u to w equals 0 if (u, w) is not in R. In the remaining cases the probability is nonzero and the sum of such probabilities gives 1.

We recall that states u of W such that $\delta_R(u) = \emptyset$ are called *dead* (or *terminal*) *states*; the remaining states are called *nonterminal*. The direct transition from a terminal state to any other state is excluded. If u is terminal, then $p^R(u, w) = 0$, for all states w. Condition (1.5.18) says that the sum of probabilities of direct transitions from any nonterminal state is 1. In particular, if u is a reflexive state, i.e., $\delta_R(u) = \{u\}$, then $p^R(u, u) = 1$, which means that the system stays in u with probability 1 (and with zero probability of leaving it).

In turn, to each atomic action $A \in \boldsymbol{\mathcal{A}}$ on (W, R) and to each pair $(u, w) \in W \times W$ a number $c^A(u, w) \in [0, 1]$ is assigned so that $c^A(u, w) = 0$ if and only if $w \notin f_A(u)$. The number $c^A(u, w) \cdot p^R(u, w)$ is interpreted as the probability that performing action A in the state u will move the system to the state w.

If $Y \subseteq W$, then the number

$$q^A(u, Y) := \sum \{c^A(u, w) \cdot p^R(u, w) : w \in Y\}, \qquad (1.5.19)$$

which is assumed to exist, is the *probability that performing the action A in the state u will lead the system to a state which belongs to Y*. The atomic action A is thus treated here as a perturbation of the discrete system (W, R). The coefficients $c^A(u, w)$ in the formula (1.5.19) provide the numerical measure of this perturbation in any state u. The number

$$p^A(u) := q^A\big(u, \delta_A(u)\big) \quad \big(= q^A(u, W)\big)$$

is the probability of the performability of the action A in the state u.

An interesting class of probabilistic systems is formed by normal systems satisfying the condition that for every state u the family $\{\delta_A(u) : A \in \boldsymbol{\mathcal{A}}\}$ forms a partition of $\delta_R(u)$ and $\sum_{A \in \boldsymbol{\mathcal{A}}} p^A(u) = 1$. (Some sets in the family $\{\delta_A(u) : A \in \boldsymbol{\mathcal{A}}\}$ may be empty.) Thus the sum of all probabilities of the performability of the actions of $\boldsymbol{\mathcal{A}}$ in each nonterminal state u is 1. The above urn example satisfies this condition.

It may be argued that randomness, inherent in many action systems, results from the negligence of factors that collectively form the situational envelope in which actions are immersed. To better understand the above issue, consider the well-known example of tossing a coin. (The issues concerning the role the notion of a situation plays in the theory will be elucidated in Chap. 2; here we will restrict ourselves to introductory remarks.)

The coin is fixed. We distinguish two actions A and B; both actions are treated as atomic:

A :	tossing the coin onto the table
B :	picking the coin up from the table

To facilitate the discussion we assume (for simplicity) that the space of states W consists of three elements, $W := \{u_0, Tails, Heads\}$. u_0 is the state in which the coin is tossed onto the table. The state u_0 is well-defined—the coin is tossed from a fixed point over the table in a definite direction with a constant force and at the same angle. (One may think in terms of a sort of cannon attached to the table shooting the coin with a constant force. The assumption that u_0 is well-defined is, of course, an idealization—there will be small fluctuations in tossing which will influence the score. This fact is crucial for the 'stochasticity' of the system.) The state $Tails$ is entirely determined by the position of the coin on the table, with the tails on the top. Similarly, the state $Heads$ is defined. The action A consists of two pairs:

$$A := \{(u_0, Heads), (u_0, Tails)\}$$

while

$$B := \{(Heads, u_0), (Tails, u_0)\}.$$

The relation R is the join of the two actions, $R := A \cup B$; thus the actions A and B are totally performable in the system $\boldsymbol{M} := (W, R, \{A, B\})$. Undoubtedly, the action B is fully controlled by the agent—the person tossing the coin. The performability of the action B does not depend on 'hidden' factors guiding the motion of the coin. (Distinguishing the action B in the model may seem irrelevant and a bit too purist. It is being done to clarify the reasoning and keep it in the framework of elementary action systems.)

The actions A and B can be performed many times. To represent formally this fact we introduce a new category of entities—the category of possible situations. A *possible situation* is any pair

$$s = (w, n), \tag{1.5.20}$$

where $w \in W$ and n is a natural number, $n \geqslant 0$. The pair (1.5.20) is interpreted as follows: "After tossing the coin n times, the state of the system $\boldsymbol{M}$ is equal to w." The number n is called the *label* of the situation s, while w is called the *state corresponding* to the situation s. Both actions A and B change not only the state of the system but the situation. To express the latter fact we shall consider the following triples:

$$(n, A, n+1) \tag{1.5.21}$$
$$(n, B, n), \tag{1.5.22}$$

where $n \in \omega$. The triples (1.5.21) and (1.5.22) are called the *labeled actions* A and B, respectively. The triple (1.5.21) is read: 'The action A performed in situation s with the label n moves the system so that the situation just after performing A has the label $n+1$.' To put it briefly, one can say that if the coin has been tossed n times, then the following toss increases the number of tosses to $n+1$. The labeled action (1.5.22) does not change the label (i.e., the number of times the coin has been tossed is not increased), but it changes the state of the system. More precisely, the labeled action (1.5.21) transforms a situation s_1 into a situation s_2 if and only if $s_1 = (w, n)$, $s_2 = (w, n+1)$ and $u\ A,\ R\ w$ (i.e., $u = u_0$ and $w \in \{Tails, Heads\}$). Analogously, the labeled action (1.5.22) transforms a situation s_1 into s_2 if and only if $s_1 = (u, n)$, $s_2 = (w, n)$ and $u\ B,\ R\ w$ (i.e., $u \in \{Tails, Heads\}$ and $w = u_0$).

We write $s_1\ Tr\ s_2$ to denote that one of the actions (1.5.21) or (1.5.22) transforms the situation s_1 into s_2. $(u_0, 0)$ is called the *initial* situation. A finite *run* of situations is any finite sequence

$$(s_0, s_1, \ldots, s_n) \tag{1.5.23}$$

of possible situations such that s_0 is the initial situation and $s_i\ Tr\ s_{i+1}$ for all i, $0 \leqslant i \leqslant n-1$. (Thus, if i is even, then s_i is transformed into s_{i+1} by means of the action $(i, A, i+1)$, i.e., by tossing the coin. If i is odd, then s_i is transformed into s_{i+1} by means of (1.5.22), i.e., by picking up the coin from the table.) The number n is called the *length* of the run (1.5.23).

There are no limitations on the length of runs of situations (1.5.23). But, of course, the agent of the above actions can stop the run (1.5.23) in any situation s_n with odd n. In other words, in each of the states $Tails$, $Heads$, the agent may switch to another action system, with compound actions built in as, say, 'having lunch' or 'going home'. In other words, he/she is able to perform an additional action, viz., Abort, in some situations.

Can the notion of a finite run of situations serve to define the probabilities $p_u^A(Tails)$ and $p_u^A(Heads)$ for $u := u_0$, i.e., the probabilities that performing the action A in the state u_0 will lead the system to the state $Tails$ ($Heads$, respectively)? An answer to this question would require adopting the additional assumption that all the runs (1.5.23) of length n are 'equally probable' provided that n is sufficiently large. Any attempts to explicate this assumption would result in *circulus in definiendo*.

The action A—tossing the coin—is almost tautologically performable in the state u_0. One may claim that the above notion of a possible situation (in a much abridged form) and of a finite run enable us to better understand what the 'stochasticity' of the action A means. From the purely intuitive point of view the agent of a 'stochastic' action has no influence on the run of situations, i.e., for no run $(s_0, \dots, s_n)$ the agent has the power and means of choosing a situation s_{n+1} that follows the actual situation s_n. But even in this case any attempt to give the above intuition a clear mathematical shape may lead us astray.

A digression. Ringing round from your mobile phone is another simple example of a situational action system with uncertain scores of actions. Suppose you want to call your friend. You know their number. You and your phone together form an action system. Let u be the initial state in which your phone is not activated. Let A be the action of dialing the number of your friend. Two more states are distinguished: w_1—after sending the signal you get an answer (your friend picks up the phone), w_2—after sending the signal you get no answer (the friend does not pick up the phone). w_2 takes place when your friend is far from their phone or the phone is deactivated or your friend does not want to talk to you. We assume that your friend's telephone works. The action A is totally performable in u (your phone is in good shape), but the result of performing A is uncertain. Only after a couple of seconds or more you find out what is going on—whether you are in the state w_1 or in w_2. (Usually a longer period elapses before finding that you are in the state w_2 rather than in w_1.) In the state w_1 the verbal (and compound) action of talking with your friend is undertaken, possibly followed by other actions. But in w_2 the action of call cancellation is performed, which results in the state u. After some time you may retry calling your friend, iterating the action A (maybe repeatedly and maybe successfully). The adequate description of this action system requires introducing the time parameter as a situational component.

The conclusion we want to draw is the following: the formalism of the action theory adopted in this book, even supplemented with the notion of a situational envelope of an action system, does not allow a clear-cut distinction between probabilistic and non-probabilistic systems. Introducing probabilities of transitions between states to the formalism in the way presented in this chapter does not change the matter—it is not self-evident *how* to attach, in definite cases, probabilities $p_u^A(Y)$ to particular actions A and pairs (u, Y).

There is no doubt that speaking of the probability $p_u^A(Y)$ needs satisfying the postulate of the repeatability of the action A. It means that the situational envelope of the action system has to guarantee, for every pair (u, w) such that $u\ A, R\ w$ holds, the possibility of the reverse transition of the system—from the state w to the state u. This latter transition should be accomplished by means of a non-stochastic action (like the action B in the example of tossing a coin). The opposition 'stochastic—non-stochastic' is understood here intuitively, otherwise we shall end up in a vicious circle.

We believe that for practical reasons, for any action A and any state u, the probability $p_u^A(\cdot)$ has to satisfy a condition which is difficult to formulate precisely and which is associated with the following intuition that concerns the number $p_u^A(Y)$. To put it simply, let us assume that the action A is performable in the state u with probability 1, i.e., the agent of A declaring and being about to perform the action A in a state u is always able to do it. Tossing a coin, shooting an arrow from a bow, casting a dice, etc., are examples of such actions. Suppose a long, 'random' series of n repetitions of A

in the state u has been performed (the series can simply be identified with a suitable run of situations defined as in the case of tossing the coin). If, as a result, performing the action A has led the system m times to a Y-state and $n - m$ times to a state lying outside Y, we can expect that the fraction m/n will be close $p_u^A(Y)$ as n grows large. In other words, with an unlimited number of repetitions of the action A in the state u, the frequency of transitions of the system from the state u to a Y-state should approach the probability $p_u^A(Y)$ that performing the action A in u will move the system to a state belonging to Y. ($p_u^A(Y)$ is viewed here as in the first formal description of probability, i.e., the function $p_u^A(\cdot)$ satisfies the formulas (1.5.7)–(1.5.10).) Attempts at explicating the above intuition encounter, however, the well-known logical difficulties concerning randomness. Axiomatic probability theory avoids random sequences, although various approaches to defining random sequences have been proposed (the frequency/measure approach by von Mises–Church, Kolmogorov complexity or the predictability approach). Probabilistic automata, invented by Rabin, employ coin tosses to decide which state transition to take. For example, the command 'Perform at random one of the actions A or B', known from dynamic logic, can be reconstructed within the above framework as performing the compound operation consisting of two sequences of actions T, A and T, B where T is the action of flipping a coin. If after tossing, the state $Heads$ is obtained, the action A is performed; otherwise, with $Tails$ as a result, the action B is performed. (To simplify matters, we assume that A and B are atomic actions.) Formally, the construction of the relevant action system requires enriching the initial system $\boldsymbol{M} = (W, R, \boldsymbol{\mathcal{A}})$ with three new states $u_0, Heads, Tails$, and then extending the relation R of $\boldsymbol{M}$ to a relation R' on $W' := W \cup \{u_0, Heads, Tails\}$, where $R' := R \cup \{(u_0, Heads), (u_0, Tails)\} \cup \{(Heads, w) : w \in Dom(A)\} \cup \{(Tails, w) : w \in Dom(B)\}$. As before T is identified with the set $\{(u_0, Heads), (u_0, Tails)\}$. T is thus totally performable in the extended system but the result of its performing is uncertain. (Strictly speaking, one should add two more atomic actions: $C := \{(Heads, w) : w \in Dom(A)\}$ and $D := \{(Tails, w) : w \in Dom(B)\}$. If $Heads$ is tossed, the agent initiates in the action A in the state $Heads$ by selecting a state $w \in Dom(A)$ (but not at random, because A is not stochastic). Similarly, if $Tails$ is tossed, the agent initiates the action B in the state $Tails$ by selecting a state $w \in Dom(B)$.)

1.6 Languages, Automata, and Elementary Action Systems

A 'symbol' is treated here as an abstract entity which is not defined formally. This notion has a similar status as 'point' or 'line' in geometry. Letters and digits are examples of frequently used symbols.

A *string* (or *word*) is a finite sequence (with repetitions) of juxtaposed symbols. Words are written without parentheses and commas between particular symbols, e.g., $abca$ instead of (a, b, c, a). The length of a string x, denoted by $|x|$, is the number of all occurrences of symbols in the word. For example, aab has length 3 and $abcb$

has length 4. The empty word, denoted by ε, is the string consisting of zero symbols. Its length $|\varepsilon|$ is equal to 0.

An *alphabet* is a nonempty finite set Σ of symbols. Σ^* denotes the set of all words (with the empty word included) over the alphabet Σ. Thus, a word x is in Σ^* if and only if all the symbols occurring in x belong to Σ. Σ^+ denotes the set of all nonempty words over Σ; thus, $\Sigma^+ = \Sigma^* \setminus \{\varepsilon\}$.

Let $x = a_1 a_2 \dots a_m$ and $y = b_1 b_2 \dots b_n$ be two words over Σ. The *composition* (or *concatenation*) of the words x and y is the word, denoted by xy, which is obtained by writing, first, the successive symbols of the word x, and then the successive symbols of the word y:

$$xy := a_1 \dots a_m b_1 \dots b_n.$$

For example, if $x := abca$ and $y := acb$ then $xy = abcaacb$.

The composition of any word with the empty word always yields the same word, i.e., for any word x, $x\varepsilon = \varepsilon x = x$.

(The set Σ^* equipped with the binary operation of concatenation and the empty word as an algebraic constant has thus the structure of a semigroup; ε is the unit of the semigroup. In fact Σ^* is a free semigroup; the symbols of Σ are its free generators.)

For any word x the nth *power* of x for $n = 0, 1, 2, \dots$ is defined inductively as follows:

$$x^0 := \varepsilon$$
$$x^{n+1} := xx^n \quad \text{for} \quad n = 0, 1, 2, \dots$$

Thus, $x^n = x \dots x$ with x occurring n times.

Induction on the length of a word enables to define the one-argument operation of the mirror reflection (or the *inverse*) of a word:

$$\varepsilon^{-1} = \varepsilon$$
$$(ax)^{-1} := x^{-1}a \quad \text{for} \quad a \in \Sigma \quad \text{and} \quad x \in \Sigma^*.$$

It follows from the definition that if $x = a_1 \dots a_m$, then $x^{-1} = a_m \dots a_1$.

Let $x, y \in \Sigma^*$. We say that x is a *subword* of the word y, symbolically: $x \sqsubseteq y$, if there exist words $z, w \in \Sigma^*$ such that $y = zxw$. If $zw \neq \varepsilon$, then x is a *proper* subword of y; if $z = \varepsilon$, then x is an *initial* subword of y; and if $w = \varepsilon$, then we say that x is a *terminal* subword of y.

A formal *language* L (over an alphabet Σ) is a set of strings of symbols from the alphabet.

Since languages are sets, the well-known set-theoretic operations such as union, intersection, difference, etc. can be applied to them. But also some special operations, derived from the operations on words, are performable on languages. For example, the *composition* of two languages L_1 and L_2 is defined to be the language

$$L_1 L_2 := \{xy : x \in L_1 \text{ and } y \in L_2\}.$$

The *power* of a language L is defined inductively, in a similar manner as for words:

$$L^0 := \{\varepsilon\}$$
$$L^{n+1} := LL^n \text{ for } n = 0, 1, 2, \ldots$$

The *closure* of a language L, denoted by L^*, is defined as the union of the consecutive powers of L:

$$L^* := \bigcup \{L^n : n \in \mathbb{N}\}.$$

The *positive closure* of L, denoted by L^+, is the union of all positive powers of L:

$$L^+ := \bigcup \{L^n : n = 1, 2, \ldots\}.$$

The only difference between L^* and L^+ consists in that L^* always contains an empty word while L^+ contains ε if and only if L does.

The notations L^* and L^+ agree with the definitions of the sets Σ^* and Σ^+ introduced earlier as the totalities of all strings (of all nonempty strings, respectively) over Σ. As a matter of fact, the set of all words over Σ is equal to the one-element set $\{\varepsilon\}$ plus the set of all words of length 1 (i.e., the set Σ) plus the set of all words of length 2 (the set Σ^2), etc.

The *inverse* of a language is the language

$$L^{-1} := \{x^{-1} : x \in L\}.$$

Regular languages are the best-examined class of formal languages. The regularity of this class of languages manifests itself in the fact that the class is recursively defined by means of simple set-theoretic and algebraic conditions: the family REG(Σ) of *regular languages* over Σ is defined as the least class $\mathfrak{R}$ of Σ-languages which satisfies the following properties:

(i) $\emptyset$, the empty language, belongs to $\mathfrak{R}$;
(ii) $\{\varepsilon\}$, the language consisting of the empty word, belongs to $\mathfrak{R}$;
(iii) for every letter $a \in \Sigma$, $\{a\}$ belongs to $\mathfrak{R}$;
(iv) if $L_1 \in \mathfrak{R}$ and $L_2 \in \mathfrak{R}$ then $L_1 \cup L_2$ and L_1L_2 belong to $\mathfrak{R}$;
(v) if $L \in \mathfrak{R}$ then $L^* \in \mathfrak{R}$.

It follows from the above definition that every finite set $L \subset \Sigma^*$ is a regular language.

REG$^+(\Sigma)$ denotes the class of all regular languages over Σ which do not contain the empty word ε. Thus, REG$^+(\Sigma) := \{L \in \text{REG}(\Sigma) : \varepsilon \notin L\}$.

The class REG$^+(\Sigma)$ can be inductively defined as the least family $\mathfrak{R}^+$ of Σ-languages which satisfies the following closure properties:

(i)$^+$ the empty language $\emptyset$ belongs to $\mathfrak{R}^+$
(ii)$^+$ for every $a \in \Sigma$, $\{a\}$ belongs to $\mathfrak{R}^+$

(iii)$^+$ if $L_1 \in \mathfrak{R}^+$ and $L^2 \in \mathfrak{R}^+$, then $L_1 \cup L_2$ and $L_1 L_2$ belong to $\mathfrak{R}^+$
(iv)$^+$ if $L \in \mathfrak{R}^+$ then $L^+ \in \mathfrak{R}^+$.

The definition of a regular language also makes sense for an *infinite* set of symbols Σ (though the latter set does not qualify as an alphabet in the strict sense): assuming that Σ is infinite we define the families REG(Σ) and REG$^+(\Sigma)$ in the same way as above, i.e., by means of clauses (i)–(v) and (i)$^+$–(iv)$^+$, respectively. Denoting by $\Sigma(L)$ the set of all symbols of Σ that occur in the words of L, one easily shows that for every $L \in$ REG(Σ), the set $\Sigma(L)$ is finite. Thus, each member of REG(Σ) is a regular language over the alphabet $\Sigma(L)$.

In the paragraphs below we shall devote much space to compound (or composite) actions on elementary action systems $\boldsymbol{M} = (W, R, \boldsymbol{\mathcal{A}})$. A *compound action* (on $\boldsymbol{M}$) is defined as a set of finite strings of atomic actions of the system $\boldsymbol{M}$. Treating the elements of $\boldsymbol{\mathcal{A}}$ (i.e., the atomic actions) as primitive symbols, we see that each compound action can be viewed as a 'language' over the 'alphabet' $\boldsymbol{\mathcal{A}}$. (Such a formulation of compound actions is thus convergent with the ideas of Nowakowska.) In particular, we will be interested in examining the properties of the class REG$^+(\boldsymbol{\mathcal{A}})$ of regular compound actions over $\boldsymbol{\mathcal{A}}$ that do not contain the empty string of actions. We must remember, however, that the set $\boldsymbol{\mathcal{A}}$ need not be finite.

The main task of the theory of formal languages is to provide methods for the description of infinite languages. It is required at the same time that the methods should satisfy the following two conditions Blikle (1973):

1. *The condition of analytical effectiveness*: the method by means of which a language L over an alphabet is described determines a universal algorithm for deciding in a finite number of steps, for each word x over Σ, if x belongs to L or it does not.
2. *The condition of synthetic effectiveness*: the method by means of which a language L over Σ is described provides an algorithm for producing all and only the words of L; at the same time for each word x of L it is possible to estimate the number of steps of the algorithm necessary to produce the word x.

The above postulates result from pragmatic aspects of the processes of communication among the language users. The analytical effectiveness of the method is necessary for the language user to receive information encoded in the language. We mean here, *inter alia*, ensuring the possibility of the grammatical parsing of expressions (sentences) of a given language, of its translation into other languages, and most importantly of all, of understanding expressions (sentences) of the language. The condition of synthetic effectiveness, in turn, is indispensable in the process of sending information encoded in the language, e.g., in speaking, writing, etc.

Besides abstract automata, which will be discussed later, combinatorial grammars are a fundamental tool for the investigation of formal languages which satisfies, under certain assumptions, the postulate of synthetic effectiveness: a combinatorial grammar describes a language providing at the same time a method of producing all and only the words of the language.

Every combinatorial grammar (over Σ) contains apart from the alphabet Σ a finite set V of auxiliary symbols, a finite set of productions and the start symbol, which is always an element of V. More formally, a *combinatorial grammar* is a quadruple

$$\mathbf{G} = (\Sigma, V, \mathscr{P}, \alpha),$$

where Σ is the given alphabet, also called the *terminal alphabet*, V is an *auxiliary alphabet* (the members of V are called *nonterminals* or *variables* or *syntactic categories*), $\mathscr{P}$ is a finite subset of $(\Sigma \cup V)^* \times (\Sigma \cup V)^*$ called the *list of productions* of the grammar; the fact that $(x, y) \in \mathscr{P}$ is written as $x \to y$. x is the *predecessor* and y is the *successor* of the production $x \to y$. α is a distinguished element of V called the *start symbol.*

We now formally define the language generated by a grammar $\mathbf{G} = (\Sigma, V, \mathscr{P}, \alpha)$. To do so, we first define two relations $\Rightarrow_{\mathbf{G}}$ and $\Rightarrow^*_{\mathbf{G}}$ between strings in $(\Sigma \cup V)^*$. Let x and y be arbitrary words over the alphabet $\Sigma \cup V$. We say that the word x *directly derives* y in the grammar $\mathbf{G}$, and write $x \Rightarrow^*_{\mathbf{G}} y$, if and only if there are words $z_1, z_2 \in (\Sigma \cup V)^*$ and a production $w_1 \to w_2$ in $\mathscr{P}$ such that $x = z_1 w_1 z_2$ and $y = z_1 w_2 z_2$. Thus, two strings are related by $\Rightarrow_{\mathbf{G}}$ just when the second is obtained from the first by one application of some production.

The relation $\Rightarrow^*_{\mathbf{G}}$ is the reflexive and transitive closure of $\Rightarrow_{\mathbf{G}}$. Thus, $x \Rightarrow^*_{\mathbf{G}} y$ if and only if there is a (possibly empty) sequence of strings $z_1, \ldots, z_n$ in $(\Sigma \cup V)^*$ such that $z_1 = x$, $z_n = y$ and $z_i \Rightarrow^*_{\mathbf{G}} z_{i+1}$ for $i = 1, 2, \ldots, n-1$. Then we say that x *derives* y in the grammar $\mathbf{G}$. The sequence $z_1, \ldots, z_n$ is called a *derivation* of y from x. Alternatively, $x \Rightarrow^*_{\mathbf{G}} y$ if y follows from x by application of zero or more productions of $\mathscr{P}$. Note that $x \Rightarrow^*_{\mathbf{G}} x$, for any word x. (We shall omit the subscript $\mathbf{G}$ in $\Rightarrow^*_{\mathbf{G}}$ when $\mathbf{G}$ is clear from context.)

The language generated by $\mathbf{G}$, denoted by $L(\mathbf{G})$, is the set $\{x : x \in \Sigma^* \text{ and } \alpha \Rightarrow^*_{\mathbf{G}} x\}$. Thus, a string x is in $L(\mathbf{G})$ if x consists solely of terminals and the string can be derived from α. The symbols of the alphabet V appear only while deriving the words of $L(\mathbf{G})$; they do not occur in the words of $L(\mathbf{G})$. Thus, V plays an auxiliary role in the process of defining the language $L(\mathbf{G})$.

Example 1.6.1 Let $\Sigma := \{A, B\}$, $V := \{\alpha, a, b\}$, and $\mathscr{P}$, the list of productions, is: $\alpha \to ab$, $a \to Aa$, $a \to A$, $b \to Bb$, $b \to B$. The following are derivations in the grammar $\mathbf{G} := (\Sigma, V, \mathscr{P}, \alpha)$:

$$\alpha \Rightarrow ab \Rightarrow Ab \Rightarrow AB$$
$$\alpha \Rightarrow ab \Rightarrow Aab \Rightarrow AAb.$$

The word AB belongs to the language $L(\mathbf{G})$. The word AAb is not in $L(\mathbf{G})$ because it contains the nonterminal b.

The language $L(\mathbf{G})$ has a very simple description: $L(\mathbf{G}) = \{A^m B^n : m, n \geqslant 1\}$. □

Right-linear grammars are the grammars in which all productions have the form $a \to xb$ or $a \to x$, where a and b are variables and x is a (possibly empty) string of terminals.

Example 1.6.2 The grammar **G** from Example 1.6.1 is not right-linear because it contains the production $\alpha \to ab$.

Let Σ and V be as above and let $\mathscr{P}'$ be the list of following productions: $\alpha \to AB$, $\alpha \to Aa$, $a \to Ab$, $b \to Bb$, $a \to AB$, $a \to Aa$, $b \to B$. $\mathbf{G}' := (\Sigma, V, \mathscr{P}', \alpha)$ is thus a right-linear grammar. The language generated by $\mathbf{G}'$ is the same as that in Example 1.6.1, i.e., $L(\mathbf{G}') = \{A^m B^n : m, n \geqslant 1\}$. □

Theorem 1.6.3 (Chomsky and Miller 1958) *Right-linear grammars characterize regular languages in the sense that a language is regular if and only if it is generated by a right-linear grammar.*

The proof of this theorem can be found, e.g., in Hopcroft and Ullman (1979). □

Abstract automata are a fundamental tool for the investigation of formal languages which satisfies the postulate of analytical effectiveness. We shall shortly present the most important facts concerning *finite automata.* These are the simplest automata. Their theory is presented, from a much broader perspective, in specialized textbooks on formal linguistics, see in Hopcroft and Ullman (1979) or *The Handbook of Theoretical Computer Science* (see van Leeuwen 1990). Here we only record the basic facts that will be employed in our action theory.

A finite automaton consists of a finite set of states and a set of transitions from a state to a state that occur on the input of symbols chosen from an alphabet Σ. One state, denoted by w_0, is singled out and is called the initial state, in which the automaton starts. Some states are designated as final or accepting states.

Formally, a (nondeterministic) *finite automaton* is a quintuple

$$\mathbf{A} := (W, \Sigma, \delta, w_0, F), \tag{1.6.1}$$

where

(i) W is a finite set of *states* of the automaton
(ii) Σ is an alphabet, whose elements are called *input symbols*
(iii) $w_0 \in W$ is the *initial state*
(iv) $F \subseteq W$ is the set of *final states*,
(v) δ is the *transition* function mapping the Cartesian product $W \times \Sigma$ into the power set $\wp(W)$.

(The set $\delta(u, a)$ is interpreted as the set of all states w such that there is a transition labeled by a from u to w.)

Since the alphabet Σ is a fixed component of the automaton (1.6.1), we shall often say that **A** is an automaton *over* the alphabet Σ.

The function δ can be extended to a function δ^* mapping $W \times \Sigma^*$ into $\wp(W)$ and reflecting sequences of inputs as follows:

(vi) $\delta^*(u, \varepsilon) := \{u\}$.
(vii) $\delta^*(u, xa) = \{w :$ for some state v in $\delta^*(u, x)$, w is in $\delta(v, a)\}$.

Briefly,

$$\delta^*(u, xa) = \bigcup\{\delta(v, a) : v \text{ in } \delta^*(u, x)\}.$$

δ^* is indeed an extension of δ since $\delta^*(u, a) = \delta(u, a)$ for any input symbol a. Thus, we may again use δ in place of δ^*.

An automaton $\boldsymbol{A}$ is *deterministic* if the transition function δ has the property that for all $u \in W$ and $a \in \Sigma$, $\delta(u, a)$ is a one-element set.

A word x of Σ^* is said to be *accepted* by a finite automaton $\boldsymbol{A} = (W, \Sigma, \delta, w_0, F)$ if $\delta(w_0, x) \cap F \neq \emptyset$.

$L(\boldsymbol{A})$ denotes the set of all words of Σ^* accepted by $\boldsymbol{A}$. The set $L(\boldsymbol{A})$ is called the *language accepted* by $\boldsymbol{A}$.

Two finite automata $\boldsymbol{A}$ and $\boldsymbol{B}$ (over the same alphabet Σ) are *equivalent* if they accept the same languages, i.e., $L(\boldsymbol{A}) = L(\boldsymbol{B})$. It is a well-known fact (see Hopcroft and Ullman 1979, p. 22) that deterministic and nondeterministic finite automata accept the same sets of words; more specifically

Proposition 1.6.4 *Let $L \subseteq \Sigma^*$ be the language accepted by a nondeterministic finite automaton. Then there exists a deterministic finite automaton that accepts L.* □

The crucial result for the theory of regular languages is that the latter are precisely the languages accepted by finite automata.

Theorem 1.6.5 *Let Σ be a finite set. For any set $L \subseteq \Sigma^*$ the following conditions are equivalent:*

(i) $L \subseteq \mathrm{REG}(\Sigma)$
(ii) $L = L(\boldsymbol{A})$ *for some finite automaton* $\boldsymbol{A}$ *over* Σ.

A proof of this theorem can be found in Hopcroft and Ullman (1979). □

Theorem 1.6.5 has interesting applications. We shall mention one.

Theorem 1.6.6 *The class* $\mathrm{REG}^*(\Sigma)$ *is closed under complementation. That is, if $L \subseteq \Sigma^*$ is regular, then $\Sigma^* \setminus L$ is regular as well.*

Proof Let $L = L(\boldsymbol{A})$ for some finite automaton $\boldsymbol{A} = (W, \Sigma_1, \delta, w_0, F)$. In view of Proposition 1.6.4 we may assume without loss of generality that M is deterministic. Furthermore, we may assume that $\Sigma_1 = \Sigma$, for if $\Sigma_1 \setminus \Sigma$ is nonempty, one can delete all transitions $\delta(u, a)$ of $\boldsymbol{A}$ on symbols a not in Σ. The fact that $L \subseteq \Sigma^*$ assures that the language accepted by $\boldsymbol{A}$ is not thereby changed. In turn, if $\Sigma \setminus \Sigma_1$ is nonempty, then none of the symbols of $\Sigma \setminus \Sigma_1$ appears in the words of L. We may therefore introduce a 'dead' state d into $\boldsymbol{A}$ with

$$\delta(d, a) = d \quad \text{for all } a \in \Sigma, \quad \text{and}$$
$$\delta(w, a) = d \quad \text{for all } w \in W \quad \text{and } a \in \Sigma \setminus \Sigma_1.$$

In order to construct an automaton over Σ which accepts the set $\Sigma^* \setminus L$, we complement the set of final states of $\boldsymbol{A} = (W, \Sigma, \delta, w_0, F)$; we define

$$\boldsymbol{A}' := (W, \Sigma, \delta, w_0, W \setminus F)$$

Then $\boldsymbol{A}'$ accepts a word x if and only if $\delta(w_0, x)$ is in $W \setminus F$, that is, x is in $\Sigma^* \setminus L$. □

Corollary 1.6.7 *The regular sets are closed under intersection.*

Proof Let L_1, L_2 belong to $\mathrm{REG}^*(\Sigma)$. Then $L_1 \cap L_2 = \Sigma^* \setminus \big((\Sigma^* \setminus L_1) \cup (\Sigma^* \setminus L_2)\big)$. Closure under intersection then follows from closure under union and complementation. □

Let

$$\boldsymbol{A} = (W, \Sigma, \delta, w_0, F) \tag{1.6.2}$$

be a (non-deterministic) finite automaton. The following elementary action system

$$\boldsymbol{M}(\boldsymbol{A}) := (W, R_{\boldsymbol{A}}, \boldsymbol{\mathcal{A}}_{\Sigma,\delta}), \tag{1.6.3}$$

is assigned to $\boldsymbol{A}$, where

(i) W coincides with the set of states of $\boldsymbol{A}$
(ii) the transition relation $R_{\boldsymbol{A}}$ is defined as follows:
$(u, w) \in R_{\boldsymbol{A}}$ if and only if $w \in \delta(a, u)$ for some $a \in \Sigma$
(iii) $\boldsymbol{\mathcal{A}}_{\Sigma,\delta} := \{A_a : a \in \Sigma\}$, where each A_a is the binary relation on W defined by means of the equivalence: $A_a(u, w)$ if and only if $w \in \delta(a, u)$.

(1.6.3) is a normal action system because $R_{\boldsymbol{\mathcal{A}}} = \bigcup\{A_a : a \in \Sigma\}$. Moreover, if $\boldsymbol{A}$ is deterministic, A_a is a function whose domain is equal to W, i.e., for any $u \in W$ there exists a unique state $w \in W$ such that $A_a(u, w)$ holds. This is a direct consequence of the fact that $\delta(a, u)$ is a singleton, for all $a \in A$, $u \in W$.

The correspondence $a \to A_a$, $a \in \Sigma$, need not be injective. The above mapping is one-to-one if and only if for every pair $a, b \in \Sigma$ it is the case that

$$a = b \text{ if and only if } (\forall u \in W)\, \delta(a, u) = \delta(b, u). \tag{1.6.4}$$

The pair $(\{w_0\}, F)$ is a task for the action system (1.6.3), where w_0 is the initial state of $\boldsymbol{A}$ and F is the set of final states.

Conversely, let

$$\boldsymbol{M} = (W, R, \boldsymbol{\mathcal{A}}) \tag{1.6.5}$$

be a *finite* elementary action system, and suppose a task (Φ, Ψ) for $\boldsymbol{M}$ has been singled out so that Φ is a one-element subset of W, $\Phi = \{w_0\}$. We wish to define an automaton assigned to $\boldsymbol{M}$, which will be called the *automaton corresponding to the system* $\boldsymbol{M}$ *and the task* (Φ, Ψ).

There are several ways of defining the transition function in the finite automaton corresponding to the action system $\boldsymbol{M}$. We present two options here. The first option identifies the transition function in the automaton with realizable performances of actions. Accordingly:

$$\delta_{\boldsymbol{M}}(u, A) := \{w \in W : u\ A, R\ w\}, \tag{1.6.6}$$

for all $u \in W$, $A \in \boldsymbol{\mathcal{A}}$.

The second possible definition of the transition function abstracts from the relation R. In this case we put:

$$\delta_{\boldsymbol{M}}(u, A) := \{w \in W : A(u, w)\}, \tag{1.6.7}$$

for all $u \in W$, $A \in \boldsymbol{\mathcal{A}}$.

In both cases the quintuple

$$\boldsymbol{A}(\boldsymbol{M}) := (W, \boldsymbol{\mathcal{A}}, \delta_{\boldsymbol{M}}, w_0, \Psi) \tag{1.6.8}$$

is a finite (non-deterministic) automaton in which $\boldsymbol{\mathcal{A}}$, the set of atomic actions of $\boldsymbol{M}$, is the input alphabet. For any $A, B \in \boldsymbol{\mathcal{A}}$, the right-hand side of formula (1.6.4), i.e., the condition

$$(\forall u, w \in W)\big(w \in \delta_{\boldsymbol{M}}(u, A) \Leftrightarrow w \in \delta_{\boldsymbol{M}}(u, B)\big)$$

is equivalent to $A \cap R = B \cap R$ if $\delta_{\boldsymbol{M}}$ is defined by formula (1.6.6), or it is equivalent to $A = B$ if $\delta_{\boldsymbol{M}}$ is given by (1.6.7). The equation $A \cap R = A \cap R$ need not imply the identity $A = B$. However, when $\boldsymbol{M}$ is normal, then the definitions (1.6.6) and (1.6.7) are equivalent, and consequently, the equivalence

$$A = B \quad \text{if and only if} \quad (\forall u \in W)\ \delta_{\boldsymbol{M}}(u, A) = \delta_{\boldsymbol{M}}(u, B)$$

holds for all $A, B \in \boldsymbol{\mathcal{A}}$ in both the meanings of $\delta_{\boldsymbol{M}}$.

Having given a finite automaton (1.6.2), we define the action system $\boldsymbol{M}(\boldsymbol{A})$. Proceeding a step further, we can then define the automaton $\boldsymbol{A}(\boldsymbol{M}(\boldsymbol{A}))$ in accordance with the formulas (1.6.6) and (1.6.8). The two automata $\boldsymbol{A}$ and $\boldsymbol{A}(\boldsymbol{M}(\boldsymbol{A}))$ are not identical as they operate on different input alphabets—$\boldsymbol{A}$ is defined over the alphabet Σ while $\boldsymbol{A}(\boldsymbol{M}(\boldsymbol{A}))$ has the set $\{A_a : a \in \Sigma\}$ as the input alphabet. The transition functions δ and $\delta_{\boldsymbol{M}(\boldsymbol{A})}$ in the respective automata satisfy the identity: $\delta_{\boldsymbol{M}(\boldsymbol{A})}(u, A_a) = \delta(u, A)$, for all $a \in \Sigma$, $u \in W$. If the automaton $\boldsymbol{A}$ satisfies condition (1.6.4), the automata $\boldsymbol{A}$ and $\boldsymbol{A}(\boldsymbol{M}(\boldsymbol{A}))$ can be identified because the map $a \to A_a$ establishes a one-to-one correspondence between the sets Σ and $\{A_a : a \in \Sigma\}$.

Similarly, having given an action system (1.6.5) with the task $(\{w_0\}, \Psi)$ distinguished, we define the transition function $\delta_{\boldsymbol{M}}$ and the automaton $\boldsymbol{A}(\boldsymbol{M})$, according to the formulas (1.6.6) and (1.6.8); then we pass to the action system $\boldsymbol{M}(\boldsymbol{A}(\boldsymbol{M}))$ which is defined by means of the formula (1.6.3). If $\boldsymbol{M}$ is normal, the relationship between the systems $\boldsymbol{M}$ and $\boldsymbol{M}(\boldsymbol{A}(\boldsymbol{M}))$ is straightforward—the two systems are identical. Since $\boldsymbol{M}$

is normal, the direct transition relation R (in $\boldsymbol{M}$) coincides with the direct transition relation $R_{A(\boldsymbol{M})}$ in the action system $\boldsymbol{M}(\boldsymbol{A}(\boldsymbol{M}))$. It is also easy to see that the sets of atomic actions of $\boldsymbol{M}$ and $\boldsymbol{M}(\boldsymbol{A}(\boldsymbol{M}))$ are the same.

1.7 Compound Actions

Let us return for a while to the example with the washing machine. In order to wash linen a sequence of atomic actions such as putting the linen into the machine, setting the wash program, turning on the machine, etc., has to be performed. In a more sophisticated model, the sequence of actions could also include spinning, starching, and so on. Thus, washing is a set of sequences of atomic actions.

Everyday life provides an abundance of similar examples of compound actions such as the baking of bread or the manufacture of cars. A compound action is then a certain set of finite sequences of atomic actions. (We could also allow for infinite strings of atomic actions as well; we shall not discuss this issue thoroughly in this book.) The above remarks lead to the following definition:

Definition 1.7.1 Let $\boldsymbol{M} = (W, R, \mathcal{A})$ be an elementary action system. A *compound action* of $\boldsymbol{M}$ is any set of finite sequences of atomic actions of $\mathcal{A}$. □

The term "composite action" will also be occasionally used hereafter and treated as being synonymous to "compound action." We will often identify any atomic action $A \in \mathcal{A}$ with the composite action $\{(A)\}$, so each atomic action qualifies as a compound one. In order to simplify the notation we adopt here the convention that any finite sequence $(A_1, \ldots, A_n)$ of atomic actions is written by juxtaposing them as $A_1 \ldots A_n$. In particular, instead of (A) we simply write A. This is in accordance with the notation for formal languages adopted in Sect. 1.6.

Infinite compound actions are allowed. The approach to action presented here is finitary in the sense that sets consisting of *infinite* strings of atomic actions are not investigated. However, in Sect. 2.6 infinite sequences of actions are occasionally mentioned in the context of infinite games.

Example There are 4 recycling bins for waste materials near your house: for paper, plastic, white glass, and colored glass. Almost everyday you perform the compound action of emptying your wastebin of plastic, paper, and glass. Taking out the trash from the wastebin is a set of finite sequences of four atomic actions: A—putting the paper from your wastebin into the paper bin, B—putting the plastic materials into the plastic bin, C—putting the white glass into the white glass bin, and D—putting the colored glass into the colored glass bin. The strings of these four actions are not uniquely determined, because you may perform them in various orders. The initial states are those in which your wastebin is full and the final one is when it is empty. More accurately, assuming that we live in the mathematized world, your bin contains today n pieces of waste and they are distributed among the four categories that add up to n. Therefore your string of actions has length n, because you put each piece separately into the appropriate recycling bin (you are very concerned about the planet and so you take special care). While performing this string of actions, i.e., taking out the trash, your action system (the cleared wastebin) passes successively through $n+1$ states $u_0, \ldots, u_n$, where u_0 is the initial state and u_n—the final one. □

$\mathcal{A}^*$ is the set of all finite sequences of atomic actions of $\mathcal{A}$. In particular the empty string ε is a member of $\mathcal{A}^*$. ε is identified with the empty set, $\varepsilon = \emptyset$. $\mathcal{A}^*$ is the largest compound action of $\boldsymbol{M}$.

$C\mathcal{A}$ denotes the family of all compound actions of the system $\boldsymbol{M} = (W, R, \mathcal{A})$. (The members of $C\mathcal{A}$ will be often referred to as compound actions on $\boldsymbol{M}$.)

We shall first distinguish two special members of $C\mathcal{A}$. $\boldsymbol{\varepsilon}$ is the compound action, whose only element is the empty sequence of atomic actions, $\boldsymbol{\varepsilon} = \{\varepsilon\} = \{\emptyset\}$. In turn, $\boldsymbol{\emptyset}$ denotes the empty compound action.

$C^+\mathcal{A}$ is the family of all compound actions which do not involve the empty string of atomic actions. Thus,

$$C^+\mathcal{A} = \{\mathbf{A} \in C\mathcal{A} : \varepsilon \notin \mathbf{A}\}.$$

Note that the empty action $\boldsymbol{\emptyset}$ belongs to $C^+\mathcal{A}$. Since the members of $C\mathcal{A}$ can be treated as languages over the (possibly infinite) alphabet $\mathcal{A}$, all the definitions adopted for formal languages can be applied to compound actions. In particular for $\mathbf{B}, \mathbf{C} \in C\mathcal{A}$ we define:

$\mathbf{B} \cup \mathbf{C} := \{A_1 \ldots A_n : A_1 \ldots A_n \in \mathbf{B} \text{ or } A_1 \ldots A_n \in \mathbf{C}\}$,

$\mathbf{B} \cup \mathbf{C}$ is the *union* of $\mathbf{B}$ and $\mathbf{C}$;

$\mathbf{B} \cap \mathbf{C} := \{A_1 \ldots A_n : A_1 \ldots A_n \in \mathbf{B} \text{ and } A_1 \ldots A_n \in \mathbf{C}\}$,

$\mathbf{B} \cap \mathbf{C}$ is the *meet* of $\mathbf{B}$ and $\mathbf{C}$;

$\sim \mathbf{B} := \{A_1 \ldots A_n \in \mathcal{A}^* : A_1 \ldots A_n \notin \mathbf{B}\}$,

$\sim \mathbf{B}$ is the *complement* of $\mathbf{B}$;

$\mathbf{B} \circ \mathbf{C} := \{A_1 \ldots A_m B_1 \ldots B_n : A_1 \ldots A_m \in \mathbf{B} \text{ and } B_1 \ldots B_n \in \mathbf{C}\}$,

$\mathbf{B} \circ \mathbf{C}$ is called the *composition* of $\mathbf{B}$ and $\mathbf{C}$;

$\mathbf{B}^* := \bigcup\{\mathbf{B}^n : n \geqslant 0\}$, where $\mathbf{B}^0 := \boldsymbol{\varepsilon}$ and $\mathbf{B}^{n+1} := \mathbf{B}^n \circ \mathbf{B}$,

$\mathbf{B}^*$ is the *iterative closure* of $\mathbf{B}$;

$\mathbf{B}^+ := \bigcup\{\mathbf{B}^n : n \geqslant 1\}$ is called the *positive iterative closure* of $\mathbf{B}$;

$\mathbf{B}^{-1} := \{A_1 \ldots A_n : A_n \ldots A_1 \in \mathbf{B}\}$,

$\mathbf{B}^{-1}$ is called the *inverse* of $\mathbf{B}$.

Note that $\mathbf{A}^* = \mathbf{A}^+ \cup \boldsymbol{\varepsilon}$, for any $\mathbf{A}$. In particular, $\boldsymbol{\varepsilon}^* = \boldsymbol{\varepsilon}^+ = \boldsymbol{\varepsilon}$ and $\boldsymbol{\emptyset}^+ = \boldsymbol{\emptyset}$, $\boldsymbol{\emptyset}^* = \boldsymbol{\varepsilon}$. The family of compound actions on $\boldsymbol{M}$ exhibits a definite algebraic structure represented by the algebra

$$(C\mathcal{A}; \cup, \cap, \sim, \circ, ^*, ^{-1}). \tag{1.7.1}$$

As $C^{+}\mathcal{A} = C\mathcal{A} \setminus \{\varepsilon\}$, the system

$$(C^{+}\mathcal{A}; \cup, \cap, \sim, \circ, {}^{+}, {}^{-1}). \tag{1.7.2}$$

is an algebra as well. (But in (1.7.2) the complement $\sim$ is taken with respect to $\mathcal{A}^{+}$, i.e., $\sim \mathbf{B} = \mathcal{A}^{+} \setminus \mathbf{B}$, for all $\mathbf{B} \in C^{+}\mathbf{A}$.) Both algebras (1.7.1) and (1.7.2) are called the *algebras of compound actions* of the system $\boldsymbol{M}$.

The algebras (1.7.1) and (1.7.2) are examples of one-sorted algebras—each operation on the algebra assigns a compound action to a single compound action or to a pair of actions. We shall further enrich the structure of (1.7.1) and (1.7.2) by adjoining to them operations which assign compound actions to single propositions. (Any proposition is identified with a subset of W.) We shall also consider operations which assign actions or propositions to pairs of type (action, proposition).

To each compound action $\mathbf{A} \in C\mathcal{A}$ a binary relation $Res\ \mathbf{A}$ on W is assigned. $Res\ \mathbf{A}$ is defined as follows:

$$Res\ \mathbf{A} := \bigcup \big\{(A_1 \cap R) \circ \ldots \circ (A_n \cap R) : n \geqslant 0,\ A_1 \ldots A_n \in \mathbf{A}\big\}, \tag{1.7.3}$$

where $P \circ Q$ is the ordinary composition of the relations P and Q. $Res\ \mathbf{A}$ is called the *resultant relation* of the compound action $\mathbf{A}$. (If $A_1 \ldots A_n$ is the empty string ε, then $(A_1 \cap R) \circ \ldots \circ (A_n \cap R)$ is equal to $E_W \cap R$, where E_W is the diagonal of W.)

If $\mathbf{A}$ is an atomic action, i.e., $\mathbf{A} = \{A\}$ for some $A \in \mathcal{A}$, then the resultant relation of $\mathbf{A}$ is equal to $A \cap R$. It follows from the definition of the reach $\boldsymbol{Re_M}$ of the system $\boldsymbol{M}$ (Definition 1.2.2) that for any compound action $\mathbf{A} \in C\mathcal{A}$, $Res\ \mathbf{A}$ is a subrelation of $\boldsymbol{Re_M}$. (We assume that $Res\ \boldsymbol{\varepsilon} = E_W \cap R$. Consequently, if the empty sequence ε belongs to $\mathbf{A}$, then $E_W \cap R \subseteq Res\ \mathbf{A}$.)

Lemma 1.7.2 *For any compound actions* $\mathbf{B}, \mathbf{C} \in C\mathcal{A}$:

(i) $Res\ (\mathbf{B} \cup \mathbf{C}) = Res\ \mathbf{B} \cup Res\ \mathbf{C}$,
(ii) $Res\ (\mathbf{B} \circ \mathbf{C}) = Res\ \mathbf{B} \circ Res\ \mathbf{C}$,
(iii) $Res\ (\mathbf{B}^{+}) = (Res\ \mathbf{B})^{+}$,
(iv) $Res\ (\mathbf{B} \cap \mathbf{C}) \subseteq Res\ \mathbf{B} \cap Res\ \mathbf{C}$,
(v) $Res\ (\boldsymbol{\varepsilon}) = E_W \cap R$ *and* $Res\ (\boldsymbol{\emptyset}) = \emptyset$.

The proof is easy and is omitted. □

Note that $Res\ (\mathbf{A}^{*}) \neq (Res\ \mathbf{A})^{*}$ because $Res\ (\mathbf{A}^{*}) = Res\ (\mathbf{A}^{+}) \cup Res(\boldsymbol{\varepsilon}) = Res\ (\mathbf{A}^{+}) \cup (E_W \cap R)$ while $(Res\ \mathbf{A})^{*}$, the transitive and reflexive closure of $Res\ \mathbf{A}$, is equal to $Res\ (\mathbf{A}^{+}) \cup E_W$. But if the relation R is reflexive, then $Res\ (\mathbf{A}^{*}) = (Res\ \mathbf{A})^{*}$.

It follows from Lemma 1.7.2 that the mapping $A \to Res\ \mathbf{A}$, $\mathbf{A} \in C^{+}\mathcal{A}$, is a homomorphism of the reduct $(C^{+}\mathcal{A}; \cup, \circ, {}^{+})$ of the algebra $(1)^{+}$ into the relational algebra $(\wp(W \times W), \cup, \circ, {}^{+})$ of binary relations on W. Moreover, if R is reflexive, then $\mathbf{A} \to Res\ \mathbf{A}$, $\mathbf{A} \in C\mathcal{A}$, is a homomorphism of the reduct $(C\mathcal{A}; \cup, \circ, {}^{*})$ of the algebra (1) into the relational algebra $(\wp(W \times W), \cup, \circ, {}^{*})$.

A *possible performance* of a compound action $\mathbf{A} \in C\mathcal{A}$ is any *nonzero* finite string $(u_0, \ldots, u_n)$ of states of W such that $u_0\ A_1\ u_1 \ldots u_{n-1}\ A_n\ u_n$ for some sequence $A_1 \ldots A_n \in \mathbf{A}$.

$|\mathbf{A}|$ is the set of all all possible performances of $\mathbf{A}$. The set $|\mathbf{A}|$ is called the *intension* of $\mathbf{A}$.

As each atomic action $A \in \mathcal{A}$ is identified with $\{A\}$, we see that $|A|$, the intension of A, is equal to A itself.

It is assumed that the diagonal E_W is the set of all possible performances of the composite action $\boldsymbol{\varepsilon} = \{\varepsilon\}$. Therefore $|\boldsymbol{\varepsilon}| = E_W$. Thus, if a compound action $\mathbf{A}$ contains the empty string of atomic actions, $|\mathbf{A}|$ is reflexive. The set of possible performances of $\boldsymbol{\emptyset}$ is empty, $|\boldsymbol{\emptyset}| = \emptyset$.

A possible performance

$$(u_0, \ldots, u_n) \tag{1.7.4}$$

of a compound action $\mathbf{A} \in C\mathcal{A}$ is said to be *realizable* if and only if $u_0\ R\ u_1 \ldots u_{n-1}\ R\ u_n$. In this case we also say that the indirect transition $u_0\ R\ u_1 \ldots u_{n-1}\ R\ u_n$ is accomplished by the action $\mathbf{A}$.

$|\mathbf{A}|_R$ is the set of all realizable performances of $\mathbf{A}$. In particular, $|\boldsymbol{\varepsilon}|_R = E_W \cap R$, $|\boldsymbol{\emptyset}|_R = \emptyset$, and $|A|_R = A \cap R$ for any atomic action A.

If the possible performance (1.7.4) of $\mathbf{A}$ is realizable then $(u_0, u_n) \in Res\mathbf{A}$; and conversely, if $(u, w) \in Res\mathbf{A}$, then there exists a realizable performance (1.7.4) of $\mathbf{A}$ such that $u_0 = u$ and $u_n = w$.

We now define the notion of the performability of a composite action.

Definition 1.7.3 Let $\boldsymbol{M} = (W, R, \mathcal{A})$ be an elementary action system, let $\mathbf{A} \in C\mathcal{A}$ be a compound action, and let $u \in W$.

(i) $\mathbf{A}$ is *performable* in u if and only if there exists a realizable performance $(u_0, \ldots, u_n)$ of $\mathbf{A}$ such that $u_0 = u$,

(ii) $\mathbf{A}$ is *totally performable* in u if and only $\mathbf{A}$ is performable in u and every possible performance $(u_0, \ldots, u_n)$ of $\mathbf{A}$ such that $u_0 = u$ is realizable,

(iii) $\mathbf{A}$ is *performable in the system* $\boldsymbol{M}$ if and only if $\mathcal{A}$ is performable in every state $u \in W$ such that there exists a possible performance $(u_0, \ldots, u_n)$ of $\mathbf{A}$ with $u_0 = u$,

(iv) $\mathbf{A}$ is *totally performable in* $\boldsymbol{M}$ if and only if every possible performance of $\mathbf{A}$ is realized. □

Thus, according to (i), $\mathbf{A}$ is performable in u if and only if there exists a sequence $A_1 \ldots A_n \in \mathbf{A}$ and a (nonzero) sequence of states $(u_0, \ldots, u_n)$ such that $u_0 = u$ and $u_0\ A_1\ R\ u_1 \ldots u_{n-1}\ A_n\ R\ u_n$. The intuitive sense of the above definitions is clear. Definition 1.7.3 of the performability of a compound action is an extension of Definitions 1.4.1 and 1.4.3. (The latter definitions are valid only for atomic actions.) If an action $\mathbf{A} \in C\mathcal{A}$ is atomic, i.e., $\mathbf{A} = \{A\}$ for some $A \in \mathcal{A}$, and $u \in W$, then $\mathbf{A}$ is performable (totally performable) in u if and only if A is performable (totally performable) in u in the sense of Definition 1.4.1.

The empty compound action $\boldsymbol{\emptyset}$ is performable in no state of $\boldsymbol{M}$. Note however that, paradoxically, $\boldsymbol{\emptyset}$ is totally performable in $\boldsymbol{M}$ because (i) is vacuously satisfied for $\boldsymbol{\emptyset}$. The action $\boldsymbol{\varepsilon} = \{\varepsilon\}$ is performable in u if and only if $u\ R\ u$. (It follows that then $\boldsymbol{\varepsilon}$ is also totally performable in every reflexive state u.) Consequently, $\boldsymbol{\varepsilon}$ is performable in the system $\boldsymbol{M}$ if and only if R is reflexive (if and only if $\boldsymbol{\varepsilon}$ is totally performable in the system $\boldsymbol{M}$).

The above definitions state that a compound action $\mathbf{A}$ is performable (at a given state) if all strings of atomic actions belonging to $\mathbf{A}$ are performable at consecutive states of $(u_0, \ldots, u_n)$. We may therefore say that the performability of a compound action $\mathbf{A}$ is a function of the performability of the constituents of $\mathbf{A}$. One may therefore argue that Definitions 1.7.3 is a form of the compositionality principle for the performability of actions.

Let us also note that $\mathbf{A} \in C\boldsymbol{\mathcal{A}}$ is totally performable in $\boldsymbol{M}$ if and only if it is totally performable in every state $u \in W$ such that there exists a (possible) performance $(u_0, \ldots, u_n)$ of $\mathbf{A}$ with $u_0 = u$.

Lemma 1.7.4 *Let $\boldsymbol{M} = (W, R, \boldsymbol{\mathcal{A}})$ be an action system and $\mathbf{A} \in C\boldsymbol{\mathcal{A}}$ a compound action. If $\mathbf{A}$ is totally performable in $\boldsymbol{M}$, then for all $A_1 \ldots A_n \in \mathbf{A}$, it is the case that $A_1 \circ \ldots \circ A_n \in R^n$.*

The proof is easy and is omitted. □

The family of all totally performable compound actions has an interesting algebraic structure.

Theorem 1.7.5 *Let $\boldsymbol{M} = (W, R, \boldsymbol{\mathcal{A}})$ be an elementary action system and let $\boldsymbol{TP}$ be the family of all compound actions totally performable in the system $\boldsymbol{M}$. Then:*

(i) $\boldsymbol{\emptyset} \in \boldsymbol{TP}$,
(ii) *if* $\mathbf{A} \in \boldsymbol{TP}$ *and* $\mathbf{B} \in \boldsymbol{TP}$, *then* $\mathbf{A} \cup \mathbf{B} \in \boldsymbol{TP}$ *and* $\mathbf{A} \circ \mathbf{B} \in \boldsymbol{TP}$,
(iii) *if* $\mathbf{A} \in \boldsymbol{TP}$, *then* $\mathbf{A}^+ \in \boldsymbol{TP}$,
(iv) *if* $\mathbf{A} \in \boldsymbol{TP}$ *and* $\mathbf{B} \subseteq \mathbf{A}$ *then* $\mathbf{B} \in \boldsymbol{TP}$.
(v) *If R is reflexive, then* $\boldsymbol{\varepsilon} \in \boldsymbol{TP}$ *and hence* $\mathbf{A}^* \in \boldsymbol{TP}$, *for any compound* $\mathbf{A}$.

Proof (ii). Let $\mathbf{A}, \mathbf{B} \in \boldsymbol{TP}$ and let $(u_0, \ldots, u_n)$ be a possible performance of $\mathbf{A} \cup \mathbf{B}$. Hence, there exists a sequence $A_1 \ldots A_n \in \mathbf{A} \cup \mathbf{B}$ such that $u_0\ A_1\ u_1 \ldots u_{n-1}\ A_n\ u_n$. Since both $\mathbf{A}$ and $\mathbf{B}$ are totally performable, we have that $u_0\ R\ u_1 \ldots u_{n-1}\ R\ u_n$. So $\mathbf{A} \cup \mathbf{B}$ is totally performable.

Now let $(u_0, \ldots, u_n)$ be a possible performance of $\mathbf{A} \circ \mathbf{B}$. Hence, there exist sequences $A_1 \ldots A_k \in \mathbf{A}$, $B_1 \ldots B_l \in \mathbf{B}$, where $l = n - k$, such that $u_0\ A_1\ u_1 \ldots\ldots u_{k-1}\ A_k\ u_k\ B_1\ u_{k+1} \ldots u_{n-1}\ B_l\ u_n$. Since $\mathbf{A}$ and $\mathbf{B}$ are totally performable, both the sequences $(u_0, \ldots, u_k)$, $(u_{k+1}, \ldots, u_n)$ are realizable. Hence $u_0\ R\ u_1 \ldots u_k\ R\ u_{k+1} \ldots\ldots u_{n-1}\ R\ u_n$ and so $\mathbf{A} \circ \mathbf{B}$ is totally performable.

(iii). Let $\mathbf{A} \in \boldsymbol{TP}$. It follows from (i) that $\mathbf{A}^n \in \boldsymbol{TP}$, for all $n \geqslant 1$. To prove that $\mathbf{A}^+$ is totally performable, assume that $(u_0, \ldots, u_m)$ is a possible performance of $\bigcup\{\mathbf{A}^n : n \geqslant 1\}$. Hence, for some $n \geqslant 1$, there exists a sequence $A_1 \ldots A_m \in \mathbf{A}^n$ such that $u_0\ A_1\ u_1 \ldots u_{m-1}\ A_m\ u_m$. Since $\mathbf{A}^n$ is totally performable, we have that $u_0\ R\ u_1 \ldots u_{m-1}\ R\ u_m$.

(iv) is immediate. As $\mathbf{A}^* = \mathbf{A}^+ \cup \boldsymbol{\varepsilon}$, (v) follows. □

If the system $\boldsymbol{M}$ is normal, the family $\boldsymbol{TP}$ of all totally performable compound actions has a simple characterization:

Proposition 1.7.6 *Let $\boldsymbol{M} = (W, R, \mathcal{A})$ be a normal elementary action system, i.e., $A \subseteq R$, for all $A \in \mathcal{A}$. Then every compound action $\mathbf{A} \in C^{+}\mathcal{A}$ is totally performable in $\boldsymbol{M}$. Moreover, if R is reflexive, then every action $\mathbf{A} \in C\mathcal{A}$ is totally performable in $\boldsymbol{M}$.*

Proof Let $(u_0, \ldots, u_n)$ be a possible performance of a compound action $\mathbf{A} \in C^{+}\mathcal{A}$. Then $u_0\ A_1\ u_1 \ldots u_{n-1}\ A_n\ u_n$ for some nonempty string $A_1 \ldots A_n \in \mathbf{A}$. As $\boldsymbol{M}$ is normal, we have that $u_0\ R\ u_1 \ldots u_{n-1}\ R\ u_n$. So $(u_0, \ldots, u_n)$ is realizable.

If R is reflexive, then $\boldsymbol{\varepsilon}$ is totally performable and hence every action in $C\mathcal{A}$ is totally performable in $\boldsymbol{M}$. □

For any compound action $\mathbf{A} \in C\mathcal{A}$ we define two subsets of W—the *domain*, $Dom\mathbf{A}$, and the *codomain* (alias *range*), $CDom\mathbf{A}$, of the action $\mathbf{A}$:

$$Dom\mathbf{A} := \{u \in W : (\exists u_0, \ldots, u_n \in W)\ u = u_0 \text{ and } (u_0, \ldots, u_n) \text{ is a possible performance of } \mathbf{A}\},$$
$$CDom\mathbf{A} := \{w \in W : (\exists u_0, \ldots, u_n \in W)\ w = u_n \text{ and } (u_0, \ldots, u_n) \text{ is a possible performance of } \mathbf{A}\}$$

We also define:

$$Dom_R\mathbf{A} := \{u \in W : (\exists u_0, \ldots, u_n \in W)\ u = u_0 \text{ and } (u_0, \ldots, u_n) \text{ is a realizable performance of } \mathbf{A}\},$$
$$CDom_R\mathbf{A} := \{w \in W : (\exists u_0, \ldots, u_n \in W)\ w = u_n \text{ and } (u_0, \ldots, u_n) \text{ is a realizable performance of } \mathbf{A}\}.$$

It follows that

$$Dom_R\mathbf{A} = \{u \in W : (\exists w \in W)(u, w) \in Res\ \mathbf{A}\}$$

and

$$CDom_R\mathbf{A} = \{w \in W : (\exists u \in W)(u, w) \in Res\ \mathbf{A}\}.$$

In particular, the above definitions make sense for any atomic action A. The intended readings of the proposition $Dom\mathbf{A}$ are: '$\mathbf{A}$ possibly starts' or '$\mathbf{A}$ possibly begins', while for $CDom\mathbf{A}$ we have '$\mathbf{A}$ possibly terminates' or '$\mathbf{A}$ possibly ends'. Analogously, $Dom_R\mathbf{A}$ says '$\mathbf{A}$ realizably starts' and $CDom_R\mathbf{A}$ says '$\mathbf{A}$ realizably terminates'.

For any two sets $\Phi, \Psi \in W$ we also define the following four compound actions:

$$\Phi \text{☞}_1 \Psi := \{A_1 \dots A_n \in \boldsymbol{\mathcal{A}}^* : (\forall u \in \Phi)(\exists w \in \Psi)\ (u, w) \in A_1 \circ \dots \circ A_n\},$$
$$\Phi \text{☞}_2 \Psi := \{A_1 \dots A_n \in \boldsymbol{\mathcal{A}}^* : (\forall u \in \Phi)(\forall w \in \Psi)\ (u, w) \in A_1 \circ \dots \circ A_n\},$$
$$\Phi \text{☞}_3 \Psi := \{A_1 \dots A_n \in \boldsymbol{\mathcal{A}}^* : (\exists u \in \Phi)(\forall w \in \Psi)\ (u, w) \in A_1 \circ \dots \circ A_n\},$$
$$\Phi \text{☞}_4 \Psi := \{A_1 \dots A_n \in \boldsymbol{\mathcal{A}}^* : (\exists u \in \Phi)(\exists w \in \Psi)\ (u, w) \in A_1 \circ \dots \circ A_n\}.$$

The compound actions $\Phi \text{☞}_i \Psi, i = 1, \dots, 4$, are collectively called "From Φ to Ψ," with the subscript i if necessary. It is not excluded that $\Phi \text{☞}_i \Psi$ may be the empty action, for some $i \in \{1, \dots, 4\}$.

The relations between the actions $\Phi \text{☞}_i \Psi, i = 1, \dots, 4$, are established by the following simple lemma.

Lemma 1.7.7 *Let* $\boldsymbol{M} = (W, R, \boldsymbol{\mathcal{A}})$ *be an action system and* $\Phi, \Psi \in W$. *Then:*

(i) $\Phi \text{☞}_1 \Psi = \bigcap \{\{u\} \text{☞}_1 \Psi : u \in \Phi\}$

(ii) $\Phi \text{☞}_2 \Psi = \bigcap \{\{u\} \text{☞}_2 \Psi : u \in \Phi\} = \bigcap \{\Phi \text{☞}_2 \{w\} : w \in \Psi\}$

(iii) $\Phi \text{☞}_3 \Psi = \bigcup \{\{u\} \text{☞}_3 \Psi : u \in \Phi\}$

(iv) $\Phi \text{☞}_4 \Psi = \bigcup \{\Phi \text{☞}_4 \{w\} : w \in \Psi\} = \bigcup \{\{u\} \text{☞}_4 \Psi : u \in \Phi\}$

(v) $\Phi \text{☞}_2 \Psi = \bigcap \{\Phi \text{☞}_1 \{w\} : w \in \Psi\}$

(vi) $\Phi \text{☞}_3 \Psi = \bigcup \{\{u\} \text{☞}_2 \Psi : u \in \Phi\}$

(vii) $\Phi \text{☞}_4 \Psi = \bigcup \{\{u\} \text{☞}_1 \Psi : u \in \Phi\}$. □

Theorem 1.7.8 *Let* $\boldsymbol{M} = (W, R, \boldsymbol{\mathcal{A}})$ *be a finite elementary action system. Let* $\Phi, \Psi \in W$. *Then for* $i = 1, \dots, 4$, *the compound action* $\Phi \text{☞}_i \Psi$ *is regular (over* $\boldsymbol{\mathcal{A}}$*).*

Proof We shall first prove the above theorem for $i = 1$.

If Φ is empty, then $\Phi \text{☞}_1 \Psi$ is equal to $\boldsymbol{\mathcal{A}}^*$. So $\Phi \text{☞}_1 \Psi$ is regular.

If Φ is a singleton, $\Phi = \{w_0\}$, we proceed as follows. We assign to the system $\boldsymbol{M}$ the finite automaton $\boldsymbol{A}(\boldsymbol{M}) := (W, \boldsymbol{\mathcal{A}}, \delta, w_0, \Psi)$ over the alphabet $\boldsymbol{\mathcal{A}}$, where the transition function δ is defined as follows: $\delta(u, A) := \{w \in W : u\ A\ w\}$, for all $A \in \boldsymbol{\mathcal{A}}$ and $u \in W$.

The proof of Theorem 1.7.8 is based on the following claims.

Claim 1 *For all* $u, w \in W$ *and for any string* $A_1 \dots A_n$ *of atomic actions,*

$$A_1 \dots A_n \in \{u\} \text{☞}_1 \{w\} \quad \textit{if and only if} \quad w \in \delta(u, A_1 \dots A_n).$$

Proof of Claim 1 It immediately follows from the definition of $\{u\} \text{☞}_1 \{w\}$ that $A_1 \dots A_n \in \{u\} \text{☞}_1 \{w\}$ if and only if $(u, w) \in A_1 \circ \dots \circ A_n$.

We prove the claim by induction on the length of the sequence $A_1 \dots A_n$. The claim holds for the empty string. If $A \in \boldsymbol{\mathcal{A}}$, then $A \in \{u\} \text{☞}_1 \{w\}$ if and only if $(u, w) \in A$ if and only if $w \in \delta(u, A)$. Suppose the claim holds for all words of length $\leqslant n$, and consider a word $A_1 \dots A_n A_{n+1}$. The following conditions are equivalent:

$A_1 \dots A_n A_{n+1} \in \{u\} \text{☞}_1 \{w\}$,

$(u, w) \in A_1 \circ \dots \circ A_n \circ A_{n+1}$,

$(\exists v \in W)\big((u, v) \in A_1 \circ \dots \circ A_n \text{ and } (v, w) \in A_{n+1}\big)$, (by the induction hypothesis)

$(\exists v \in W)(v \in \delta(u, A_1 \dots A_n) \text{ and } w \in \delta(v, A_{n+1})$,

$w \in \delta(u, A_1 \dots A_n A_{n+1})$. □

Claim 2 *For every sequence $A_1 \dots A_n$ of atomic actions of $\boldsymbol{\mathcal{A}}$, the word $A_1 \dots A_n$ is accepted by $\boldsymbol{A}(\boldsymbol{M})$ if and only if $A_1 \dots A_n \in \{w_0\} \text{☞}_1 \Psi$.*

Proof of Claim 2 $A_1 \dots A_n$ is accepted by $\boldsymbol{A}(\boldsymbol{M})$ if and only if

$(\exists w)(w \in \Psi \text{ and } w \in \delta(w_0, A_1 \dots A_n)$ if and only if (by Claim 1)

$(\exists w)(w \in \Phi \text{ and } A_1 \dots A_n \in \{w_0\} \text{☞}_1 \{w\})$ if and only if

$A_1 \dots A_n \in \{w_0\} \text{☞}_1 \Psi$. □

It follows from Claim 2 and Theorem 1.6.5 that for all $w_0 \in W$ and all $\Psi \subseteq W$, the compound action $\{w_0\} \text{☞}_1 \Psi$ is regular.

If Φ is a quite arbitrary nonempty subset of W, then by Lemma 1.7.7.(i), $\Phi \text{☞}_1 \Psi = \bigcap \big\{\{u\} \text{☞}_1 \Psi : u \in \Phi\big\}$. Since Φ is finite (for $\boldsymbol{M}$ is assumed to be finite), $\Phi \text{☞}_1 \Psi$ is thus an intersection of finitely many regular compound actions. This, in view of Corollary 1.6.7, implies that $\Phi \text{☞}_1 \Psi$ is regular as well.

Recalling again Corollary 1.6.7 and Lemma 1.7.7.(v)–(vi), we easily get that $\Phi \text{☞}_2 \Psi$ and $\Phi \text{☞}_3 \Psi$ are regular.

To prove that $\Phi \text{☞}_4 \Psi$ is regular, we notice that $\{u\} \text{☞}_4 \Psi = \{u\} \text{☞}_1 \Psi$, for all $u \in W$, $\Psi \subseteq W$. By Lemma 1.7.7.(iv), $\Phi \text{☞}_4 \Psi = \bigcup \big\{\{u\} \text{☞}_4 \Psi : u \in \Phi\big\} = \bigcup \big\{\{u\} \text{☞}_1 \Psi : u \in \Phi\big\}$. Since $\bigcup \big\{\{u\} \text{☞}_1 \Psi : u \in \Phi\big\}$ is the union of finitely many regular actions, we infer that $\Phi \text{☞}_4 \Psi$ is also regular.

This completes the proof of the theorem. □

If the system $\boldsymbol{M}$ is normal, Proposition 1.7.6 implies that the actions $\Phi \text{☞}_i \Psi$, $i = 1, \dots, 4$, are totally performable in $\boldsymbol{M}$.

Apart from the compound actions $\Phi \mathrel{☞}_i \Psi$, $i = 1, \ldots, 4$, we can also define actions of the type 'From Φ to Ψ' in the stronger version, in which the relation R of direct transition is necessarily involved. Their definitions are strictly linked with the issue of reaching goals in elementary action systems.

A *task* for an action system $\boldsymbol{M}$ is any pair (Φ, Ψ) with $\Phi, \Psi \subseteq W$. Φ is the initial condition of the task while Ψ is the *terminal condition* or the *goal* of the task.

We shall now discuss three concepts of the attainability of the goal Ψ from the set Φ that differ in strength.

Definition 1.7.9 Let (Φ,Ψ) be a task for an elementary action system $\boldsymbol{M} = (W, R, \boldsymbol{\mathcal{A}})$.

(i) The goal Ψ is *weakly attainable* from Φ in $\boldsymbol{M}$ if and only if $(\Phi \times W) \cap \boldsymbol{Re_M}$ is non-empty, where $\boldsymbol{Re_M}$ is the reach of $\boldsymbol{M}$, i.e., there exist states $u \in \Phi$, $w \in \Psi$ and an operation of $\boldsymbol{M}$ with the initial state u and the terminal state w;

(ii) The goal Ψ is *attainable* from Φ in $\boldsymbol{M}$ if and only if for every $u \in \Phi$, $(\{u\} \times \Psi) \cap \boldsymbol{Re_M}$ is nonempty, i.e., from every state $u \in \Phi$, the system turns to a *certain* state $w \in \Psi$ by means of an operation of $\boldsymbol{M}$, whose initial state is u and the initial state is w;

(iii) Ψ is *strongly attainable* from Φ in $\boldsymbol{M}$ if and only if $\Phi \times \Psi \subseteq \boldsymbol{Re_M}$, i.e., it is possible to move the system from *any* state $u \in \Phi$ to any state $w \in \Psi$ through an operation of $\boldsymbol{M}$ with the initial state u and the terminal state w. □

For any task (Φ, Ψ) we define the compound action $\Phi \mathrel{☞}_i^R \Psi$, for $i = 1, \ldots, 4$, in an analogous way as the actions $\Phi \mathrel{☞}_i \Psi$:

$$\Phi \mathrel{☞}_1^R \Psi := \{A_1 \ldots A_n \in \boldsymbol{\mathcal{A}}^* : (\forall u \in \Phi)(\exists w \in \Psi)\, (u, w) \in (A_1 \cap R) \circ \cdots \circ (A_n \cap R)\},$$
$$\Phi \mathrel{☞}_2^R \Psi := \{A_1 \ldots A_n \in \boldsymbol{\mathcal{A}}^* : (\forall u \in \Phi)(\forall w \in \Psi)\, (u, w) \in (A_1 \cap R) \circ \cdots \circ (A_n \cap R)\},$$
$$\Phi \mathrel{☞}_3^R \Psi := \{A_1 \ldots A_n \in \boldsymbol{\mathcal{A}}^* : (\exists u \in \Phi)(\forall w \in \Psi)\, (u, w) \in (A_1 \cap R) \circ \cdots \circ (A_n \cap R)\},$$
$$\Phi \mathrel{☞}_4^R \Psi := \{A_1 \ldots A_n \in \boldsymbol{\mathcal{A}}^* : (\exists u \in \Phi)(\exists w \in \Psi)\, (u, w) \in (A_1 \cap R) \circ \cdots \circ (A_n \cap R)\}.$$

We recall that if the sequence $A_1 \ldots A_n \in \boldsymbol{\mathcal{A}}^*$ is empty, then $(A_1 \cap R) \circ \cdots \circ (A_n \cap R) = E_W \cap R$. If Φ and Ψ are singletons, $|\Phi| = |\Psi| = 1$, then the actions $\Phi \mathrel{☞}_i^R \Psi$, $i = 1, \ldots, 4$, coincide, i.e., $\Phi \mathrel{☞}_1^R \Psi = \cdots = \Phi \mathrel{☞}_4^R \Psi$.

Proposition 1.7.10 *Let (Φ, Ψ) be a task for $\boldsymbol{M}$. Then*

(i) Ψ is weakly attainable from Φ if and only if $\Phi \mathrel{☞}_3^R \Psi$ is nonempty;

(ii) Ψ is attainable from Φ if and only if $(\forall u \in \Phi)\ \{u\} \mathrel{☞}_3^R \Psi \neq \emptyset$;

(iii) Ψ is strongly attainable from Φ if and only if $(\forall u \in \Phi)(\forall w \in \Psi)$ $\{u\} \mathrel{☞}_3^R \{w\} \neq \emptyset$. □

It is not difficult to show that the actions $\Phi \,☞_i^R\, \Psi$, $i = 1, \ldots, 4$, are regular in every finite elementary action system $\boldsymbol{M}$. (To prove this, one defines the automaton $\boldsymbol{A}(\boldsymbol{M})$ in which the transition function δ is given by the formula: $\delta(u, A) := \{w \in W : u\, A, R\, w\}$; the proof then goes as in Theorem 1.7.8.)

If $\Phi \,☞_i^R\, \Psi$ is non-empty, where $i = 1, 2$, then it is performable in every state $u \in \Phi$. (If $\Phi \,☞_2^R\, \Psi \neq \emptyset$, we can say more: for every pair $u \in \Phi$, $w \in \Psi$ there exists a realizable performance $(u_0, \ldots, u_n)$ of $\Phi \,☞_2^R\, \Psi$ such that $u_0 = u$ and $u_n = w$.)

For any $\Phi \subseteq W$ and $\mathbf{B} \in C\mathcal{A}$ we also define the following propositions.

$$[\mathbf{B}]\Phi := \{u \in W : (\forall w \in W)(Res\, \mathbf{B}(u, w) \text{ implies } w \in \Psi)\},$$
$$\langle \mathbf{B} \rangle \Phi := \{u \in W : (\exists w \in W)(Res\, \mathbf{B}(u, w) \text{ implies } w \in \Phi)\}.$$

$[\mathbf{B}]\Phi$ is the set of all states u such that *every* realizable performance of $\mathbf{B}$ in u leads the system to a state belonging to Φ. Analogously, $\langle \mathbf{B} \rangle \Phi$ is the set of all states u such that *some* realizable performance of $\mathbf{B}$ in the state u carries the system to a Φ-state.

To each pair $\mathbf{B}, \mathbf{C}$ of compound actions in $C\mathcal{A}$ the proposition $\mathbf{B} \Rightarrow_R \mathbf{C}$ is assigned, where

$$\mathbf{B} \Rightarrow_R \mathbf{C} := \big\{u \in W : \big(\forall w \in W\big)\big(Res\, \mathbf{B}(u, w) \text{ implies } Res\, \mathbf{C}(u, w)\big)\big\}.$$

The proposition $\mathbf{B} \Rightarrow_R \mathbf{C}$ expresses the fact that the resultant relation of $\mathbf{B}$ is a subrelation of the resultant relation of $\mathbf{C}$. Indeed, $\mathbf{B} \Rightarrow_R \mathbf{C} = W$ if and only if $Res\, \mathbf{B} \subseteq Res\, \mathbf{C}$.

We also define the proposition $\mathbf{B} \Leftrightarrow_R \mathbf{C}$, where

$$\mathbf{B} \Leftrightarrow_R \mathbf{C} := (\mathbf{B} \Rightarrow_R \mathbf{C}) \cap (\mathbf{C} \Rightarrow_R \mathbf{B}).$$

The proposition $\mathbf{B} \Leftrightarrow_R \mathbf{C}$ expresses the fact that the compound actions $\mathbf{B}$ and $\mathbf{C}$ are equivalent in the sense that their resultant relations are equal. Indeed, $\mathbf{B} \Leftrightarrow_R \mathbf{C} = W$ if and only if $Res\, \mathbf{B} = Res\, \mathbf{C}$.

We close this section with the definition of the $\boldsymbol{\delta}$-operator. This operator, examined by Segerberg (1989) in a somewhat different context, assigns to each proposition $\Phi \subseteq W$ a certain (compound) action $\boldsymbol{\delta}\Phi$ for which he suggested the reading "the bringing about of Φ" or, more briefly, "doing Φ." This action should be thought of as the maximally reliable way of bringing about the state-of-affairs represented by Φ. The $\boldsymbol{\delta}$-operator is a primitive notion in the Segerberg's approach. Here it can be defined as follows:

$$\boldsymbol{\delta}\Phi := \{A_1 \ldots A_n \in \mathcal{A}^* : CDom(A_1 \circ \cdots \circ A_n) \subseteq \Phi\}.$$

($CDom(P)$ is the co-domain (the range) of the relation P, $CDom(P) := \{w \in W : (\exists u \in W)(u, w) \in P\}$.) Note that the action $\boldsymbol{\delta}\Phi$ may be empty.

Denoting by $[\delta\Phi]\Psi$ the proposition

$$\{u \in W : (\forall w \in W)((u, w) \in Res\ \delta\Phi \Rightarrow w \in \Psi)\}$$

(i.e., $[\delta\Phi]\Psi = [Res\ \delta\Phi]\Psi$), we see that δ satisfies Segerberg's conditions:

$$[\delta\Phi]\Psi = W \quad \text{and} \quad [\delta\Phi]\Psi \cap [\delta\Psi]\Lambda \subseteq [\delta\Phi]\Lambda.$$

1.8 Programs and Actions

Definition 1.8.1 Let $\boldsymbol{M} = (W, R, \boldsymbol{\mathcal{A}})$ be an elementary action system. A (finitistic) *program* P over $\boldsymbol{M}$ is any set of possible operations of $\boldsymbol{M}$; that is, P is a set of finite sequences

$$u_1, A_1, u_2, A_3, u_2, \ldots u_m, A_m, u_{m+1}, \tag{1.8.1}$$

where $u_1, u_2, \ldots, u_m, u_{m+1}$ is a nonempty sequence of states of W and $A_1, A_2, \ldots, A_m$ is a sequence of atomic actions of $\boldsymbol{\mathcal{A}}$ such that $(u_i, u_{i+1}) \in A_i$ for $i = 1, 2, \ldots, m$.

As is customary, the sequence (1.8.1) is written down in a more compact form as

$$u_1\ A_1\ u_2\ A_2\ u_3 \ldots u_m\ A_m\ u_{m+1}. \tag{1.8.2}$$

This notation indicates that $(u_i, u_{i+1}) \in A_i$ for $i = 1, 2, \ldots, m$.

Any sequence (1.8.1) belonging to P is interchangeably called a *run*, an *execution* or an *operation* of the program P. □

Definition 1.8.1 provides a semantic concept of a program. In logic, there is a clear division line between the syntactic notion of propositional formula and the semantic notion of a proposition. In computer science, programs are viewed as syntactic entities. Here programs are treated semantically; the semantic counterpart of a program is constituted by the set of all possible runs (executions) of a syntactically-defined program in a definite space of states. We shall use the word *program* with both two meanings, leaving the disambiguation to context.

Definition 1.8.2 Let $\boldsymbol{M} = (W, R, \boldsymbol{\mathcal{A}})$ be an elementary action system.

If $A \in \boldsymbol{\mathcal{A}}$ is an atomic action, then by the *program corresponding to* A we shall understand the set $Pr(A)$ of all triples u_1, A, u_2 with $(u_1, u_2) \in A$.

In many contexts the program $Pr(A)$ will be simply identified with the action A itself.

If $\mathbf{A}$ is a compound action over $\boldsymbol{M}$, then the *program* $Pr(\mathbf{A})$ *corresponding to* $\mathbf{A}$ is the set of all sequences (1.8.1) such that $A_1, A_2, \ldots, A_m \in \mathbf{A}$.

Each sequence (1.8.1) belonging to $Pr(A)$ is called a *possible run* or an *execution* of the action $\mathbf{A}$. □

From the set-theoretic viewpoint, possible *executions* of a compound action **A** are *not* the same objects as possible *performances* of **A**, because the latter are sequences of *states* $u_1, u_2, u_3, \ldots, u_m, u_{m+1}$ such that $u_1 A_1 u_2 A_2 u_3 \ldots u_m A_m u_{m+1}$ for some sequence $A_1, A_2, A_2, \ldots, A_m \in \mathbf{A}$.

According to the above convention, $Pr(\mathbf{A})$ is identified with the set of all sequences (1.8.2) for which $A_1, A_2, \ldots, A_m \in \mathbf{A}$.

Two Examples

(A). Let A be an atomic action of $\boldsymbol{M} = (W, R, \boldsymbol{\mathcal{A}})$ and Φ a subset of W. The program "**while** Φ **do** A" defines the way the action A is to be performed: whenever the system remains in a state belonging to Φ, iterate performing the action A until the system moves outside Φ. (It is assumed here for simplicity that A is atomic; the definition of "**while – do**" programs also makes sense for composite actions as well.) The program "**while** Φ **do** A" is therefore identified with the set of all sequences of the form

$$u_0 \; A \; u_1 \; A \; u_2 \ldots u_m \; A \; u_{m+1} \tag{1.8.3}$$

such that the states $u_0, \ldots, u_m$ belong to Φ and $u_{m+1} \notin \Phi$.

The above program also comprises infinite runs

$$u_0 \; A \; u_1 \; A \; u_2 \ldots u_m \; A \; u_{m+1} \ldots \tag{1.8.4}$$

such that $u_i \in \Phi$ for all $i \geqslant 0$. Each sequence (1.8.4) is called a *divergent run* of the program "**while** Φ **do** A" in the space W.

(B). Let Φ and Ψ be subsets of W and **A** a compound action over $\boldsymbol{M} = (W, R, \boldsymbol{\mathcal{A}})$. A *Hoare program* $\{\Phi\}\mathbf{A}\{\Psi\}$ for **A** is the set of all runs $u_1 A_1 u_2 A_2 u_3 \ldots u_m A_m u_{m+1}$ of **A** such that $u_1 \in \Phi$ and $u_{m+1} \in \Psi$.

Φ is named the *precondition* and Ψ the *postcondition*: when the precondition is met, the action **A** establishes the postcondition.

If A is an atomic action, then $\{\Phi\}A\{\Psi\}$ is defined as the set of all runs $u_1 \; A \; u_2$ of A with $u_1 \in \Phi$ and $u_2 \in \Psi$. □

Example 1.8.3 Collatz's Action System.

This example stems from number theory. We start with an arbitrary but fixed natural number n. If n is even, we divide it by 2 and get $n/2$. If n is odd, we multiply n by 3 and add 1 to obtain the even number $3n + 1$. Then repeat the procedure indefinitely. The Collatz conjecture says that no matter what number n we start with, we will always eventually reach 1.

The Collatz conjecture is an unsolved problem in number theory, named after Lothar Collatz, who proposed it in 1937.

In the framework of our theory, we may reformulate the Collatz conjecture in an equivalent way as follows.

If $n \geqslant 1$ is a natural number, then, by the Fundamental Theorem of Arithmetic, it has a unique representation as a product of powers of consecutive primes, $n = 2^{\alpha} 3^{\beta} 5^{\gamma} 7^{\delta} \ldots$, where all but finitely many exponents $\alpha, \beta, \gamma, \delta, \ldots$ are equal to 0.

Let $\mathbb{N}^+$ be the set of positive integers $\{1, 2, 3, \ldots\}$. We define two atomic actions A and B on $\mathbb{N}^+$. For $m, n \in \mathbb{N}^+$ we put:

$$A(m, n) \Leftrightarrow_{df} m = 2^{\alpha}3^{\beta}5^{\gamma}7^{\delta}\ldots, \text{ with } \alpha \geqslant 1 \text{ and } n = 3^{\beta}5^{\gamma}7^{\delta}\ldots (= 2^{0}3^{\beta}5^{\gamma}7^{\delta}\ldots).$$

Thus, $A(m, n)$ if and only if m is even and n is the (odd) number being the result of dividing m by the largest possible power of 2.

$$B(m, n) \quad \Leftrightarrow_{df} \quad m = 3^{\beta}5^{\gamma}7^{\delta}\ldots \text{ and } n = 3m + 1.$$

In other words, $B(m, n)$ if and only if m is odd and n is the (even) number obtained by multiplying m by 3 and then adding 1. The prime number representation of n may differ much from that of m but it always contains 2 with a positive exponent.

We put: $R := A \cup B$. $\boldsymbol{M} := (W, R, \{A, B\})$ is a normal and deterministic elementary action system.

The compound action $\mathbf{A} := \{AB\}^+ \cup \{BA\}^+$ is called here the *Collatz action* and the system $\boldsymbol{M}$ itself is referred to as the *Collatz system*. $\mathbf{A}$ is performable in every state u of the system. In fact, if u is even, then every finite string of the form $ABABAB\ldots AB$ is performable in u. If u is odd, then every finite string of the form $BABABA\ldots BA$ is performable in u. Thus, every state u gives rise to only one path of states beginning with u and being a possible performance of $\mathbf{A}$.

The program $Pr(\mathbf{A})$ corresponding to $\mathbf{A}$ is the union of two sets of runs:

$$u_1 \; A \; u_2 \; B \; u_3 \; A \; u_4 \; B \ldots u_{2m-1} \; A \; u_{2m} \; B \; u_{2m+1}, \tag{1.8.5}$$

where u_1 is an even number, and

$$u_1 \; B \; u_2 \; A \; u_3 \; B \; u_4 \; A \ldots u_{2m-1} \; B \; u_{2m} \; A \; u_{2m+1}, \tag{1.8.6}$$

where u_1 is odd, and $m \geqslant 1$.

Since A and B are deterministic actions, for each even number u and each m there exists exactly one run (1.8.5) such that $u = u_1$. Similarly, for each odd u and each m there exists exactly one run (1.8.6) such that $u = u_1$.

The *Collatz Conjecture* formulated in the framework of the action system $\boldsymbol{M}$ states that no matter what number u_1 is taken as the beginning of run (1.8.5) or (1.8.6) of $Pr(\mathbf{A})$, the run will always eventually reach 1 for sufficiently large m. In other words, for every u there exists a (sufficiently big) natural number m such that the (unique) run of $Pr(\mathbf{A})$ with $u_1 = u$ terminates with 1, i.e., $u_{2m+1} = 1$. □

Chapter 2
Situational Action Systems

Abstract Situational aspects of action are discussed. The presented approach emphasizes the role of situational contexts in which actions are performed. These contexts influence the course of an action; they are determined not only by the current state of the system but also shaped by other factors as time, the previously undertaken actions and their succession, the agents of actions and so on. The distinction between states and situations is explored from the perspective of action systems. The notion of a situational action system is introduced and its theory is expounded. Numerous examples illustrate the reach of the theory.

In this chapter, structures more complex than elementary action systems are investigated. These are *situational action systems*. Before presenting, in a systematic way, the theory of these structures, we shall illustrate the basic ideas by means of two examples.

2.1 Examples—Two Games

2.1.1 Noughts and Crosses

Noughts and crosses, also called tic-tac-toe, is a game for two players, **X** and **O**, who take turns marking the spaces in a 3×3 grid. Abstracting from the purely combinatorial aspects of the game, we may identify the set of possible states of the game with the set of 3×3 matrices, in which each of the entries is a number from the set $\{0, 1, 2\}$. 0 marks blank, 1 and 2 mark placing Xs and Os on the board, respectively. (It should be noted that many of the 3^9 positions are unreachable in the game.)[1] The rules are well known and so there is no need here to repeat them in detail. The initial state is formed by the matrix with noughts only. The final states are formed by matrices, in which all the squares are filled with 1s or 2s; yet there is neither a column nor a row nor a diagonal filled solely with 1s or solely with 2s (the state of a draw). Neither are there matrices in which the state of 'three-in-a-row'

[1]One may also encode each position by a sequence of length 9 of the digits 0, 1, 2.

J. Czelakowski, *Freedom and Enforcement in Action*, Trends in Logic 42,
DOI 10.1007/978-94-017-9855-6_2

is obtained; i.e., exactly one column or one row, or one diagonal filled with 1s or 2s (the victory of either of the players). The direct transition R from one state u to w is determined by writing either a 1 or a 2 in the place of any one appearance of a nought in matrix u. There is no transition from the final states u into any state. (Relation R is therefore not total.) There are only two elementary actions: A_X and A_O. The action A_X consists in writing a 1 in a free square, whereas the action A_O consists in writing a 2 (as long as there are such possibilities). A_X and A_O are then binary relations on the set of states. The above remarks define the elementary action system $\boldsymbol{M} = (W, R, \{A_X, A_O\})$ associated with noughts and crosses.

Not all the principles of the game are encoded in the elementary system $\boldsymbol{M}$. The fact that the actions are performed alternately is of the key importance, with the provision that the action A_X is performed by the player (agent) **X** and A_O by **O**. We assume that the first move is made by **X**. We shall return to discuss this question after the next example is presented.

The game is operated by two agents **X** and **O**. The statement that somebody or something is an agent, i.e., the doer/performer of a given action, says something which is difficult (generally) to explicate. What matters here is that we show the intimate relations occurring between the agent and the action. If we know that the computer is the agent of, for example, the action A_X (or A_O), then this sentence entails something more: in the wake of it there follows a structure of orders, i.e., a certain program that the computer carries out. It is not until such a programme is established that we can speak, in a legitimate way, of the agency as the special bond between the computer as an agent and the action being performed. When, on the other hand, it is a *person* that is a player (i.e., he is the agent of one of the above-mentioned actions), then in order to establish this bond it is sufficient, obviously, to know that this person understands what he is doing, knows the rules of the game (which he can communicate to us himself), and is actually taking part in the game.

The game of noughts and crosses can also be viewed from the perspective of the theory of automata because the game has well-determined components such as: a finite set of states, a two-element alphabet composed of atomic actions A_X, A_O, the initial state, the set of terminal states, and the indeterministic transition function between states. In consequence, we obtain a finite automaton accepting (certain) words of the form $(A_X A_O)^n$ or $(A_X A_O)^n A_O$, where $n \geqslant 1$. However, as is easily noticed, the above perspective is useless in view of the goals of the game. We are not interested here in *what* words are accepted by the above-described automaton (since it is well-known that these are actions A_X, A_O which are performed alternatingly), but *how* the successive actions should be carried out so that the victory (or at least a draw) is secured.

The details above concerning the game noughts and crosses are fully understood by human beings. One can say even more: such detailed knowledge of the mathematical representation of the game is not necessary. It suffices just to have a pencil and a sheet of paper to start the game. Still, if one of the players is to be a computer (a defective creature) seeing that it is not equipped with senses, the above rules are not sufficient and must be further expanded on in programming language of the computer. It is not hard, though, as a competent programmer can easily write a suitable program on his own.

2.1.2 Chess Playing

We shall use the standard **algebraic notation** (**AN**) (but in a rather rudimentary way) for recording and describing the moves in the game of chess. Each square of the chessboard is identified by a unique coordinate pair consisting of a letter and a number. The *vertical* rows of squares (called *files*) from White's left (the queenside) to his right (the kingside) are labeled $\boldsymbol{a}$ through $\boldsymbol{h}$. The *horizontal* rows of squares (called *ranks*) are numbered **1** to **8** starting from White's side of the board. Thus, each square has a unique identification of the file letter followed by the rank number.

The Cartesian product of the two sets $X := \{\boldsymbol{a}, \boldsymbol{b}, \boldsymbol{c}, \boldsymbol{d}, \boldsymbol{e}, \boldsymbol{f}, \boldsymbol{g}, \boldsymbol{h}\}$ and $Y := \{\mathbf{1}, \mathbf{2}, \mathbf{3}, \mathbf{4}, \mathbf{5}, \mathbf{6}, \mathbf{7}, \mathbf{8}\}$, i.e., the set $X \times Y$, is called the *chessboard*. The elements of $X \times Y$ are called *squares*. The *black squares* are the elements of the set

$$\{\boldsymbol{a}, \boldsymbol{c}, \boldsymbol{e}, \boldsymbol{g}\} \times \{\mathbf{1}, \mathbf{3}, \mathbf{5}, \mathbf{7}\} \cup \{\boldsymbol{b}, \boldsymbol{d}, \boldsymbol{f}, \boldsymbol{h}\} \times \{\mathbf{2}, \mathbf{4}, \mathbf{6}, \mathbf{8}\},$$

and the *white squares* are the elements of the set

$$\{\boldsymbol{a}, \boldsymbol{c}, \boldsymbol{e}, \boldsymbol{g}\} \times \{\mathbf{2}, \mathbf{4}, \mathbf{6}, \mathbf{8}\} \cup \{\boldsymbol{b}, \boldsymbol{d}, \boldsymbol{f}, \boldsymbol{h}\} \times \{\mathbf{1}, \mathbf{3}, \mathbf{5}, \mathbf{7}\}.$$

The game is played by two players (agents): *White* and *Black*. Each of them has 16 pieces (chessmen) at his disposal. *White* plays with the white pieces while *Black* with the black ones. Any arrangement of pieces (not necessarily all of them) on the chessboard is a *possible position*. We consider only the positions where there are at least two kings (white or black) on the chessboard. Thus, a position on the chessboard is any non-empty injective partial function from the set *White pieces* ∪ *Black pieces* into $X \times Y$ such that the two kings belong to its domain. (Not all possible positions appear during the game; some never occur in any game.)

Let W be the set of all possible positions. Instead of the word 'position' we will also be using the term 'chessboard state' or 'configuration'. (The drawings showing the chessboard states are customarily called diagrams.)

The relation R of the direct transition between the chessboard states is determined by the rules of the chessmen's movements. Thus, the two states u and w are in the relation R, i.e., $u\,R\,w$, if and only if in the position u one of the players moves a chessman of a proper colour, and w is the position just after the move. The move is made in accordance with the rules. The position w may have as many chessmen as u, or fewer in the case when some chessmen have been taken. So, a chessman's move that is followed by the taking of an opposite colour chessman is recognized as one move.

The first simplification we will is to reject the possibility of both big and small castling. Accommodating castling is something hampered by certain conditions which will not be analyzed here. The second simplification is to omit the replacement of pawns with chessmen on the change line. (A pawn, having moved across the whole board to the change line must be exchanged for the following chessmen: the queen, a castle, a bishop or a knight from the set of spare chessmen.)

The transition $u\ R\ w$ can be made by the *White* or *Black* player. In a fixed position u, $\{w \in W : u\ R\ w\}$ is then the set of all possible configurations on the chessboard which one arrives at by any side's move undertaken in the state u. The relation R is the join of two relations: B^{Black} and R^{White}, where $R^{White}(u, w)$ takes place if and only if in the position u the *White* player moves, according to the regulations, a white piece which is in the configuration u, and w is the position just following the move. The relation R^{Black} is defined analogously. Thus, $R = R^{White} \cup R^{Black}$. For a given configuration u, the set of all pairs (u, w) such that $R^{White}(u, w)$ takes place ($R^{Black}(u, w)$ takes place, respectively) can be called the set of all moves in the state u admissible for the *White* player (admissible for the *Black* player, respectively).

We adopt here a restrictive definition of the direct transition relation between configurations: in checking situations only the transitions resulting in releasing the opponent's king (if such exist) are admissible. Checking puts the king under the threat of death. When the king is threatened, the danger must be averted by the king's being moved to a square not in any line of attack of any of the opponent's pieces. In other words, in any configuration u, in which the king of a given colour is checked only the transitions from u to the states w, in which the king is released from check are admissible. So, if u is a configuration where, for example, the white king is checked, then $R^{White}(u, w)$ takes place if and only if there exists a white piece's move from u to the position w and the white king is not checked in the position w, i.e., it is not in a position where it could be taken by a black piece. The transition $R^{Black}(u, w)$ is similarly defined in the position u, where the black king is checked. Thus, if one of the kings is checked in the state u, then $u\ R\ w$ must be a move that releases the king from check.

There are various ways of selecting atomic actions in the game of chess. One option is to assign to each piece a certain atomic action. Thus, there are as many atomic actions as there are pieces. Each player can thus perform 16 actions with particular pieces. The above division of atomic actions is based on the tacit assumption that pieces retain their individuality in the course of the game. We will be more parsimonious and distinguish only six atomic actions to be performed by each player. Consequently, the atomic actions we shall distinguish refer to the *types* of pieces, and not to individual pieces. According to **AN**, each type of piece (other than pawns) is identified by an uppercase letter. English-speaking players use **K** for king, **Q** for queen, **R** for rook, **B** for bishop, and **N** for knight (since **K** is already used). Pawns are not indicated by a letter, but rather by the absence of any letter. This is due to the fact that it is not necessary to distinguish between pawns for moves, since only one pawn can move to a given square. (Pawn captures are an exception and indicated differently.)

Here is the list of atomic actions performed by the *White* player:

$$\boldsymbol{K}^{White}, \boldsymbol{Q}^{White}, \boldsymbol{R}^{White}, \boldsymbol{B}^{White}, \boldsymbol{N}^{White}, \boldsymbol{P}^{White}.$$

$\boldsymbol{K}^{White}$ is the action any performance of which is a single move of the white knight, $\boldsymbol{Q}^{White}$ are the white queen's moves, $\boldsymbol{R}^{White}$ are the white rooks' moves, $\boldsymbol{B}^{White}$ the bishops' moves, $\boldsymbol{N}^{White}$ the knights' moves, and $\boldsymbol{P}^{White}$ the action any performance of which is a move of an *arbitrary* white pawn on the chessboard.

A similar division of pieces and atomic actions is also adopted for the *Black* player:

$$\boldsymbol{K}^{Black}, \boldsymbol{Q}^{Black}, \boldsymbol{R}^{Black}, \boldsymbol{B}^{Black}, \boldsymbol{N}^{Black}, \boldsymbol{P}^{Black}.$$

A move with a piece includes taking the opponent's piece when performing this move.

Actions are conceived of extensionally. A given atomic action is identified with the set of its possible performances. Thus, if $A \in \{\boldsymbol{K}^{White}, \boldsymbol{Q}^{White}, \boldsymbol{R}^{White}, \boldsymbol{B}^{White}, \boldsymbol{N}^{White}, \boldsymbol{P}^{White}\}$, e.g. $A = \boldsymbol{K}^{White}$, and $u, w \in W$, then $A(u, w)$ takes place if and only if in the position u the *White* player moves the knight (according to the movement regulations for the knight), and w is the chessboard position just after the move. (According to the algebraic notation, each move of a piece is indicated by the piece's uppercase letter, plus the coordinate of the destination square. For example, ***Be*5** (move a bishop to ***e*5**), ***c*5** (move a pawn to ***c*5**—no letter in the case of pawn moves, remember).

The system

$$\Big(W, R, \{\boldsymbol{K}^{White}, \boldsymbol{Q}^{White}, \boldsymbol{R}^{White}, \boldsymbol{B}^{White}, \boldsymbol{N}^{White}, \boldsymbol{P}^{White}\} \cup \\ \{\boldsymbol{K}^{Black}, \boldsymbol{Q}^{Black}, \boldsymbol{R}^{Black}, \boldsymbol{B}^{Black}, \boldsymbol{N}^{Black}, \boldsymbol{P}^{Black}\}\Big), \tag{2.1.1}$$

where $R = R^{White} \cup R^{Black}$ is an elementary action system operated by two agents *White* and *Black*. The system is not normal. The reason is in the fact that in the positions u in which the king of a given colour is checked, e.g. the white one, the set of all possible performances of the actions $\boldsymbol{K}^{White}, \boldsymbol{Q}^{White}, \boldsymbol{R}^{White}, \boldsymbol{B}^{White}, \boldsymbol{N}^{White}, \boldsymbol{P}^{White}$ in the state u is, in general, larger than the set of admissible transitions from the state u to others, in which the king is no longer under threat. In other words, R^{White} is a proper subset of the union the relations $\boldsymbol{K}^{White} \cup \boldsymbol{Q}^{White} \cup \boldsymbol{R}^{White} \cup \boldsymbol{B}^{White} \cup \boldsymbol{N}^{White} \cup \boldsymbol{P}^{White}$. Analogously, R^{Black} is a proper subset of $\boldsymbol{K}^{Black} \cup \boldsymbol{Q}^{Black} \cup \boldsymbol{R}^{Black} \cup \boldsymbol{B}^{Black} \cup \boldsymbol{N}^{Black} \cup \boldsymbol{P}^{Black}$. Thus, the actions from the above list are not totally performable in some states.

The system (2.1.1) is complete. Although certain configurations u occur in none of the chess games, each direct transition $u\,R\,w$ is made by means of a certain atomic action from the above list, i.e., there exists an A such that uA, Rw. From this remark the completeness of the system follows.

u_0 is the initial configuration accepted in any chess game, i.e., the arrangement of pieces on the chessboard at the start of each game. The proposition $\Phi := \{u_0\}$ is thus the initial condition of each game.

In the game of chess, a few types of final conditions are distinguished. The most two important are where the black king is checkmated and where the white king is checkmated.

Let Ψ_0^{Black} be the set of all possible positions u on the chessboard in which the black king is checked by some white piece. The proposition Ψ_0^{Black} expresses the fact that the black king is checked (but not the fact that the next move must be performed by the *Black* player). Analogously, the set Ψ_0^{White} is defined as the set of positions u in which the white king is checked.

The king is checkmated when he is in a configuration in which he is checked *and* when there is no way of escape nor cannot he in any other way be protected against the threat of being taken. Checkmate results in the end of the game.

So, the checkmate of the black king is the set Ψ^{Black} of all positions u in which the black king is checkmated and none of the actions $\boldsymbol{K}^{Black}$, $\boldsymbol{Q}^{Black}$, $\boldsymbol{R}^{Black}$, $\boldsymbol{B}^{Black}$, $\boldsymbol{N}^{Black}$, $\boldsymbol{P}^{Black}$ can be undertaken in order to protect the black king. (As noted above, we are omitting here the possibility of castling as a way to escape checkmate.) This fact can be simply expressed in terms of the relation R^{Black} as follows:

$$\Psi^{Black} := \{u \in \Psi_0^{Black} : \delta_R^{Black}(u) = \emptyset\}.$$

(δ_R^{Black} is the graph of the relation R^{Black}.)

In the analogous way we define the proposition Ψ^{White} which expresses the checkmate of the white king:

$$\Psi^{White} := \{u \in \Psi_0^{White} : \delta_R^{White}(u) = \emptyset\}.$$

Thus, in the game of chess, we distinguish two tasks: (Φ_0, Ψ^{Black}) and (Φ_0, Ψ^{White}). The task (Φ_0, Ψ^{Black}) is taken by the *White* player. He aims to checkmate the black king. The states belonging to the proposition Ψ^{Black} define then the goal the *White* player wants to achieve. Analogously, (Φ_0, Ψ^{White}) is the task for the *Black* player. His goal is to checkmate the white king.

A stalemate is a position in which one of the players, whose turn comes, cannot move either the king or any other chessman; at the same time the king is not checked. When there is a stalemate on the chessboard, the game ends in a draw.

Let Λ^{Black} denote the set of stalemate positions of the black king, i.e.,

$$\Lambda^{Black} := \{u \in W : u \notin \Psi_0^{Black} \text{ and } \delta_R^{Black}(u) = \emptyset\}.$$

Similarly, Λ^{White} denotes the set of stalemate positions of the white king:

$$\Lambda^{White} := \{u \in W : u \notin \Psi_0^{White} \text{ and } \delta_R^{White}(u) = \emptyset\}.$$

If during the game the position $u \in \Lambda^{Black}$ is reached and the *Black* player is to make a move in this position, the game ends in a draw; similarly for when $u \in \Lambda^{White}$ and the *White* player is to move.

The stalemate positions do not exhaust the set of configurations which give rise to the game ending. A game ends in a draw when both players know that there exist possibilities for continuing the game indefinitely which do not lead to positions of $\Psi^{Black} \cup \Psi^{White}$. In practice, a purely pragmatic criterion is applied which says that the game is drawn after the same moves have been repeated in sequence three times; the game is also limited by the rule that in 50 moves a pawn must have been moved.

In spite of simplifications made in the chess tournament, we have not explored all the rules of the game. We know what actions the players perform but we have not added that they can only do them alternately. The same player is not allowed to make successively two or more moves (except the case of castling). Moreover, the game is started only by the *White* player in the strictly defined initial position. These facts indicate that the description of the game of chess, expressed by means of formula (2.1.1), is oversimplified and it does not adequately reflect the real course of the game. We shall return to this issue in the next paragraph.

The above situational description of the game of chess does not have, obviously, any greater value as regards its usability. For obvious reasons neither the best of players nor any computer is able to take on a victor's strategy through searching all possible configurations on the chessboard that are continuations of the given situation. The phenomenon of combinatorial explosion blocks further calculations (we get very large numbers, even for a small number of steps). At the time of writing, the best computer can search at the most three moves ahead in each successive configuration (a move is one made by *White* and the successive move by *Black*). It is in this fact, as well as the rate at which computers operate, that the advantage the computer has over the human manifests itself. One obvious piece of advice and the most reasonable suggestion offered to the human player is to select a strategy referring based on his experience as a chess-player and the knowledge gathered over the centuries, where this includes a description of the finite set (larger but not too large—within the limits of what a human brain cell remember) of particularly significant games.

2.2 Actions and Situations

The above examples reveal the significant role of situational contexts in which action systems function. These contexts that influence the course of an action are not determined just by the current state of the action system but also by other factors. It is often difficult or even not feasible to specify them all. Moving a black piece to an empty square may be allowed by the relation R^{Black} if only the current arrangement of chessmen is taken into account; the move however will not be permitted when the previous move was also made by the *Black* player for two successive moves by the same player does not comply with the rules of chess. A similar remark applies to tic-tac-toe.

The above examples also show that the performability of an action in a particular state u of the system may depend on the previously undertaken actions, their succession, and so on. In the light of the above remarks a distinction should be made between the notion of a state of the system and the situation of the system. In this chapter this distinction is explored from the perspective of action systems.

The set W of possible states, the relation R of a direct transition between states and the family $\mathcal{A}$ of atomic actions, fully determine the possible transformations of the atomic system, i.e., the transformations (carried out by the agents) that move the system from one state to others. The current situation of the system is in general shaped by a greater number of factors. A given situation is determined by the data from the surroundings of the system. The current state of the system is one of the elements constituting the situation but, of course, not the most important. The agents are involved in a network of mutual relations. The situation in which the agents act may depend on certain principles of cooperation (or hostility) as well. Acceptance of certain norms of action is an example of such interdependencies. Thus, each state of the system is immersed in certain wider situational contexts. The context is specified, among others, by factors, not directly bound up with the state of the system as: the moment when a particular action is undertaken, the place where the system or its part is located at the moment the action is started or completed, the previously performed action (or actions) and their agents, the strategies available to the agents, etc. Not all of the mentioned factors are needed to make up a given situation—it depends on the depth of the system description and the principles of its functioning. These elements constitute the situational setting of a given state.

In a game of chess the relation R of direct transition between configurations on the chessboard defined in Sect. 2.1 as the union $R = R^{White} \cup R^{Black}$ is insensitive to the situational context the players are involved in. The atomic actions $\boldsymbol{K}^{White}, \boldsymbol{Q}^{White}, \boldsymbol{R}^{White}, \boldsymbol{B}^{White}, \boldsymbol{N}^{White}, \boldsymbol{P}^{White}$ or $\boldsymbol{K}^{Black}, \boldsymbol{Q}^{Black}, \boldsymbol{R}^{Black}, \boldsymbol{B}^{Black}, \boldsymbol{N}^{Black}, \boldsymbol{P}^{Black}$ transform some configurations into others. The above purely extensional, 'input—output' description of atomic actions abstracts from some situational factors which influence the course of the game. The performability of an atomic action depends here exclusively on the state of the system in which the action is undertaken (Definition 1.4.1). Thus performability is viewed here as a context-free property, devoid of many situational aspects which may be relevant in the description of an action.

We shall outline here a situational concept of action performability, according to which the fact that an action is performable in a given state of the system depends not only on the state and the relation R of direct transition between states but also on certain external factors—the situational context the system is set in.

Performing an action changes the state of the system and at the same time it creates a new situation. A move made by a player in a game of chess changes the arrangement of chessmen on the board. It also changes the player's situation: the next move will be made by his opponent unless the game is finished. It does not mean that the notion of an action should be revised—as before we identify (atomic) actions with binary relations on the set W of possible states of the system. We shall, however, extend the notion of an elementary action system by enlarging it with new components: the set S of possible situations, the relation Tr of a direct transition between possible situations, and a map f which to every possible situation s assigns a state $f(s) \in W$. The state $f(s)$ is a part of the situation s—if s occurs then $f(s)$ is the state of the system corresponding to s. These components enable one to articulate a new, 'situational' definition of the performability of an action. Thus, we will speak of action performability in a given state of the system with respect to a definite situational context, or shortly, the *performability of an action in a given situation*.

Situations are not investigated here in depth as a separate category. That is, we shall not investigate and develop an ontology of situations but rather limit ourselves to some general comments. The focus is rather on illustrating the role of situations in the action theory in various contexts than outlining a general account of structured situations. The notion of a situation we shall use here is built out of elements taken from automata theory, theory of algorithms, games and even physics and does not fully agree with the notions that occur in the literature.

We shall present here a simplified, 'labeled' theory of situations. This theory is convergent with the early formal approaches to logical pragmatics (see e.g. Montague 1970; Scott 1970). On the other hand, this concept is based on some ideas which directly stem from the theory of algorithms.

The ontology of actions adopted in this book shows similarities with the situation calculus applied in logic programming. The situation calculus is a formalism designed for representing and reasoning about dynamic entities (McCarthy 1963; Reiter 1991). It is based on three key ingredients: the actions that can be performed in the world, the situations, and the fluents that describe the state of the world. In Reiter's approach situations are finite sequences of actions. Here, situations are isolated and form a separate ontological category.

Let S be a set, whose elements will be called *possible* (or *conceivable*) situations.[2] Each situation $s \in S$ is determined by a system of factors. Their specification depends on the ways of organization and functioning of the action system. A possible situation can include the following components: the state of the system, the location of the system (or its distinguished parts), the agent of each atomic action, the action currently performed on the system, the previously performed action and its agent, and so on. We can roughly characterize any situation s as a sequence of entities

$$s = (w, t, x, \ldots), \tag{2.2.1}$$

where $w \in W$ is a possible state of the system, t—time, x—location of the system, etc. To each possible situation (2.2.1) a unique state w is assigned—the first component of the above sequence—which is called the *state of the system in the situation* s. The values of all the other parameters (and strictly speaking the names of the values of

[2]This sentence is a convenient conceptual metaphor. Situations are constituents of the real world. In this chapter situations are viewed as *mathematical* (or rather *set-theoretical*) representations of these constituents. We therefore speak of sets of conceivable situations.

Of course, any representation of the world of situations and actions does not involve mentioning all of them individually but it rather classifies them as uniform sorts or types. But token situations or actions can be individuated, such as the stabbing of Julius Caesar.

these parameters) defining a given situation form a subsequence of s which is called the *label* of the situation s. In the case of situation (2.2.1), its label is equal to the sequence $(t, x, \ldots)$. Thus, each possible situation s can be represented by an ordered pair

$$s = (w, a),$$

where $w \in W$ is a state of the system and a is a label.

The labels are assumed to form a set which is denoted by V. (It is not assumed that the set V is finite.) Thus, the set S of possible situations is equal to the Cartesian product $W \times V$, i.e.,

$$S := W \times V.$$

Apart from the set S there is a transition relation Tr between situations. If $Tr(s_1, s_2)$ holds, then s_2 is the situation immediately occurring after the situation s_1; we also say that the situation s_1 *directly turns* into the situation s_2. The fact that $Tr(s_1, s_2)$ holds does not prejudge the occurrence of the situation s_1.

The above remarks enable us to articulate the definition of a situational action system.

Definition 2.2.1 A *situational action system* is a six-tuple

$$\boldsymbol{M}^s := (W, R, \boldsymbol{\mathcal{A}}, S, Tr, f),$$

where

(1) the reduct $\boldsymbol{M} := (W, R, \boldsymbol{\mathcal{A}})$ is an elementary action system in the sense of Definition 1.2.1;

(2) S is a non-empty set called the set of *possible situations* the action system $\boldsymbol{M}$ is set in. The set S is also called the *situational envelope* of the action system $\boldsymbol{M}$;

(3) Tr is a binary relation on S, called the *direct transition relation* between possible situations;

(4) $f : S \to W$ is a mapping which to each situation $s \in S$ assigns a state $f(s) \in W$ of the action system $\boldsymbol{M}$. $f(s)$ is called the *state of the action system* $\boldsymbol{M}$ *corresponding to the situation* s, or simply, the *state of the system in the situation* s. It is therefore unique, for each situation s;

(5) the relation R of direct transition between states of the action system $\boldsymbol{M} = (W, R, \boldsymbol{\mathcal{A}})$ is compatible with Tr, i.e., for every pair $s_1, s_2 \in S$ of situations, if $s_1 \, Tr \, s_2$ then $f(s_1) \, R \, f(s_2)$. □

The changes in the situational envelope take place according to the relation Tr. Condition (5) says that the evolution of situations in the envelope is compatible with transformations between the states of the system $\boldsymbol{M}$, defined by the relation R. (In the 'labeled' setting of situations, that is, when $S = W \times V$, the function f is defined as the projection of $W \times V$ onto W, i.e., $f(s) = w$, for any situation $s = (w, a) \in S$.)

It follows from (5) that the set

$$\big\{(u, w) \in W \times W : \big(\exists s, t \in S\big)\big(s \, Tr \, t \;\; \& \;\; f(s) = u \;\; \& \;\; f(t) = w\big)\big\}$$

is contained in the relation R. The two sets need not be equal. In other words, it is not postulated that for any $s_1 \in S$, $w \in W$ the condition $f(s_1)\, R\, w$ implies that $s_1\, Tr\, s_2$ for some $s_2 \in S$ such that $w = f(s_2)$. (For example, this implication fails to hold in the situational model of chess playing defined below.) This shows that, in general, the relation $R \subseteq W \times W$ cannot be eliminated from the description of situational action systems and replaced by the relation Tr).

In the simplest case, possible situations of S are identified with states of the action system $\boldsymbol{M} = (W, R, \boldsymbol{\mathcal{A}})$, that is, $S = W$ (the label of each situation is the empty sequence), f is the identity map, and $Tr = R$, and the situational action system $\boldsymbol{M}^s$ reduces to the elementary system $\boldsymbol{M}$.

Definition 2.2.1 is illustrated by means of so-called *iterative* action systems; the latter form a subclass of situational systems. Iterative action systems function according to simple algorithms that define the order in which particular atomic actions are performed. A general scheme which defines these systems can be briefly described in the following way. Let $\boldsymbol{M} = (W, R, \boldsymbol{\mathcal{A}})$ be an elementary action system. It is assumed that S, the totality of possible situations, is equal to the Cartesian product $W \times V$, where V is a fixed set of labels. The function f that assigns to each situation s the corresponding state of the system $\boldsymbol{M}$ is the projection from $W \times V$ onto W. In order to define the relation Tr of a direct transition between situations we proceed as follows. An action performed in a definite situation changes the state of the system $\boldsymbol{M}$ and also results in a change of the situational context. The triples

$$(a, A, b), \tag{2.2.2}$$

where a and b are labels in V and $A \in \boldsymbol{\mathcal{A}}$ is an atomic action, are called *labeled actions*. The action (2.2.2) has the following interpretation: the atomic action A performed in a situation with label a changes the state of the system $\boldsymbol{M}$ in such a way that the situation just after performing this action is labeled by b. Thus, (a, A, b) leads from situations labeled by a to ones with label b while the states of $\boldsymbol{M}$ transform themselves according to the action A and the relation R.

A non-empty set $\boldsymbol{LA}$ of labeled actions is singled out. $\boldsymbol{LA}$ need not contain all the triples (2.2.2) nor need be a finite set. We say that a labeled action $(a, A, b) \in \boldsymbol{LA}$ *transforms* a situation s_1 into s_2 if and only if $s_1 = (u, a)$, $s_2 = (w, b)$ and, moreover, it is the case that $u\, A,\, R\, w$.

The set $\boldsymbol{LA}$ makes it possible to define the transition relation $Tr(\boldsymbol{LA})$ between situations. Let $s_1 = (u, a)$, $s_2 = (w, b) \in S$. Then we put:

$$(s_1, s_2) \in Tr(\boldsymbol{LA}) \quad \text{if and only if} \quad (\exists A \in \boldsymbol{\mathcal{A}})\big((a, A, b) \in \boldsymbol{LA}\ \&\ uA, Rw\big). \tag{2.2.3}$$

It follows from (2.2.3) that $(s_1, s_2) \in Tr(\boldsymbol{LA})$ implies that $f(s_1)\, R\, f(s_2)$, i.e., R is compatible with Tr and that the conditions (1)–(5) of Definition 2.2.1 are met. The six-tuple

$$(W, R, \boldsymbol{\mathcal{A}}, S, Tr(\boldsymbol{LA}), f)$$

is a situational action system.

A finite *run* of situations is as any finite sequence

$$(s_0, s_1, \ldots, s_n)$$

of situations such that $(s_i, s_{i+1}) \in Tr(\boldsymbol{LA})$ for every $i, 0 \leqslant i \leqslant n-1$. The situation s_0 is called the *beginning* of the run.

We may also consider infinite (or *divergent*) runs of situations as well. These are infinite sequences of situations

$$(s_0, s_1, \ldots, s_n, \ldots)$$

such that $(s_i, s_{i+1}) \in Tr(\boldsymbol{LA})$, for all i. The problem of divergent runs, interesting from the viewpoint of computability theory, is not discussed in this book.

In tic-tac-toe two labels are distinguished: X and O. Thus, $V := \{\mathrm{X}, \mathrm{O}\}$. A possible situation is a pair $s = (w, a)$, where $a \in V$ and $w \in W$ is a configuration on the board (each configuration being identified with an appropriate 3×3 matrix). The pair $s = (w, \mathrm{X})$ is read: w is the current configuration on board and **X** is to make a move. One interprets the pair (w, O) in a similar way.

The set of labeled actions has two elements: $(\mathrm{X}, A_{\mathrm{X}}, \mathrm{O})$ and $(\mathrm{O}, A_{\mathrm{O}}, \mathrm{X})$.[3] The action $(\mathrm{X}, A_{\mathrm{X}}, \mathrm{O})$ transforms a situation of the form $s_1 = (u, \mathrm{X})$ into a situation $s_2 = (w, \mathrm{O})$. The transformation from s_1 to s_2 is accomplished by performing the action A_{X} by **X**. This action moves the system from the state u to w. The labeled action $(\mathrm{O}, A_{\mathrm{O}}, \mathrm{X})$ is read in a similar way.

The second example is similarly reconstructed as a situational action system. In a game of chess we distinguish two labels, BLACK and WHITE, that is $V := \{\mathrm{BLACK}, \mathrm{WHITE}\}$. A possible chess situation is thus any pair of the form $s = (w, a)$, where $a \in V$ and $w \in W$ is a configuration on the chessboard. The pair $s = (w, \mathrm{WHITE})$ is interpreted as follows: w is the current configuration on the chessboard and *White* is to make a move. The pair (w, BLACK) is read analogously.

The set $\boldsymbol{LA}$ of labeled atomic actions consists of the following triples:

$$(\mathrm{WHITE}, A^{White}, \mathrm{BLACK}) \tag{2.2.4}$$

where $A^{White} \in \{\mathbf{K}^{White}, \mathbf{Q}^{White}, \mathbf{R}^{White}, \mathbf{B}^{White}, \mathbf{N}^{White}, \mathbf{P}^{White}\}$ and

$$(\mathrm{BLACK}, A^{Black}, \mathrm{WHITE}) \tag{2.2.5}$$

where $A^{Black} \in \{\mathbf{K}^{Black}, \mathbf{Q}^{Black}, \mathbf{R}^{Black}, \mathbf{B}^{Black}, \mathbf{N}^{Black}, \mathbf{P}^{Black}\}$.

The labeled action (2.2.4) thus transforms situations of the type $s_1 = (u, \mathrm{WHITE})$ into situations $s_2 = (w, \mathrm{BLACK})$. The transformation is accomplished by

---

[3] Indeed, instead of the triples $(\mathrm{X}, A_{\mathrm{X}}, \mathrm{O})$ and $(\mathrm{O}, A_{\mathrm{O}}, \mathrm{X})$ it suffices to take the pairs $(A_{\mathrm{X}}, \mathrm{O})$ and $(A_{\mathrm{O}}, \mathrm{X})$, since the actions A_{X} and A_{O} are already labeled by X and O, respectively (that is, they possess their own names). Defining labeled actions as triples is justified by the fact that elements of the set of atomic actions in elementary action systems are not basically linguistically distinguished through investing them with special names.

performing in the state u an atomic action by *White* so that w is the configuration of the chessboard just after the move. The labeled action (2.2.5) is interpreted in a similar way.

Having defined the set $\boldsymbol{LA}$, we then define the relation $Tr(\boldsymbol{LA})$ according to formula (2.2.3). The function $f : S \to W$ is defined, as expected, as the projection of S onto $W : f\big((w, a)\big) := w$, for all $(w, a) \in S$.

There is only one initial situation $s_0 := (u_0, \text{WHITE})$, where u_0 is the configuration at the outset of the game. Thus, a *game of chess* is a finite sequence of situations $(s_0, s_1, \ldots, s_n)$, $n \geqslant 0$, such that $s_0 = (u_0, \text{WHITE})$ is the initial situation and $(s_i, s_{i+1}) \in Tr(\boldsymbol{LA})$, for all $i, 0 \leqslant i \leqslant n-1$. It follows from the definition of $Tr(\boldsymbol{LA})$ that successive moves are alternately performed by the players *Black* and *White*. The transition from s_0 to s_1 is accomplished by *White* who starts the game. In turn, the transition from s_1 to s_2 is made by *Black*, and so on. Thus, a game of chess can be represented as an iterative action system, taking into account the simplifications we have made.

A few words about terminal situations. In the light of the above definitions, a game of chess may be *any* long sequence $(s_0, \ldots, s_n)$ of a situation starting with the initial situation. A game thus defined need not respect the rules of chess. First, there exist time limitations that do not permit game to go on for too long. (They can be inserted into the scheme of situation presented here by adjoining an additional time parameter.) However, there is a more important reason. A game of chess *has* to finish in the case of checkmating the adversary. The above definition of a game does not take this factor into consideration. To resolve the matter we will define a certain subclass of the class of possible situations. First, we will extend the set of labels with a new label ω, which will be called the *terminal* label. Thus

$$V' := \{\text{BLACK}, \text{WHITE}, \omega\}.$$

We also say that

$$S' := W \times V'.$$

A possible *terminal* situation is any situation of the form (w, ω), where w is a configuration belonging to the set $\Psi^{Black} \cup \Psi^{White}$ i.e., w is a configuration in which either the black or the white king is checkmated.

The set $\boldsymbol{LA}$ of labeled atomic actions defined by means of the formulas (2.2.4) and (2.2.5) is also extended to the set $\boldsymbol{LA}'$ by augmenting it with the following actions:

$$(\text{BLACK}, A^{White}, \omega) \tag{2.2.6}$$

where A^{White} is an atomic action assigned to *White*, and

$$(\text{WHITE}, A^{Black}, \omega) \tag{2.2.7}$$

where A^{Black} is an atomic action of *Black*.

Note One should carefully distinguish between linguistic interpretations of a game and situational runs of a game. Here ω, BLACK, WHITE are linguistic entities while A^{White} and A^{Black} are binary relation on the set of states. □

Let *Tr*(***LA***$'$) be the transition relation between situations of S' determined by the labeled actions of ***LA***$'$. As is easy to check, the terminal situations defined as above are indeed terminal—if s is terminal then there exists no situation s' such that *Tr*(***LA***$'$)(s, s'). The notion of a game of chess is defined similarly as above; that is, as a finite sequence of situations $(s_0, s_1, \ldots, s_n)$, $n \geqslant 0$, such that $s_0 = (u_0, \text{WHITE})$ is the initial situation and $(s_i, s_{i+1}) \in$ *Tr*(***LA***$'$), for all i. A game is concluded if the situation s_n is terminal. (The definition of a concluded game does not exhaust the set of all situations in which the game is actually concluded. A game can be concluded through being a draw. To take these situations into account, we would have to further extend the notion of a situation and the transition relation between situations.)

The distinction between the meanings of 'state' and 'situation' is not absolute. When speaking about elementary action systems, we do not always have in mind sharply-distinguished material objects subject to the forces exerted by the agents. The definition of an elementary system distinguishes only certain states of affairs and relations between them; in particular the relation of a direct transition. The selection of one or other set of states and binary relations representing atomic actions greatly depends on the 'world perspective' and on a definite perception of the analyzed actions in particular. We may figuratively say that elementary action systems define *what* actions are performed while situational action systems also take into account *how* they are performed. The borderline between the two concepts is fluid. For example, one may modify the scheme of chess presented above so as to include the players in the notion of a state (i.e. a configuration of pieces). Thus, e.g., the fact that *Black* is to make a move would be a component of the current state of the game. Such a step is, of course, possible; it would however complicate the description of the game.

Suppose $\boldsymbol{M} = (W, R, \boldsymbol{\mathcal{A}})$ is a quite arbitrary elementary action system. If one wants to restrict uniformly all actions in the system to sequences of states of length at most n, it suffices to introduce n labels, being e.g. the consecutive natural numbers (or numerals) $0, 1, \ldots, n$, and to define the set of possible situations as the product $S := W \times \{0, 1, \ldots, n\}$. The function f assigning to each situation s its unique state is the projection onto the first axis, i.e., $f(s) := w$ for any situation $s = (w, k)$.

A given state may therefore receive $n + 1$ different labels. The (direct) transition relation *Tr* between situations is defined as follows: for $s = (w, k)$ and $s' = (w', k')$,

$$s\,Tr\,s' \quad \Leftrightarrow_{df} \quad w\,R\,w' \;\wedge\; w \neq w' \;\wedge\; k' = k + 1.$$

Each situational transition changes states and increases the label from k to $k + 1$. (The second conjunct excludes reflexive points of R but does not exclude loops of states of longer length from situational transitions. If we want to exclude loops $u\,R\,u_1\,R\,\ldots u_l\,R\,u$ altogether as 'futile' computations, the problem is much more complicated and not analysed here.)

$\boldsymbol{M}^s = (W, R, \mathcal{A}, Tr, f)$ is a well-defined situational action system in which $\boldsymbol{M} = (W, R, \mathcal{A})$ is contained.

The situations of the form $s = (w, n)$ are terminal, where w is an arbitrary state. This means that for $s = (w, n)$ there is no situation s' such that $s\,Tr\,s'$. Therefore the systems halts at $s = (w, n)$ and the work of the action system $\boldsymbol{M}$ ceases as being subject to the organization of $\boldsymbol{M}^s$. In turn, situations of the form $(w, 0)$ may be treated as initial ones.

Since loops in runs of situations are not excluded, it may happen that there is a sequence of transitions between situations of length $\geqslant 3$, say $s\,Tr\,s_1\,Tr \dots s_l\,Tr\,s$. In this case the system starts at s and finishes at s (in the same state). But of course we may declare at the outset that R does not admit loops.

The width of the situational envelope of an elementary action system depends on how the system functions. Let us take a look at Example 1.3.3. W_T is here the set of all proofs carried out from T with the help of the rules from a set Θ. Let $w = (\phi_0, \dots, \phi_{n-1}, \phi_n)$ be a fixed proof in W_T. In the simplest case, the situational context of the proof w is constituted by the way the formula ϕ_n, i.e., the last element of the proof w, is adjoined to the shorter proof $u = (\phi_0, \phi_1, \dots, \phi_{n-1})$. The proof w is a result of applying a definite rule $r \in \Theta$ to the some 'prior' formulas $\{\phi_j : j \in J\}$, where J is a subset of $\{0, \dots, n-1\}$; that is, to some formulas occurring in u. The situational context of w is therefore represented by means of the triple (u, r, J), which encodes the above fact. Denoting the pair (r, J) by a, we see that the above situation can be identified with the pair $s = (u, a)$. A deeper description of possible situations would take into account not only the way the proposition ϕ_n was affixed to u forming the proof w, but also, for example, the ways all or some of the sub-proofs of u were formed. Along with changing the width and the depth of the description of possible situations, the description of the relation Tr of a direct transition between possible situations would be modified too.

The second remark concerns current situations of action systems. An action system always finds itself in a definite situation—the *current situation* of the system. The state corresponding to this situation (the state in which the system is) is called the *current state* of the system. It is not possible to single out the separate category of current situations by means of linguistic procedures though it is possible to provide an exhaustive list of attributes determining these situations. Hybrid logic introduces nominals which are objects that signify only one situation, namely the one which actually happens. For example, one can always describe the configuration of pieces which are *now* placed on the chessboard but no linguistic operation is able to fully render the meaning of the word 'now'. The terms 'the current situation' and 'the current state' are demonstratives—their proper understanding always requires knowledge of extralinguistic factors.

The third remark concerns the relationship between situations and (possibly infinite) runs of situations. Suppose

$$\boldsymbol{M}^s = (W, R, \mathcal{A}, S, Tr, f)$$

is a situational system. A *possible* (or *hypothetical*) *run* of situations (in $\boldsymbol{M}^s$) is any function mapping an interval of integers into S. There are thus *a priori* four possible *types* of runs. A run is of type $(-\infty, +\infty)$ if it is indexed by the set of all integers, i.e., it is represented as an infinite sequence without end points

$$(\ldots, s_{-m}, s_{-m+1}, \ldots, s_0, \ldots, s_n, s_{n+1}, \ldots), \tag{2.2.8}$$

where $s_k \; Tr \; s_{k+1}$, for every integer k.

A run is of type $(-\infty, 0)$ if it is indexed by the non-positive integers, i.e., it is of the form

$$(\ldots, s{-m}, s_{-m+1}, \ldots, s_0), \tag{2.2.9}$$

where $s_k \; Tr \; s_{k+1}$ for every negative integer k and there does not exist a situation $s \in S$ such that $s_0 \; Tr \; s$.

A run is of type $(0, +\infty)$ if it is of the form

$$(s_0, \ldots, s_n, s_{n+1}, \ldots), \tag{2.2.10}$$

where $s_k \; Tr \; s_{k+1}$ for every natural number $k \in \omega$ and there does not exist a situation $s \in S$ such that $s \; Tr \; s_0$.

Finally, a run is *finite* (and terminated in both directions) if it is of the form $(s_0, \ldots, s_n)$ for some natural number n and there do not exist situations $s, s' \in S$ such that $s \; Tr \; s_0$ or $s_n \; Tr \; s'$.

Whenever we speak of a run of situations, we mean *any* run falling into one of the above four categories. If a situation s occurs in a run, then the situation s' in the run that directly follows s is called the *successor* of s; the situation s is then called the *predecessor* of s'.

Proposition 2.2.2 *Let* $\boldsymbol{M}^s = (W, R, \boldsymbol{\mathcal{A}}, S, Tr, f)$ *be a situational action system. Let* s, s' *be two possible situations in* S*. Then* $s \; Tr \; s'$ *holds if and only if there exists a run (of one of the above types) such that* s *and* s' *occur in the run and* s' *is the successor of* s.

The proof is simple and is omitted. □

To each situational system $\boldsymbol{M}^s$ the class $\boldsymbol{R}$ of all possible runs of situations in the system is assigned. It may happen that $\boldsymbol{R}$ includes runs of all 4 types. This property gives rise to a certain classification of situational systems. For example, a system $\boldsymbol{M}^s$ would be of category 1 if $\boldsymbol{R}$ contained only finite runs. There are $2^4 - 1$ $(=15)$ possible categories of situational action systems.

The relation Tr of direct transitions between situations can be unambiguously described in terms of runs of situations. More precisely, any situational action system can be equivalently characterized as a quintuple

$$(W, R, \boldsymbol{\mathcal{A}}, S, f) \tag{2.2.11}$$

satisfying the earlier conditions imposed on $(W, R, \mathcal{A})$ and S, f, for which additionally the class $\boldsymbol{R}$ of runs of situations from S is singled out. $\boldsymbol{R}$ is postulated to be the class of all runs of situations admissible for the system (2.2.11). The relation *Tr* of direct transitions between situations can be then *defined* by the right-hand side of the statements of Proposition 2.2.2.

We conclude this section with remarks on the performability of actions in situational systems. Suppose we are given a situational action system $\boldsymbol{M}^s$ in the sense of Definition 2.2.1. The fact that A is performable in a definite situation s is fully determined by the relations *Tr* and R. Accordingly:

If s is a situation and $A \in \mathcal{A}$ is an atomic action, then the act of performing the action A in this situation turns s into a situation s' such that $s\, Tr\, s'$ and $f(s)\, A\, f(s')$.

Immediately after performing the action A the system $\boldsymbol{M} = (W, R, \mathcal{A})$ is in the state $f(s')$. This is due to the fact that R is compatible with *Tr*, i.e., $s\, Tr\, s'$ implies that $f(s)\, R\, f(s')$.

The above remarks give rise to the following definition:

Definition 2.2.3 Let $\boldsymbol{M}^s = (W, R, \mathcal{A}, S, Tr, f)$ be a situational action system and let $s \in S$ and $A \in \mathcal{A}$.

(i) The atomic action A is *performable in the situation* s if and only if there exists a situation $s' \in S$ such that $s\, Tr\, s'$ and $f(s)\, A\, f(s')$; otherwise A is *unperformable* in s.
(ii) The action A is *totally performable in the situation* s if and only if it is performable in s and for every state $w \in W$, if $f(s)\, A\, w$, then $s\, Tr\, s'$ for some situation s' such that $w = f(s')$. □

The performability of an action A in a situation s is thus tantamount to the existence of the direct *Tr*-transition from s to another situation s' such that the pair $\big(f(s), f(s')\big)$ is a possible performance of A (i.e., A is accounted for the transition $s\, Tr\, s'$). The second conjunct of the definition of total performability of A in s states that *every* possible performance of A in the state $f(s)$ results in a new situation s' such that $s\, Tr\, s'$ (i.e., the relation *Tr* imposes no limitations on the possible performances of A in $f(s)$).

Similar to the case of elementary action systems, total performability always implies performability. The converse holds for deterministic actions, i.e. actions A which are partial functions on W.

The concept of the situational performability of an atomic action is extended onto non-empty compound actions. Let $\boldsymbol{M}^s$ be a situational system and $\mathbf{A} \in C\mathcal{A}$ a compound action. $\mathbf{A}$ is *performable in a situation* s if and only if there exists a non-empty string $(s_0, \ldots, s_n)$ of situations and a string of atomic actions $A_1 \ldots A_n \in \mathbf{A}$ such that $s = s_0\, Tr\, s_1 \ldots s_{n-1}\, Tr\, s_n$ and $f(s_0)\, A_1\, f(s_1) \ldots f(s_{n-1})\, A_n\, f(s_n)$ (i.e., the transition from s to s_n is effected by means of consecutive performances of the actions $A_1, \ldots, A_n$).

The action $\mathbf{A}$ is *totally performable in* s if and only if $\mathbf{A}$ is performable in s and for every possible performance $(u_0, \ldots, u_n)$ of $\mathbf{A}$ such that $u_0 = f(s)$ there exist situations $s_0, \ldots, s_n$ with the following properties: $s_0 = s$, $u_i = f(s_i)$ for $i = 0, \ldots, n$, and $u_i \, Tr \, u_{i+1}$ for all $i \leqslant n-1$.

The second conjunct of the above definition states that for every performance $(u_0, \ldots, u_n)$ of **A** starting with $u_0 = f(s)$ there exists a run of situations $(s_0, \ldots, s_n)$ with $s_0 = s$ such that $u_0, \ldots, u_n$ are the states corresponding to $s_0, \ldots, s_n$.

The existence of a situational envelope of an elementary action system radically restricts the possibilities of performing compound actions. In any normal elementary action system $\boldsymbol{M}$, *every* compound action not containing the empty string ε is totally performable (Proposition 1.7.6). However, if the elementary system $\boldsymbol{M}$ is a part of some situational action system $\boldsymbol{M}^s = (W, R, \mathcal{A}, S, Tr, f)$, that is, the reduct $(W, R, \mathcal{A})$ of $\boldsymbol{M}^s$ coincides with $\boldsymbol{M}$, then the above result is no longer true (if performability is taken in the sense of $\boldsymbol{M}^s$).

2.2.1 An Example. Thomson Lamp

Intuitively, a supertask is an activity consisting of an infinite number of steps but taken, as a whole, during a finite time. Examples of modern supertasks resemble ancient paradoxes posed by Zeno of Elea (e.g. Achilles and the tortoise). Supertasks may be identified with *infinite* strings of atomic actions performed in a finite time period.

A Thomson lamp is a device consisting of a lamp and a switch set on an electrical circuit. If the switch is on, then the lamp is lit, and if the switch is off, then the lamp is off. A Thomson lamp is therefore modelled as a simple elementary action system $\boldsymbol{M}$ with two states and two atomic action A : 'Switching on the lamp' and B : 'Switching off the lamp'. The picture gets more complicated if $\boldsymbol{M}$ is contained in a situational envelope involving only one additional situational parameter—time. Suppose that:

1. At time $t = 0$ the switch is on.
2. At time $t = 1/2$ the switch is off.
3. At time $t = 3/4$ the switch is on.
4. At time $t = 7/8$ the switch is off.
5. At time $t = 15/16$ the switch is on, etc.

What is the state of the lamp at time $t = 1$? Is it lit or not? We see that the above infinite sequence of alternate atomic actions gives rise to a formulation of non-trivial questions, depending on the selection of the situational envelope.[4] Solving them requires introducing genuinely infinitistic components into the picture of action theory presented here. These components are introduced in different ways. One option is to endow the *set of states* with a topology or with some order-continuous properties. This option requires infinite, ordered sets of states exhibiting various forms of order-completeness. The other option is to embed the atomic system in a

[4]The above example is taken from Jerzy Pogonowski's essay *Entertaining Math Puzzles* which can be found at www.glli.uni.opole.pl.

situational envelope endowed with various continuity properties. (This is the case with a Thomson lamp.) A mixture of these two approaches is also conceivable. In Chap. 3 the first option is elaborated in the context of an ordered action system.

2.3 Iterative Algorithms

Let $\boldsymbol{M} = (W, R, \boldsymbol{\mathcal{A}})$ be an elementary action system. An *iterative algorithm for* $\boldsymbol{M}$ is a quintuple

$$D := (W, V, \boldsymbol{\alpha}, \boldsymbol{\omega}, \boldsymbol{LA}), \tag{2.3.1}$$

where W is the set of states of $\boldsymbol{M}$, V is a finite set called the set of *labels* of the algorithm, $\boldsymbol{\alpha}$ and $\boldsymbol{\omega}$ are designated elements of V, called the *initial* and the *terminal* (or *end*) label of the algorithm, respectively, and $\boldsymbol{LA}$ is a finite subset of the Cartesian product

$$V \setminus \{\boldsymbol{\omega}\} \times \boldsymbol{\mathcal{A}} \times V \setminus \{\boldsymbol{\alpha}\}.$$

The members of $\boldsymbol{LA}$ are called *labeled atomic actions* of the algorithm D. Thus, the labeled actions are triples (a, A, b), where a and b are labels, called respectively the *input* and the *output* label, and A is an atomic action. Since $\boldsymbol{LA}$ involves only finitely many atomic actions of $\boldsymbol{\mathcal{A}}$, the members of the set

$$\boldsymbol{\mathcal{A}}_D := \{A \in \boldsymbol{\mathcal{A}} : (\exists a, b \in V)\ (a, A, b) \in \boldsymbol{LA}\}$$

are called the *atomic actions of the algorithm* D. $\boldsymbol{LA}$ may be empty.

The elements of the set $S_D := W \times V$ are called the *possible situations* of the algorithm. If $s = (w, a) \in S_D$, then w is the *state corresponding* to s, and a is the *label* of s. The situations with label $\boldsymbol{\alpha}$, i.e., the members of $W \times \{\boldsymbol{\alpha}\}$ are (possible) *initial situations* of the algorithm while the members of $W \times \{\boldsymbol{\omega}\}$, i.e., the situations labeled by ω are (possible) *terminal situations* of the algorithm D.

The relation Tr_D of *direct transition* in the algorithm is defined as follows: if $s = (u, a)$, $t = (w, b)$ are possible situations of D, then

$$Tr_D(s, t) \quad \text{if and only if} \quad (\exists A)\ (a, A, b) \in \boldsymbol{LA}\ \ \&\ \ uA, Rw.$$

Thus, if s is an initial situation, then $Tr(s', s)$ for no situation s' of D. Similarly, if s is terminal, then there does not exist a situation $s' \in S$ such that $Tr(s, s')$.

The relation Tr_D^*, the transitive and reflexive closure of Tr_D, is called the *transition relation* in D. Thus,

$$(s, t) \in Tr_D^* \quad \text{if and only if} \quad (\exists n \geqslant 0)\ (s, t) \in (Tr_D)^n.$$

(Thus, in particular, $(Tr_D)^0 = \{(s, s) : s \in S\}$.) *Finite runs* of the algorithm D are sequences of situations

$$(s_0, s_1, \ldots, s_n) \tag{2.3.2}$$

such that s_0 is an initial situation of D and $(s_i, s_{i+1}) \in Tr_D$ for all i, $0 \leqslant i \leqslant n-1$. The situations s_0 and s_n are called the *beginning* and the *end* of the run (2.3.2). The run (2.3.2) is *terminated* in D if and only if s_n is a terminal situation of D.

It may happen that for some run (2.3.2) there does not exist a situation s with the property that $Tr_D(s_n, s)$; i.e., there is no possibility of continuation of the run (2.3.2), even if the situation s_n is *not* terminal in D. It is said then that the algorithm D is stuck in the situation s_n. A run with this property is not qualified as terminated in D.

The fact that an algorithm may get jammed in some situations leads us to isolate the so-called integral iterative algorithms. A situation $s = (w, a)$ is *non-terminal* if $a \neq \boldsymbol{\omega}$. An iterative algorithm D is *integral* if the domain of the relation Tr_D coincides with the set of all non-terminal situations; that is, if for every non-terminal situation s there exists a situation s' such that $Tr_D(s, s')$.

If the system $\boldsymbol{M} = (W, R, \boldsymbol{\mathcal{A}})$ is normal and every atomic action $A \in \boldsymbol{\mathcal{A}}$ is a total function (with domain W) and the iterative algorithm D for $\boldsymbol{M}$ has the property that for every triple $(a, A, b) \in \boldsymbol{LA}$ with $b \neq \boldsymbol{\omega}$ there exists a triple $(c, B, d) \in \boldsymbol{LA}$ such that $b = c$, then D is integral.

Infinite (or *divergent*) *runs* of D are defined as infinite sequences of situations

$$(s_0, s_1, \ldots, s_n, \ldots)$$

such that s_0 is an initial situation of D and $(s_i, s_{i+1}) \in Tr_D$ for all i. Thus, no terminal situation occurs in a divergent run.

Every elementary action system $\boldsymbol{M} = (W, R, \boldsymbol{\mathcal{A}})$ with a distinguished iterative algorithm D for $\boldsymbol{M}$ forms a situational action system. For let

$$\boldsymbol{M}^s := (W, R, \boldsymbol{\mathcal{A}}, S_D, Tr_D, f),$$

where S_D $(= W \times V)$ is the set of all possible situations of D, Tr_D is the transition relation in D, and $f : W \times V \to W$ is the projection onto W. $\boldsymbol{M}^s$ is a situational action system in the sense of Definition 2.2.1.

Each state of W corresponds to a certain initial situation of $\boldsymbol{M}^s$. This is not the case in the iterative system associated with chess playing (Sect. 2.2). In a game of chess there is only one initial situation.

To each iterative algorithm (2.3.1) a binary relation Res_D on the set W of states is assigned and called the *resultant relation* of the algorithm D:

$$(u, w) \in Res_D \qquad \text{if and only if} \quad \big((u, \boldsymbol{\alpha}), (w, \boldsymbol{\omega})\big) \in Tr_D^*.$$

Equivalently, $(u, w) \in Res_D$ if and only if there exists a terminated run $(s_0, \ldots, s_n)$ such that $s_0 = (u, \boldsymbol{\alpha})$ and $s_n = (w, \boldsymbol{\omega})$.

The resultant relation Res_D is a sub-relation of the reach of the elementary action system $\boldsymbol{M}$.

Example Let $\boldsymbol{M}$ be an arbitrary elementary action system, and let $(A_1, \ldots, A_n)$ be a fixed non-empty sequence of atomic actions of $\mathcal{A}$. Let $V := \{0, 1, \ldots, n\}$, $\boldsymbol{\alpha} := 0$, $\boldsymbol{\omega} := n$, and $\boldsymbol{LA} := \{(0, A_1, 1), \ldots, (n-1, A_n, n)\}$. Then $D := (W, V, \boldsymbol{\alpha}, \boldsymbol{\omega}, \boldsymbol{LA})$ is an algorithm for $\boldsymbol{M}$. The resultant relation of D is equal to $(A_1 \cap R) \circ \ldots \circ (A_n \cap R)$. □

An iterative algorithm (2.3.1) is *well-designed* if and only if for every labeled action $(a, A, b) \in \boldsymbol{LA}$ there exists a finite string $A_1, \ldots, A_n$ of atomic actions of D and a finite sequence $a_0, a_1, \ldots, a_n$ of labels of V such that $a_0 = \boldsymbol{\alpha}$, $a_n = \boldsymbol{\omega}$, $(a_i, A_{i+1}, a_{i+1}) \in \boldsymbol{LA}$ for all $i \leqslant n-1$, and (a, A, b) is equal to a triple (a_i, A_{i+1}, a_{i+1}) for some $i \leqslant n-1$. Intuitively, D is well-designed if every labeled action of $\boldsymbol{LA}$ is employed in some terminated run of D.

Well-designed algorithms are singled out for technical reasons. Each algorithm D can be transformed into a well-designed algorithm by deleting from $\boldsymbol{LA}$ those labeled actions (a, A, b) which do not satisfy the above condition. This operation does not change the resultant relation of the algorithm.

We recall that REG($\mathcal{A}$) denotes the family of all regular compound actions (over $\mathcal{A}$), and $\mathrm{REG}^{+}(\mathcal{A}) := \{\mathbf{B} \in \mathrm{REG}(\mathcal{A}) : \varepsilon \notin \mathbf{B}\}$. The properties of the family REG($\mathcal{A}$) are independent of the internal structure of an action system $\boldsymbol{M} = (W, R, \mathcal{A})$; the cardinality of $\mathcal{A}$ is the only factor that matters. This fact lead us to distinguish, for each $\mathbf{B} \in C\mathcal{A}$, the set $\mathcal{A}_{\mathbf{B}}$ of atomic actions of $\mathcal{A}$ occurring in the strings of $\mathbf{B}$.

Lemma 2.3.1 *If* $\mathbf{B} \in \mathrm{REG}(\mathcal{A})$, *then* $\mathcal{A}_{\mathbf{B}}$ *is finite.*

The lemma follows from the definition of a regular language over a (possibly infinite) alphabet (see Sect. 1.6). □

Let $\boldsymbol{A}$ be a family of binary relations on a set W. The *positive Kleene closure* of $\boldsymbol{A}$, denoted by $Cl^{+}(\boldsymbol{A})$, is the least family $\boldsymbol{B}$ of binary relations on W which includes $\boldsymbol{A}$ as a subfamily and satisfies the following conditions:

(i) $\emptyset \in \boldsymbol{B}$
(ii) if $\{P, Q\} \subseteq \boldsymbol{B}$, then $\{P \circ Q, P \cup Q, P^{+}\} \subseteq \boldsymbol{B}$.

The *Kleene closure* of $\boldsymbol{A}$, denoted by $Cl^{*}(\boldsymbol{A})$, is the least family $\boldsymbol{B}$ of binary relations on W which includes $\boldsymbol{A}$ as a subfamily and satisfies

(i) $\emptyset, E_W \in \boldsymbol{B}$
(ii) if $\{P, Q\} \subseteq \boldsymbol{B}$, then $\{P \circ Q, P \cup Q, P^{*}\} \subseteq \boldsymbol{B}$.

The purpose of this section is to determine the principles of the functioning of elementary action systems $\boldsymbol{M}$ which are regulated by appropriate automata. This will give rise to a class of situational action systems erected on atomic systems.

We henceforth will work with action systems $\boldsymbol{M} = (W, R, \mathcal{A})$ which are normal and in which the relation R is reflexive. These assumptions imply the total performability of all non-empty compound actions of $C\mathcal{A}$ in $\boldsymbol{M}$, and of the action $\boldsymbol{\varepsilon}$ in particular. (If the reflexivity of R is dropped, one should rather work with the compound actions of $C^{+}\mathcal{A}$ rather than those of $C\mathcal{A}$ and with the positive Kleene closure Cl^{+}.)

Proposition 2.3.2 *Let $\boldsymbol{M} = (W, R, \mathcal{A})$ be a normal elementary action system with reflexive R. Then the following conditions hold:*

(i) For every regular action $\mathbf{B} \in \mathrm{REG}(\mathbf{A})$, $Res\,\mathbf{B}$, the resultant relation of $\mathbf{B}$ defined as in the formula (1.7.3) of Sect. 1.7, belongs to $Cl(\mathcal{A}_{\mathbf{B}})$;
(ii) Conversely, for every binary relation $Q \in Cl(\mathcal{A})$ there exists a regular action $\mathbf{B} \in \mathrm{REG}(\mathcal{A})$ such that $Res\,\mathbf{B} = Q$ and $Q \in Cl(\mathcal{A}_{\mathbf{B}})$.

Proof (i) We define:

$$\mathcal{P} := \{\mathbf{B} \in C\mathcal{A} : Res\,\mathbf{B} \in Cl(\mathcal{A}_{\mathbf{B}})\}.$$

($\mathcal{A}_{\mathbf{B}}$ is a set of binary relations on W, viz, the set of atomic actions occurring in the compound action $\mathbf{B}$. As R is reflexive, $Res\,\mathbf{B}$ is reflexive, for all non-empty $\mathbf{B}$.)

$\mathcal{P}$ has the following properties:

(a) *If $\mathbf{B} \in C\mathcal{A}$ is finite then $\mathbf{B} \in \mathcal{P}$; in particular $\boldsymbol{\emptyset} \in \mathcal{P}$;*
(b) *If $\mathbf{B}, \mathbf{C} \in \mathcal{P}$, then $\mathbf{B} \cup \mathbf{C}, \mathbf{B} \circ \mathbf{C}$ and $\mathbf{B}^*$ belong to $\mathcal{P}$ as well.*

((b) directly follows from Lemma 1.7.2 and the remarks following it.) Thus, every regular action of $\mathrm{REG}(\mathcal{A})$ belongs to $\mathcal{P}$.

(ii) We define:

$$\boldsymbol{L} := \big\{Q \subseteq W \times W : (\exists \mathbf{B})\ \ \mathbf{B} \in \mathrm{REG}(\mathcal{A})\ \ \&\ \ Q = Res\,\mathbf{B}\big\}.$$

We claim that $Cl(\mathcal{A}) \subseteq \boldsymbol{L}$.

$\emptyset \in \boldsymbol{L}$ and $E_W \in \boldsymbol{L}$, because $\emptyset = Res\,\boldsymbol{\emptyset}$ and $E_W = Res\,\boldsymbol{\varepsilon}$. For every atomic action $Q \in \mathcal{A}$, $\mathbf{B} := \{Q\}$ is a regular action on $\boldsymbol{M}$ and, since $\boldsymbol{M}$ is normal, $Res\,\mathbf{B} = Q$. Hence, $\mathcal{A} \subseteq \boldsymbol{L}$. Now assume that $P, Q \in \boldsymbol{L}$. We shall show that $\{P \circ Q, P \cup Q, P^*\} \subseteq \boldsymbol{L}$. We have that $P = Res\,\mathbf{B}$ and $Q = Res\,\mathbf{C}$ for some regular compound actions $\mathbf{B}$ and $\mathbf{C}$ on $\boldsymbol{M}$. Then, by Lemma 1.7.2, $P \circ Q = Res\,\mathbf{B} \circ Res\,\mathbf{C} = Res\,(\mathbf{B} \circ \mathbf{C})$, $P \cup Q = Res\,\mathbf{B} \cup Res\,\mathbf{C} = Res\,(\mathbf{B} \cup \mathbf{C})$, and, as R is reflexive, $(Res\,\mathbf{B})^* = Res\,(\mathbf{B}^*)$. As $\mathbf{B} \circ \mathbf{C}$, $\mathbf{B} \cup \mathbf{C}$, and $\mathbf{B}^*$ are regular, this proves that $Cl(\mathcal{A}) \subseteq \boldsymbol{L}$.

If $Q = Res\,\mathbf{B}$ for some regular $\mathbf{B}$, then by (i), $Res\,\mathbf{B} \in Cl(\mathcal{A}_{\mathbf{B}})$, and hence $Q \in Cl(\mathcal{A}_{\mathbf{B}})$. So (ii) holds. □

An iterative algorithm $D = (W, V, \boldsymbol{\alpha}, \boldsymbol{\omega}, \boldsymbol{LA})$ for $\boldsymbol{M}$ is said to *accept a string* $A_1 \ldots A_n$ of atomic actions of $\boldsymbol{M}$ if there exists a sequence $a_0, \ldots, a_n$ of labels of V such that $a_0 = \boldsymbol{\alpha}$, $a_n = \boldsymbol{\omega}$ and $(a_i, A_{i+1}, a_{i+1}) \in \boldsymbol{LA}$, for all $i \leqslant n - 1$. (Thus D accepts the empty string ε if and only if $\boldsymbol{\alpha} = \boldsymbol{\omega}$. D accepts no strings if and only if $\boldsymbol{LA}$ is empty and $\boldsymbol{\alpha} \neq \boldsymbol{\omega}$.)

The *total* (compound) *action* of the algorithm D, denoted by $\mathbf{A}(D)$, is defined as the set of all strings of atomic actions accepted by D.

In particular, for a given algorithm $D = (W, V, \boldsymbol{\alpha}, \boldsymbol{\omega}, \boldsymbol{LA})$, if $\boldsymbol{LA}$ is empty and $\boldsymbol{\alpha} \neq \boldsymbol{\omega}$, then the total action of D is the empty compound action $\boldsymbol{\emptyset}$. If $\boldsymbol{LA}$ is empty and $\boldsymbol{\alpha} = \boldsymbol{\omega}$, then the total action of D equals $\boldsymbol{\varepsilon}$.

Thus, D is well-designed if and only $\mathcal{A}_D = \mathcal{A}_{\mathbf{A}(D)}$, i.e., every atomic action of D occurs in some sequence belonging to $\mathbf{A}(D)$.

Proposition 2.3.3 *Let D be an iterative algorithm for an elementary action system $\boldsymbol{M} = (W, R, \mathcal{A})$. Then:*

(i) the total action $\mathbf{A}(D)$ of D is regular, and
(ii) the resultant relation of the total action $\mathbf{A}(D)$ (in the sense of formula (1.7.3) of Sect. 1.7) coincides with the resultant relation Res_D of D.

Proof To prove (i) we shall employ some facts from the theory of formal grammars. Let $D = (W, V, \boldsymbol{\alpha}, \boldsymbol{\omega}, \boldsymbol{LA})$. We shall treat the sets V and $\mathcal{A}_D$ (the set of atomic actions of D) as disjoint alphabets. V will be the auxiliary alphabet (variables) while $\mathcal{A}_D$ will be the terminal alphabet (terminals). The initial label will be regarded as the start symbol. The set $\boldsymbol{LA}$ of labeled actions of D specifies, in turn, a certain finite set $\mathscr{P}$ of productions:

$$\mathscr{P} := \{a \to Ab : (a, A, b) \in \boldsymbol{LA}\} \ \cup \ \{a \to A : (\exists a \in V) \ (a, A, \boldsymbol{\omega}) \in \boldsymbol{LA}\},$$

where $\boldsymbol{\omega}$ is the terminal label. The system $\mathbf{G}_D := (\mathcal{A}_D, V, \mathscr{P}, \boldsymbol{\alpha})$ is, thus, a combinatorial grammar.

Lemma 2.3.4 *The language generated by the grammar $\mathbf{G}_D$ coincides with the total action $\mathbf{A}(D)$ of the algorithm D.*

Proof of the lemma. We first show that $\mathbf{A}(D) \subseteq L(\mathbf{G}_D)$. Let $A_1 \dots A_n \in \mathbf{A}(D)$. Then, for some $a_0, \dots, a_n \in V$ with $a_0 = \boldsymbol{\alpha}$ and $a_n = \boldsymbol{\omega}$, we have that $(a_i, A_{i+1}, a_{i+1}) \in \boldsymbol{LA}$ for all $i \leqslant n - 1$. This implies that

$$\boldsymbol{\alpha} \Rightarrow A_1a_1 \Rightarrow A_1A_2a_2 \Rightarrow \dots \Rightarrow A_1 \dots A_{n-1}a_{a-1} \Rightarrow A_1 \dots A_{n-1}A_n$$

is a terminated derivation in $\mathbf{G}_D$. Hence $A_1 \dots A_n \in L(\mathbf{G}_D)$.

To prove the reverse inclusion $L(\mathbf{G}_D) \subseteq \mathbf{A}(D)$, suppose that $A_1 \dots A_n \in L(\mathbf{G}_D)$ and let $z_0 \Rightarrow z_1 \Rightarrow \dots \Rightarrow z_{m-1} \Rightarrow z_m$ be a derivation of $A_1 \dots A_n$ in $\mathbf{G}_D$, where $z_0 = \boldsymbol{\alpha}$ and $z_m = A_1 \dots A_n$. We show by induction for $i = 1, \dots, n-1$ that $z_i = A_1 \dots A_i a_i$ for some $a_i \in V$. This holds for $i = 1$. Let $i \geqslant 2$ and assume that $z_j = A_1 \dots A_j a_j$ for all $j \leqslant i - 1$. In particular, $z_{i-1} = A_1 \dots A_{i-1}a_{i-1}$. Since $i \leqslant n - 2$, no production of the form $a \to A$ can be applied to z_{i-1} (for otherwise we would get a word of the form $A_1 \dots A_{i-1}A$ which is different from $A_1 \dots A$; and the string $A_1 \dots A_{i-1}A$ blocks further derivations). So the word z_{i-1} derives z_i by means of a production $a_{i-1} \to Ab$ in which $A = A_i$.

Since $z_{n-1} = A_1 \dots A_{n-1}a_{n-1}$, we see that z_n must be equal to $A_1 \dots A_n$. So $m = n$, $z_m = z_n$, and the word z_{n-1} derives z_n by means of the production $a_{n-1} \to A_n$.

It follows from the definition of $\mathbf{G}_D$ that the sequence $a_0 := \boldsymbol{\alpha}, a_1, \dots, a_{n-1}$, $a_n := \boldsymbol{\omega}$ has the property that $(a_i, A_{i+1}, a_{i+1}) \in \boldsymbol{LA}$ for all $i \leqslant n-1$. So $A_1 \dots A_n \in \mathbf{A}(D)$. This concludes the proof of the lemma. □

$\mathbf{G}_D$ is a right linear-grammar. This fact, in view of Theorem 1.6.3, implies that the language $L(\mathbf{G}_D)$ $(= \mathbf{A}(D))$ is regular. So (i) holds.

(ii) We have to show that $Res_D = Res\,\mathbf{A}(D)$. Let $(u, w) \in Res_D$, i.e., $\big((u, \boldsymbol{\alpha}), (w, \boldsymbol{\omega})\big) \in (Tr_D)^+$. Hence there exists a finite sequence of situations $s_0, \ldots, s_n$, where $s_i = (u_i, a_i)$, for all $i \leqslant n$, such that $s_0 = (u, \boldsymbol{\alpha})$, $s_n = (w, \boldsymbol{\omega})$ and, for every $i \leqslant n-1$, there exists an action $A_{i+1} \in \boldsymbol{A}(D)$ with the property that $(a_i, A_{i+1}, a_{i+1}) \in \mathscr{P}$ and $u_i\ A_{i+1},\ R\ u_{i+1}$. It follows that the sequence $A_1 \ldots A_n$ belongs to $\mathbf{A}(D)$ and $(u, w) \in (A_1 \cap R) \circ \ldots \circ (A_n \cap R)$, i.e., $(u, w) \in Res\,\mathbf{A}(D)$.

Conversely, let $(u, w) \in Res\,\mathbf{A}(D)$. Hence there exists a sequence $A_1 \ldots A_n \in \mathbf{A}(D)$ and a string of states $u_0, \ldots, u_n$ such that $u_0 = u$, $u_n = w$, and $u_i\ A_{i+1},\ R\ u_{i+1}$ for all $i \leqslant n-1$. Since $A_1 \ldots A_n \in \mathbf{A}(D)$, there exists also a sequence of labels $a_0, \ldots, a_n$ such that $a_0 = \boldsymbol{\alpha}$, $a_n = \boldsymbol{\omega}$, and $(a_i, A_{i+1}, a_{i+1}) \in \boldsymbol{LA}$, for all $i \leqslant n-1$, which means that $\big((u_i, a_i), (u_{i+1}, a_{i+1})\big) \in Tr_D$ for every $i \leqslant n-1$. Hence $(u, w) \in Res_D$. This proves that $Res_D = Res\,\mathbf{A}(D)$. □

Corollary 2.3.5 *Let $D = (W, V, \boldsymbol{\alpha}, \boldsymbol{\omega}, \boldsymbol{LA})$ be an iterative algorithm for a normal system $\boldsymbol{M} = (W, R, \boldsymbol{\mathcal{A}})$ with reflexive R. Then Res_D belongs to the Kleene closure of $\boldsymbol{\mathcal{A}}_D$.*

Proof Use Propositions 2.3.2 and 2.3.3. □

Regular languages are those accepted by finite automata. Regular compound actions, however, can be also conveniently characterized in terms of total actions of iterative algorithms.

Theorem 2.3.6 *Let $\boldsymbol{M} = (W, R, \boldsymbol{\mathcal{A}})$ be an elementary action system with reflexive R. For every compound action $\mathbf{B} \in C\boldsymbol{\mathcal{A}}$ the following conditions are equivalent:*

(i) **B** *is regular.*
(ii) There exists a well-designed iterative algorithm D for $\boldsymbol{M}$ such that the total action of D is equal to **B** *and $\boldsymbol{\mathcal{A}}_D = \boldsymbol{\mathcal{A}}_{\mathbf{B}}$.*

Proof (ii) ⇒ (i) This is an immediate consequence of Proposition 2.3.3.

(i) ⇒ (ii) The proof of this part of the theorem consists in, for every regular action $\mathbf{B} \in C\boldsymbol{\mathcal{A}}$, the effective construction of an iterative algorithm D satisfying the conditions mentioned in clause (ii). The proof is by induction on complexity of the regular action **B**.

Suppose first that the action **B** is finite and non-empty. Let the initial and the terminal labels $\boldsymbol{\alpha}$ and $\boldsymbol{\omega}$ be fixed. For each sequence $\underline{A} = A_1 \ldots A_n \in \mathbf{B}$ we select a string of distinct labels $a_1, \ldots, a_{n-1}$, and define:

$$\boldsymbol{P}_{\underline{A}} := \{(\boldsymbol{\alpha}, A_1, a_1), \ldots, (a_{i-1}, A_i, a_i), \ldots, (a_{n-1}, A_n, \boldsymbol{\omega})\},$$
$$\boldsymbol{LA} := \bigcup \{\boldsymbol{P}_{\underline{A}} : \underline{A} \in \mathbf{B}\}.$$

Since **B** is assumed to be finite, $\boldsymbol{LA}$ is finite as well.

Let V be the collection of all the labels occurring in the labeled actions of $\boldsymbol{LA}$. Then $D = (W, V, \boldsymbol{\alpha}, \boldsymbol{\omega}, \boldsymbol{LA})$ is a well-designed iterative algorithm for $\boldsymbol{M}$. Moreover, it follows from the construction of D that the compound action of D is equal to $\mathbf{B}$ and that $\mathcal{A}_D = \mathcal{A}_{\mathbf{B}}$.

If $\mathbf{B}$ is empty, it is assumed that $\boldsymbol{\alpha} \neq \boldsymbol{\omega}$ and $\boldsymbol{LA}$ is empty. It follows that the transition relation Tr_D is empty. Consequently, the action of D is equal to $\emptyset$.

For the compound action $\boldsymbol{\varepsilon}$, it is assumed that $\boldsymbol{\alpha} = \boldsymbol{\omega}$ and $\boldsymbol{LA}$ is empty. This implies that the transition relation Tr_D is equal to E_W. Consequently, the action of D is equal to $\boldsymbol{\varepsilon}$.

Now suppose $\mathbf{B}$ and $\mathbf{C}$ are compound actions on $\boldsymbol{M}$ such that well-designed algorithms D_1 and D_2 for $\boldsymbol{M}$ have been constructed so that the total action of D_1 is equal to $\mathbf{B}$ and the total action of D_2 is equal to $\mathbf{C}$. We shall build an algorithm D for $\boldsymbol{M}$ such that the total action of D is equal to $\mathbf{B} \cup \boldsymbol{C}$. Let $D_i = (W_i, V_i, \boldsymbol{\alpha}_i, \boldsymbol{\omega}_i, \boldsymbol{LA}_i)$ for $i = 1, 2$. Without loss of generality we can assume that $\boldsymbol{\alpha}_1 = \boldsymbol{\alpha}_2 = \boldsymbol{\alpha}$ and $\boldsymbol{\omega}_1 = \boldsymbol{\omega}_2 = \boldsymbol{\omega}$, and that the sets $V_1 \setminus \{\boldsymbol{\alpha}, \boldsymbol{\omega}\}$ and $V_2 \Sigma \{\boldsymbol{\alpha}, \boldsymbol{\omega}\}$ are disjoint. (Otherwise one can simply rename the labels of V_1 and V_2.) We then put:

$$D := (W, V_1 \cup V_2, \boldsymbol{\alpha}, \boldsymbol{\omega}, \boldsymbol{LA}_1 \cup \boldsymbol{LA}_2).$$

It is easy to see that the total action of D is equal to $\mathbf{B} \cup \mathbf{C}$ and $\mathcal{A}_D = \mathcal{A}_{\mathbf{B} \cup \mathbf{C}}$. Moreover D is well-designed.

We shall now construct an algorithm D° for $\boldsymbol{M}$ such that the total action of D° is equal to $\mathbf{B} \circ \mathbf{C}$. We put:

$$D^\circ := (W, V_1 \cup V_2 \cup \{c\}, \boldsymbol{\alpha}, \boldsymbol{\omega}, \boldsymbol{LA}^\circ),$$

where c is a new label adjoined to $V_1 \cup V_2$, and $\boldsymbol{LA}$ is the 'concatenation' of $\boldsymbol{LA}_1$ and $\boldsymbol{LA}_2$. To define $\boldsymbol{LA}^\circ$ formally, we need an auxiliary notion.

A *terminated* sequence of labeled atomic actions of an algorithm $D = (W, V, \boldsymbol{\alpha}, \boldsymbol{\omega}, \boldsymbol{LA})$ is any finite sequence of triples

$$(a_0, A_1, a_1), (a_1, A_2, a_2), \ldots, (a_{n-2}, A_{n-1}, a_{n-1}), (a_{n-1}, A_n, a_n) \qquad (2.3.3)$$

of $\boldsymbol{LA}$ such that $a_0 = \boldsymbol{\alpha}$ and $a_n = \boldsymbol{\omega}$.

For any terminated sequence (2.3.3) of elements of $\boldsymbol{LA}_1$ and any terminated sequence

$$(b_0, B_1, b_1), (b_1, B_2, b_2), \ldots, (b_{m-2}, B_{m-1}, b_{m-1}), (b_{m-1}, B_m, b_m)$$

of elements of $\boldsymbol{LA}_2$, where $b_0 = \boldsymbol{\alpha}$ and $b_m = \boldsymbol{\omega}$, we define a new sequence

$$\begin{aligned} &(\boldsymbol{\alpha}, A_1, a_1), (a_1, A_2, a_2), \ldots, (a_{n-2}, A_{n-1}, a_{n-1}), (a_{n-1}, A - n, c), \qquad (2.3.4)\\ &(c, B_1, b_1), (b_1, B_2, b_2), \ldots, (b_{m-2}, B_{m-1}, b_{m-1}), (b_{m-1}, B_m, \boldsymbol{\omega}), \end{aligned}$$

where c is the new label adjoined to $V_1 \cup V_2$.

$\boldsymbol{LA}^\circ$ is, by definition, the set of all labeled atomic actions occurring in all the sequences of the form (2.3.4).

It is easy to show that the total action of D° is equal to $\mathbf{B} \circ \mathbf{C}$, the set of atomic actions of D° is equal to $\mathcal{A}_{\mathbf{B}\circ\mathbf{C}}$, and D° is well-designed.

Having given a well-designed algorithm $D = (W, V, \boldsymbol{\alpha}, \boldsymbol{\omega}, \boldsymbol{LA})$ for $\boldsymbol{M}$ such that the total action of D is equal to $\mathbf{B}$, we define a new well-designed algorithm

$$D^{\#} = (W, V^{\#}, \boldsymbol{\alpha}^{\#}, \boldsymbol{\omega}^{\#}, \boldsymbol{LA}^{\#}).$$

Here $\boldsymbol{\alpha}^{\#} = \boldsymbol{\omega}^{\#}$, $V^{\#} = (V \setminus \{\boldsymbol{\alpha}, \boldsymbol{\omega}\}) \cup \{\boldsymbol{\alpha}^{\#}, c\}$, where $\boldsymbol{\alpha}^{\#}$ and c are new labels not occurring in V. Thus, the labels $\boldsymbol{\alpha}$ and $\boldsymbol{\omega}$ are removed from V and new labels $\boldsymbol{\alpha}^{\#}$, c are adjoined. $\boldsymbol{LA}^{\#}$ is defined as follows: for every finite number, say k, of terminated non-empty sequences of labeled atomic actions of D:

$$(a_0^j, A_1^j, a_1^j), (a_1^j, A_2^j, a_2^j), \ldots, (a_{n_j-1}^j, A_{n_j}^j, a_{n_j}^j) \tag{2.3.5}$$

where $j = 1, \ldots, k$, and $a_0^j = \boldsymbol{\alpha}$, $a_{n_j}^j = \boldsymbol{\omega}$ for all j, a new sequence of labeled actions is formed:

$$\begin{array}{llll}
(\boldsymbol{\alpha}^{\#}, A_1^1, a_1^1), & (a_1^1, A_2^1, a_2^1), & \ldots, & (a_{n_1-1}^1, A_{n_1}^1, c), \\
(c, A_1^2, a_1^2), & (a_1^2, A_2^2, a_2^2), & \ldots, & (a_{n_2-1}^2, A_{n_2}^2, c), \\
 & & \vdots & \\
(c, A_1^j, a_1^j), & (a_1^j, A_2^j, a_2^j), & \ldots, & (a_{n_j-1}^j, A_{n_j}^j, c), \\
 & & \vdots & \\
(c, A_1^k, a_1^k), & (a_1^k, A_2^k, a_2^k), & \ldots, & (a_{n_k-1}^k, A_{n_k}^k, \boldsymbol{\omega}^{\#}).
\end{array} \tag{2.3.6}$$

If $k = 0$, (2.3.6) reduces to the sequence $(\boldsymbol{\alpha}^{\#}, \boldsymbol{\omega}^{\#})$, which is equal to $(\boldsymbol{\alpha}^{\#}, \boldsymbol{\alpha}^{\#})$.

$\boldsymbol{LA}^{\#}$ is the set of all labeled atomic actions occurring in the sequences (2.3.6).

The total compound action of the algorithm $D^{\#}$ is equal to $\mathbf{B}^*$, the iterative closure of $\mathbf{B}$. Moreover the set of atomic actions of $D^{\#}$ is equal to the set of all atomic actions occurring in $\mathbf{B}$.

It follows from the above constructions that for *every* action $\mathbf{B} \in \mathrm{REG}(\mathcal{A})$ there exists a well-designed iterative algorithm D such that the total action of D is equal to $\mathbf{B}$ and $\mathcal{A}_D = \mathcal{A}_{\mathbf{B}}$. This completes the proof of the theorem 2.3.6. □

The following corollary (in a slightly different but equivalent form) is due to Mazurkiewicz (1972a):

Corollary 2.3.7 *Let $\boldsymbol{M} = (W, R, \mathcal{A})$ be a normal action system with reflexive R, and let Q be a relation that belongs to the Kleene closure of $\mathcal{A}$. Then there exists an iterative algorithm D for $\boldsymbol{M}$ such that $Res_D = Q$ and $Q \in Cl(\mathcal{A}_D)$.*

Proof According to Proposition 2.3.2.(ii), there exists a regular compound action $\mathbf{B}$ on $\boldsymbol{M}$ such that $Q = Res\,\mathbf{B}$ and $Q \in Cl(\mathcal{A}_{\mathbf{B}})$. In turn, in view of Theorem 2.3.6, there exists an iterative algorithm D for $\boldsymbol{M}$, whose total action $\mathbf{A}(D)$ is equal to $\mathbf{B}$ and $\mathcal{A}_D = \mathcal{A}_{\mathbf{B}}$. Hence $Q \in Cl(\mathcal{A}_D)$. Since $\mathbf{A}(D) = \mathbf{B}$, Proposition 2.3.3 yields $Res_D = Res\,\mathbf{B} = Q$. □

Let $\boldsymbol{M} = (W, R, \mathcal{A})$ be a *finite* elementary action system. In Sect. 1.7, the compound actions Φ☞$_i\Psi$ have been defined for all $\Phi, \Psi \subseteq W$ and $i = 1, \ldots, 4$. Since these actions are regular, Theorem 2.3.6 implies that Φ☞$_i\Psi$ is 'algorithmizable' for all Φ, Ψ and $i = 1, \ldots, 4$; that is, there exists a well-designed iterative algorithm D for $\boldsymbol{M}$ such that the total action $\mathbf{A}(D)$ of D is equal to Φ☞$_i\Psi$. In particular this implies that the resultant relation of Φ☞$_i\Psi$ coincides with Res_D. Speaking figuratively, it means that the ways the task (Φ, Ψ) is implemented and the goal Ψ is reached in the finite system $\boldsymbol{M}$ is subordinated to a well-designed iterative algorithm. Thus, there exists, at least a theoretical possibility of 'automatizing' the tasks (Φ, Ψ) in the system $\boldsymbol{M}$. The practical realization of this idea requires adopting realistic assumptions as regards the cardinalities of the sets $\mathcal{A}$, W and V (the set of labels of the algorithm).

The reach $\boldsymbol{Re_M}$ of an elementary action system $\boldsymbol{M}$ is equal to the resultant relation of the action $\mathcal{A}^*$, the set of all non-empty strings of atomic actions. Since $\mathcal{A}^*$ is regular if $\mathcal{A}$ is finite, Theorem 2.3.6 and Corollary 2.3.7 imply that $\boldsymbol{Re_M}$ belongs to the Kleene closure of $\mathcal{A}$ whenever $\boldsymbol{M}$ is a finite normal elementary system with reflexive R. But this fact can be established directly by appealing to the definition of $\boldsymbol{Re_M}$.

Let $\boldsymbol{M} = (W, R, \mathcal{A})$ be an elementary action system. Two compound actions $\mathbf{B}, \mathbf{C} \in C\mathcal{A}$ are *equivalent over* $\boldsymbol{M}$ if and only if $\mathcal{A}_{\mathbf{B}} = \mathcal{A}_{\mathbf{C}}$ and $Res\,\mathbf{B} = Res\,\mathbf{C}$, i.e., $\mathbf{B}$ and $\mathbf{C}$ are 'built up' with the same atomic actions and the resultant relations of $\mathbf{B}$ and $\mathbf{C}$ are identical.

The facts we have established thus far enable us to draw the following simple corollary:

Corollary 2.3.8 *Let $\boldsymbol{M}$ be an elementary normal action system with reflexive R. Then for every action $\mathbf{B} \in C\mathcal{A}$ the following conditions are equivalent:*

(i) $\mathbf{B}$ is equivalent to a regular action;
(ii) There exists a well-designed algorithm D for $\boldsymbol{M}$ such that $\mathcal{A}_D = \mathcal{A}_{\mathbf{B}}$ and the resultant relation of $\mathbf{B}$ is equal to Res_D.

Proof (i) ⇒ (ii) Suppose $\mathbf{B}$ is equivalent to a regular action $\mathbf{C}$. By Theorem 2.3.6 there exists an iterative algorithm D for $\boldsymbol{M}$ such that $\mathcal{A}_D = \mathcal{A}_{\mathbf{C}}$ and the total action $\mathbf{A}(D)$ of D is equal to $\mathbf{C}$. Hence $\mathcal{A}_D = \mathcal{A}_{\mathbf{B}}$ and $Res\,\mathbf{B} = Res\,\mathbf{C} = Res\,\mathbf{A}(D) = Res_D$. So (ii) holds.

(ii) ⇒ (i) Assume (ii) Since D is well-designed, $\mathcal{A}_{\mathbf{B}} = \mathcal{A}_D = \mathcal{A}_{\mathbf{A}(D)}$. By Proposition 2.3.3, the action $\mathbf{A}(D)$ of D is regular. So $\mathbf{B}$ is equivalent to $\mathbf{A}(D)$ over $\boldsymbol{M}$. □

The above corollary gives rise to the problem of the preservation of regular actions by the relation of equivalence. More specifically, we ask if the following is true in normal action systems $\boldsymbol{M}$:

(∗) for any two compound actions **B**, **C** $\in C\boldsymbol{\mathcal{A}}$, if **B** is regular and equivalent to **C** over $\boldsymbol{M}$, then **C** is regular as well.

To answer the above question, we shall make use of the Myhill-Nerode Theorem. Let Σ be an alphabet. For an arbitrary language L over Σ the equivalence relation R_L on the set Σ^* for all non-empty words is defined as follows:

$x\ R_L\ y$ if and only if for each word z, either both or neither xz and yz is in L, i.e., for all z, $xz \in L$ if and only if $yz \in L$.

The Myhill-Nerode Theorem states that for an arbitrary language $L \subseteq \Sigma^*$, the relation R_L is of finite index (i.e., the number of equivalence classes of R_L is finite) if and only if L is accepted by some finite automaton (if and only if, by Theorem 1.6.5, L is regular).

Let $\Sigma := \{A, B\}$ and let $L := \{A^n B^n : n = 1, 2, \ldots\}$. The language L is not regular. To show this we define $x_n := A^n$, $y_n := B^n$ for all $n \geqslant 1$. If $m \neq n$ then $x_m x_n \notin L$ and $x_n x_n \in L$. Hence $m \neq n$ implies that $[x_m] \neq [x_n]$, where $[x]$ denotes the equivalence class of x with respect to R_L. This means that the index of R_L is infinite. So L cannot be regular.

Let $\boldsymbol{M} = (W, R, \boldsymbol{\mathcal{A}})$ be an action system, where $\boldsymbol{\mathcal{A}} = \{A, B\}$, the relations A and B are transitive, $B \subseteq A \subseteq R$ and R is reflexive. The compound action **B** is defined in the same manner as the language L as above, that is, $\mathbf{B} := \{A^n B^n : n \geqslant 1\}$.

The above argument shows that **B** is not regular. It is also easy to prove that $Res\,\mathbf{B} = A \circ B$. Hence trivially $Res\,\mathbf{B}$ belongs to the positive Kleene closure of $\boldsymbol{\mathcal{A}}_{\mathrm{B}} = \{A, B\}$ which implies, by Corollary 2.3.7, that there exists an iterative algorithm D for $\boldsymbol{M}$ such that $\boldsymbol{\mathcal{A}}_D = \boldsymbol{\mathcal{A}}$ and the resultant relation of D is equal to $Res\,\mathbf{B}$. Thus, **B** is equivalent over $\boldsymbol{M}$ to the regular action $\mathbf{A}(D)$ of the algorithm. As **B** is not regular, we see that (∗) does not hold for **B** and **C**.

The above result is not surprising because the definition of a regular compound action over an action system $\boldsymbol{M}$ does not take into account the internal set-theoretic structure of the family of atomic actions of $\boldsymbol{M}$.

2.4 Pushdown Automata and Pushdown Algorithms

A grammar $\mathbf{G} = (\Sigma, V, \mathscr{P}, \alpha)$ is *context-free* if each production of $\mathscr{P}$ is of the form $a \to x$, where a is a variable and x is a string of symbols from $(\Sigma \cup V)^*$.

The adjective 'context-free' comes from the fact that every production of the above sort can be treated as a substitution rule enabling to replace the symbol a by the word x irrespective of the context in which the symbol occurs.

A language L is *context-free* if it is generated by some context-free grammar.

Example 2.4.1 Let $L := \{A^n B^n : n = 1, 2, \ldots\}$. We showed in Sect. 2.3 that L is not regular. Let us consider the grammar $\mathbf{G} = (\Sigma, V, \mathcal{P}, \alpha)$, where $\Sigma = \{A, B\}$, $V = \{\alpha\}$ and $\mathcal{P} = \{\alpha \to A\alpha B,\ \alpha \to AB\}$. $\mathbf{G}$ is context-free. By applying the first production $n - 1$ times, followed by an application of the second production, we have

$$\alpha \Rightarrow A\alpha B \Rightarrow AA\alpha BB \Rightarrow A^3\alpha B^3 \Rightarrow \ldots \Rightarrow A^{n-1}\alpha B^{n-1} \Rightarrow A^n B^n.$$

Furthermore, induction on the length of a derivation shows that if $\alpha \Rightarrow_{\mathbf{G}} z_1 \Rightarrow_{\mathbf{G}} z_2 \ldots z_{n-1} \Rightarrow_{\mathbf{G}} z_n$ then $z_n = A^n B^n$ if z_n is a terminal and $z_n = A^n \alpha B^n$ otherwise. Thus $L = L(\mathbf{G})$. □

$\mathrm{CFL}(\Sigma)$ denotes the family of all context-free languages over Σ. $\mathrm{CFL}^+(\Sigma)$ is the class of all context-free languages over Σ which do not involve the empty word ε,

$$\mathrm{CFL}^+(\Sigma) := \{L : L \in \mathrm{CFL}(\Sigma) \text{ and } \varepsilon \notin L\}.$$

There are several ways by means of which one can restrict the format of productions without reducing the generative power of context-free grammars. If L is a non-empty context-free language and ε is not in L, then L can be generated by a context-free grammar $\mathbf{G}$ which uses productions whose right-hand sides each start with a terminal symbol followed by some (possibly empty) string of variables. This special form is called *Greibach normal form*. We shall formulate this result in a slightly sharper form.

Let $\mathbf{G}$ be context-free. If at each step in a $\mathbf{G}$-derivation a production is applied to the *leftmost* variable, then the derivation is said to be leftmost. Similarly a derivation in which the rightmost variable is replaced at each step is said to be *rightmost*.

Theorem 2.4.2 (Greibach 1965) *Every context-free language L without ε is generated by a grammar* $\mathbf{G}$ *for which every production is of the form $v \to Ax$, where v is a variable, A is a terminal and x is a (possibly empty) string of variables. Furthermore, every word of L can be derived by means of a leftmost derivation in* $\mathbf{G}$.

The proof of the above theorem is constructive—an algorithm is provided which, for every context free-grammar $\mathbf{G}$ in which no production of the form $v \to \varepsilon$ occurs, converts it to Greibach normal form. We omit the details. □

We shall define a broad class of algorithms which comprises iterative algorithms as a special, limit case.

Let $\boldsymbol{M} = (W, R, \boldsymbol{\mathcal{A}})$ be an elementary action system. A *pushdown algorithm for* $\boldsymbol{M}$ is a quadruple

$$D := (W, V, \boldsymbol{\alpha}, \boldsymbol{LA}),$$

where W is, as above, the set of states of $\boldsymbol{M}$, V is a finite set called the *stack* alphabet, $\boldsymbol{\alpha}$ is a particular stack symbol called the *start* symbol, and $\boldsymbol{LA}$ is a finite set of labeled atomic actions of $\boldsymbol{M}$, i.e., $\boldsymbol{LA}$ is a finite set of triples of the form

$$(a, A, \beta), \tag{2.4.1}$$

where $A \in \mathcal{A}$, $a \in V$, and $\beta \in V^*$ is a possibly empty string of symbols of V. (The empty string $\varepsilon \in V^*$ is called the *end* symbol.) The labels a and β of (2.4.1) are called the *input* and the *output* label of (2.4.1), respectively.

The set

$$\mathcal{A}_D := \{A \in \mathcal{A} : (\exists a, \beta)\ (a, A, \beta) \in \boldsymbol{LA}\}$$

is called the set of *atomic actions of the algorithm D*. The set is always finite.

Iterative algorithms may be regarded as a limit case of pushdown algorithms. The former are pushdown algorithms in which the terminal labels β of labeled atomic actions $(a, A, \beta) \in \boldsymbol{LA}$ have length $\leqslant 1$, i.e., $|\beta| \leqslant 1$.

The elements of the set $S_D := W \times V^*$ are called the possible *situations* of the algorithm. If $s = (w, \beta) \in S_D$, w is the state corresponding to s and β is the label of the situation s. β is also called the *stack* corresponding to the situation s. A situation s is *initial* if it is of the form (w, α) for some $w \in W$. A situation s is *terminal* if it is labeled by the empty string ε of stack symbols; i.e., the stack in this situation is empty; so terminal situations are of the form (w, ε), $w \in W$.

A labeled action (a, A, β) *transforms a situation* $s_1 = (w_1, \gamma_1)$ *into a situation* $s_2 = (w_2, \gamma_2)$ if and only if $\gamma_1 = a\sigma$, $\gamma_2 = \beta\sigma$ for some $\sigma \in V^*$, and $w_1\, A, R\, w_2$. Thus, s_1 is a situation in which a is the top symbol on the stack and s_2 is the situation which results from s_1 by replacing the top symbol a by the string β and performing the atomic action A so that the system moves from w_1 to the state w_2.

The transition relation Tr_D in a pushdown algorithm D is defined similarly as in the case of iterative algorithms:

> $(s_1, s_2) \in Tr_D$ if and only if there exists a labeled action $(a, A, \beta) \in \boldsymbol{LA}$ which transforms s_1 into s_2.

The notions of finite, infinite or terminated runs of situations in D are defined in the well-known way. The latter are finite sequences $(s_0, \ldots, s_n)$ such that s_0 is an initial situation, $Tr(s_i, s_{i+1})$ for all $i \leqslant n - 1$, and s_n is a terminal situation.

Every elementary normal action system $\boldsymbol{M} = (W, R, \mathcal{A})$ together with a distinguished pushdown algorithm D for $\boldsymbol{M}$ form, in a natural way, a situational action system. (To simplify matters, it is also assumed that R is reflexive.) We put:

$$\boldsymbol{M}^s := (W, R, \mathcal{A}, S_D, Tr_D, f)$$

where S_D and Tr_D are defined as above, and f is the projection from S_D $(= W \times V^*)$ onto W. $\boldsymbol{M}^s$ is easily seen to satisfy the conditions imposed on situational action systems.

The resultant relation Res_D of a pushdown algorithm D is defined as follows:

$$(u, w) \in Res_D \quad \text{if and only if} \quad \big((u, \boldsymbol{\alpha}), (w, \varepsilon)\big) \in (Tr_D)^*.$$

Res_D is a subrelation of the reach of $\boldsymbol{M}$.

The members of V^* represent possible contents of the stack. A string $a_1 \ldots a_n \in V^*$ is a possible state of the stack with a_1 the top symbol of the stack. The strings of V^* are called *control labels* of the algorithm. Every pair (a, β) with $a \in V$ and $\beta \in V^*$ defines the *control relation*, denoted by $a \to \beta$, on the set V^*:

$$\gamma_1(a \to \beta)\gamma_2 \quad \text{if and only if} \quad (\exists \sigma \in V^*)\ (\gamma_1 = a\sigma \ \ \&\ \ \gamma_2 = \beta\sigma). \qquad (2.4.2)$$

$a \to \beta$ is in fact a partial function, whose intended meaning is to modify control labels of situations during the work of the algorithm. The functions $a \to \beta$ enable us to express neatly possible transformations of the situations of the algorithm: a labeled action (a, A, β) transforms a situation $s_1 = (w_1, \gamma_1)$ into $s_2 = (w_2, \gamma_2)$ if and only if $\gamma_1(a \to \beta)\gamma_2$ and $w_1\, A, R\, w_2$. The functions $a \to \beta$ are also used in defining a certain compound action on the system $\boldsymbol{M}$ called the total action of the pushdown algorithm D. More specifically, a finite string $A_1 \ldots A_n$ of atomic actions of $\boldsymbol{M}$ is said to be *accepted by the algorithm D* if and only if there exists a sequence

$$(a_1, A_1, \beta_1), \ldots, (a_n, A_n, \beta_n)$$

of labeled actions of $\boldsymbol{LA}$ and a sequence of control labels $\gamma_1, \ldots, \gamma_{n+1} \in V^*$ with the following properties:

(i) $\gamma_1 = a_1 = \boldsymbol{\alpha}$ (= the start symbol), $\gamma_{n+1} = \varepsilon$ (= the end symbol)
(ii) $\gamma_i(a_i \to \beta_i)\gamma_{i+1}$ for all $i \leqslant n$.

The elements of the string $A_1 \ldots A_n$ are therefore consecutive actions of the algorithm leading from initial to terminal situations.

The *total action* of a pushdown algorithm D is defined as the set $\mathbf{A}(D)$ of all non-empty strings of atomic actions accepted by D.

Since $\mathcal{A}_D$, the set of all atomic actions the algorithm D involves, is always finite, the composite action $\mathbf{A}(D)$ is a language over the alphabet $\mathcal{A}_D$.

The resultant relation $Res\ \mathbf{A}(D)$ of the action $\mathbf{A}(D)$ coincides with the resultant relation Res_D of the algorithm D. This is an immediate consequence of the following simple observations:

(i) If $(s_0, \ldots, s_n)$ is a terminated run of situations, where $s_i = (w_i, \gamma_i)$, $i = 0, 1, \ldots, n$, then $(w_0, w_1, \ldots, w_n)$ is a realizable performance of $\mathbf{A}(D)$.
(ii) If $(w_0, w_1, \ldots, w_n)$ is a realizable performance of $\mathbf{A}(D)$, then there exist labels $\boldsymbol{\alpha} = \gamma_0, \ldots, \gamma_n = \varepsilon$ such that $(w_0, \gamma_0), \ldots, (w_n, \gamma_n)$ is a terminated run of situations.

The following theorem characterizes the compound actions of pushdown algorithms.

Theorem 2.4.3 *Let $\boldsymbol{M} = (W, R, \mathcal{A})$ be an elementary action system. Then for every non-empty compound action $\mathbf{B} \in C\mathcal{A}$ which does not contain the empty word ε the following conditions are equivalent:*

(i) $\mathbf{B}$ *is context-free,*
(ii) There exists a pushdown algorithm D for $\boldsymbol{M}$ such that $\mathbf{B}$ is the total action of D.

Proof (ii) $\Rightarrow$ (i) Assume $\mathbf{B} = \mathbf{A}(D)$ for some pushdown automaton D for $\boldsymbol{M}$. The idea of the proof is similar to the one applied in the proof of Proposition 2.3.3. The stack alphabet V is treated here as an auxiliary alphabet, while the members of $\mathcal{A}_D$ are terminals. The label $\boldsymbol{\alpha}$ is the start symbol. To the set $\boldsymbol{LA}$ of labeled actions of D the following finite list $\mathscr{P}$ of productions is assigned:

$$\mathscr{P} := \{a \to A\beta : (a, A, \beta) \in \boldsymbol{LA}\}.$$

The system

$$\mathbf{G}_D := (\mathcal{A}_D, V, \mathscr{P}, \boldsymbol{\alpha})$$

is then a grammar in Greibach normal form.

It is easy to see that the total action $\mathbf{A}(D)$ of D coincides with the set L of words over $\mathcal{A}_D$ that can be derived from $\boldsymbol{\alpha}$ by means of leftmost derivations in $\mathbf{G}_D$. Theorem 2.4.2 implies that $L = L(\mathbf{G}_D)$. Hence $\mathbf{B} = \mathbf{A}(D) = L(\mathbf{G}_D)$. Since every language generated by a Greibach grammar is context-free, (i) thus follows.

(i) $\Rightarrow$ (ii) The proof of this part is also easy. Let $\mathbf{B}$ be a context-free compound action (over the finite alphabet $\mathcal{A}_{\mathbf{B}}$). Since $\mathbf{B}$ contains only non-empty finite strings of atomic actions, the language $\mathbf{B}$ is generated by a grammar

$$\mathbf{G} = (\mathcal{A}_{\mathbf{B}}, V, \mathscr{P}, \boldsymbol{\alpha}) \tag{2.4.3}$$

in which every production of $\mathscr{P}$ is of the form $a \to A\beta$, where $a \in V$, $A \in \mathcal{A}_{\mathbf{B}}$, and β is a possibly empty string of elements of V. (This immediately follows from Greibach Normal Form Theorem 2.4.2; the auxiliary alphabet V is obviously individually suited to the language $\mathbf{B}$.)

Let $D := (W, V, \boldsymbol{\alpha}, \boldsymbol{LA})$ be the pushdown algorithm for $\boldsymbol{M}$ in which the set $\boldsymbol{LA}$ of labeled actions is defined as follows:

$$\boldsymbol{LA} := \{(a, A, \beta) : a \to A\beta \text{ is in } \mathscr{P}\}.$$

The total action of the algorithm D is easily seen to be equal to the set of words over $\mathcal{A}_{\mathbf{B}}$ obtained by means of all leftmost derivations in the grammar (2.4.3). But in view of Theorem 2.4.2, the latter set of words is equal to the language (over $\mathcal{A}_{\mathbf{B}}$) generated by the grammar (2.4.3). Thus, $\mathbf{A}(D) = L(\mathbf{G}) = \mathbf{B}$. □

Finite automata over some finite alphabet function according to a simple rule. They read consecutive symbols in a word while assuming different states. Formally, this requires assuming a specification of a finite set of states, a next-move function

δ from states and symbols to finite subsets of states, and a distinction between an initial state and the set of final states of the automaton.

Pushdown automata are much more powerful tools equipped with a restricted memory. A pushdown automaton can keep a finite stack during its reading a word. The automaton may use a separate stack alphabet. The 'next-move' function indicates, for each state, the symbol read and the top symbol of the stack, the next state of the automaton, as well as the string of stack symbols which will be entered or removed at the top of the stack.

Formally, a *pushdown automaton* (PDA) is a system

$$\boldsymbol{A} = (W, \Sigma, \Gamma, \delta, w_0, \boldsymbol{\alpha}, F) \tag{2.4.4}$$

where

(i) W is a finite set of *states* of the automaton
(ii) Σ is an alphabet called the *input alphabet*
(iii) Γ is an alphabet called the *stack alphabet*
(iv) w_0 is the *initial state* of the automaton
(v) $\boldsymbol{\alpha} \in \Gamma$ is a distinguished stack symbols called the *start symbol*
(vi) $F \subseteq W$ is the set of *final states*
(vii) δ is a mapping from $W \times \Sigma \cup \{\varepsilon\} \times \Gamma$ to finite non-empty subsets of $W \times \Gamma^*$.

The interpretation of

$$\delta(u, a, Z) = \{(w_1, \gamma_1), (w_2, \gamma_2), \ldots, (w_m, \gamma_m)\},$$

where $w_1, \ldots, w_m \in W$ and $\gamma_1, \ldots, _m \in \Gamma^*$, is that the PDA in state u, with input symbol a and Z the top symbol on the stack, can, for any i $(i = 1, \ldots, m)$, enter state w_i, replace the symbol Z by the string γ_i, and advance the input head one symbol. The convention is adopted that the leftmost symbol of γ_i will be placed highest on the stack and the rightmost symbol lowest on the stack. It is not permitted to choose the state w_i and the string γ_j for some $j \neq i$ in one move. In turn,

$$\delta(u, \varepsilon, Z) = \{(w_1, \gamma_1), (w_2, \gamma_2), \ldots, (w_m, \gamma_m)\}$$

says that $\boldsymbol{M}$ in the state u, independent of the input symbol being scanned with Z the top symbol on the stack, enters one of the states w_i and replaces Z by γ_i for $i = 1, \ldots, m$. The input head is not advanced.

To formally describe the configuration of a PDA at a given instant onc introduces the notion of an *instantaneous description* (ID). Each ID is defined as a triple (w, x, γ), where w is a state, x is a string of input symbols, and γ is a string of stack symbols. According to the terminology adopted in this chapter, each instantaneous description $s = (w, x, \gamma)$ can be regarded as a possible situation of the automaton—w is the state of the automaton in the situation s while the label (x, γ) of the situation

s defines the string x of input symbols read by the automaton and the string γ of symbols on the stack.

If $\boldsymbol{M} = (W, \Sigma, \Gamma, \delta, w_0, \boldsymbol{\alpha}, F)$ is a PDA, then we write $(u, ax, Z\gamma) \vdash_{\boldsymbol{M}} (w, x, \beta\gamma)$ if $\delta(u, a, Z)$ contains (w, β). a is an input symbol or ε. $\vdash_{\boldsymbol{M}}^{*}$ is the reflexive and transitive closure of $\vdash_{\boldsymbol{M}}$. $L(\boldsymbol{M})$ is the language defined by *final state* of $\boldsymbol{M}$. Thus,

$$L(\boldsymbol{M}) := \{x \in \Sigma^* : (w_0, x, \boldsymbol{\alpha}) \vdash_{M}^{*} (w, \varepsilon, \gamma) \text{ for some } w \in F \text{ and } \gamma \in \Gamma^*\}.$$

Theorem 2.4.4 *The class of languages accepted for by PDA's coincides with the class of context free languages.*

The book by Hopcroft and Ullman (1979) contains a proof of the above theorem and other relevant facts concerning various aspects of PDA's. □

Turing machines can be represented as situational action systems too. The transition relations between possible situations in such systems are defined in a more involved way however.

The characteristic feature of iterative and pushdown algorithms is that they provide *global* transformation rules of possible situations. In the simple case of iterative algorithms, transformation rules are defined by means of a finite set $\boldsymbol{LA}$ of labeled actions and then by distinguishing the set of all possible pairs of sequences of labels $(a_0, \ldots, a_n)$ and atomic actions $A_1 \ldots A_n$ such that $a_0 = \boldsymbol{\alpha}$, $a_n = \boldsymbol{\omega}$, and $(a_i, A_{i+1}, a_{i+1}) \in \boldsymbol{LA}$.

The conjugate sequences $(a_0, \ldots, a_n)$ and $(A_1, \ldots, A_n)$ define ways of transforming some situations into others. The transition from a situation labeled by a_i to a situation labeled by a_{i+1} is done through performing the action A_{i+1} (for $i = 0, \ldots, n-1$). The rules neither distinguish special states of the system nor do they impose any restrictions on transitions between the states corresponding to situations other than what relation R dictates. To put it briefly, they abstract from the *local* properties of the action system.

Schemes of action which are not captured by the above classes of algorithms can be easily provided. Let us fix an atomic action $A \in \boldsymbol{\mathcal{A}}$. Let $\Phi \subset W$ be a non-empty set of states. The program '**while** Φ **do** A' defines the way the action A should be performed: whenever the system is in a state belonging to Φ, iterate performing the action A until the system moves out of Φ. (We assume here for simplicity that A is atomic; the definition of '**while – do**' programs also makes sense for compound actions.) The program '**while** Φ **do** A' is identified with the set of all operations of the form

$$u_0 \, A \, u_1 \, A \, u_2 \ldots u_{n-1} \, A \, u_n \tag{2.4.5}$$

such that $u_0, \ldots, u_{n-1} \in \Phi$ and $u_n \notin \Phi$. (The above program also comprises infinite operations

$$u_0 \, A \, u_1 \, A \, u_2 \ldots u_{n-1} \, A \, u_n \ldots$$

such that $u_i \in \Phi$ for every $i \geqslant 0$.)

We claim that the above program is not determined by a pushdown algorithm. This statement requires precise formulation. To this end we define, for every pushdown algorithm $D = (W, V, \boldsymbol{\alpha}, \boldsymbol{LA})$, the notion of the *program* $Pr(D)$ corresponding to the algorithm D (see Sect. 1.8).

$Pr(D)$ is the set of all finite operations $u_0\, A_1\, u_1\, \dots\, u_{n-1}\, A_n\, u_n$ with $A_1 \dots A_n$ ranging over the total action $\mathbf{A}(D)$ of D. (As mentioned earlier, the notion of a program is usually conceived of as a certain syntactic entity. Here it is identified with its meaning; i.e., as a set of computations.)

Let $\boldsymbol{M} = (W, R, \{A\})$ be a normal elementary action system in which A is a total unary function on W. Let Φ be a proper non-empty subset of W closed with respect to A. (This means that for every pair of states $u, w \in W$, $u \in \Phi$ and $u\, A\, w$ imply that $w \in \Phi$.)

Proposition 2.4.5 *There does not exist a pushdown algorithm D for $\boldsymbol{M}$ such that*

$$Pr(D) = \textbf{while } \Phi \textbf{ do } A. \tag{2.4.6}$$

Proof It follows from the definitions of A and Φ that:

For any $n \geqslant 1$, there exist states $u_0, \dots, u_n$ such that $u_0\, A\, u_1 \dots u_{n-1}\, A\, u_n$ and $u_0, \dots, u_n \in \Phi$. (2.4.7)

We apply a *reductio ad absurdum* argument. Suppose that (2.4.6) holds for some algorithm D. It is clear that $\mathbf{A}(D)$ is a compound action over the one-element alphabet $\{A\}$. Let $A \dots A$ be a word of $\mathbf{A}(D)$ of length n. As (2.4.6) holds, we have that for *any* string $(u_0, \dots, u_n)$ of states, $u_0\, A\, u_1 \dots u_{n-1}\, A\, u_n$ implies that $u_0, \dots, u_{n-1} \in \Phi$ and $u_n \notin \Phi$. But this contradicts (2.4.7). □

The program '**while** Φ **do** A' defines a situational system $\boldsymbol{M}^s$ over $\boldsymbol{M}$. (The system $\boldsymbol{M}$ need not be normal.) The set V of labels consists of two elements, $V := \{a, \boldsymbol{\omega}\}$, where $\boldsymbol{\omega}$ is the terminal label and the label a is identified with the name of the action A.

$S := W \times V$ is the set of possible situations. A pair $(u, a) \in S$ has the following interpretation: u is the current state of the system and the action A is performed. Analogously, $(u, \boldsymbol{\omega})$ is read: u is the current state of the system and no action is performed in u. $S_0 := \Phi \times \{a\}$ is the set of initial situations, while the members of $S := (W \setminus \Phi) \times \{\boldsymbol{\omega}\}$ are terminal situations. The relation Tr of the transition between situations is defined as follows:

$Tr(s, t)$ if and only if **either** $s = (u, a), t = (w, a), u \in \Phi, w \in \Phi$ and uA, Rw **or** $s = (u, a), t = (w, \boldsymbol{\omega}), u \in \Phi, w \notin \Phi$ and uA, Rw.

Let $f : W \times V \to W$ be the projection from S onto W. The six-tuple

$$\boldsymbol{M}^s := (W, R, \{A\}, S, Tr, f)$$

is called the *situational action system associated with the program* '**while** Φ **do** A' on $\boldsymbol{M}$.

Finite runs of situations are defined in the usual way; they are sequences of situations

$$(s_0, \ldots, s_n) \tag{2.4.8}$$

such that $s_0 \in S_0$ and $Tr(s_i, s_{i+1})$ for all $i \leqslant n - 1$. The run (2.4.8) is terminated if $s_n \in S_\omega$.

Not every run (2.4.8) can be prolonged to a terminated run. It may happen that for some $s_n = (u_n, a)$ there are no states w such that $u_n\ A, R\ w$. The program gets jammed in the state u_n.

2.5 The Ideal Agent

A theory of action should distinguish between praxeological and epistemic aspects of action. (This issue is also discussed in the last chapter of this book.) In this work, we will not discuss at length such praxeological elements of action as the cost of action, available means, money, etc. These 'hard' praxeological factors are *en bloc* represented by the relation R of direct transition between states. These problems are closely connected with practical reasoning and with the studies on deliberation in particular. As Segerberg (1985, p. 195) points out: "It is well-known how crucially deliberation depends on the faculties and outlook of the agent, but it depends not only on his knowledge, beliefs and values, but also how he perceives the situation, what ways he has of acquiring new situation, his imagination, what action plans he has available and what decision he employs."

The very nature of agency and the problem of the epistemic status of agents in particular are the most difficult issues that a theory of action has to resolve. In its lexical meaning, the theory typically describes an action as behaviour caused by an *agent* in a particular *situation*. The agent's *desires* and *beliefs* (e.g. my wanting a glass of water and believing the clear liquid in the cup in front of me is water) lead to bodily behavior (e.g. reaching over for the glass). According to Davidson (1963), the desire and belief jointly cause the action. But Bratman (1999) has raised problems for such a view and has argued that one should take the concept of intention as basic and as not analyzable into beliefs and desires (the Belief-Desire-Intention model of action).[5]

[5] An interesting example which might be discussed with the problem of agency is that of micromanagement where pathological relations hold between individual and collective agents. Roughly, micromanagement is defined as "attention to small details in management: control of a person or a situation by paying extreme attention to small details". The notion of micromanagement in the wider sense describes social situations where one person (the micromanager) has an intense degree of control and influence over the members of a group. Often, this excessive obsession with the most minute of details causes a direct management failure in the ability to focus on the major details.

We make in this section the assumption that the agents operating a situational action system

$$M^s = (W, R, \mathcal{A}, S, Tr, f)$$

are omniscient in the generous sense. To be more specific, suppose $a_1, \ldots, a_n$ is a list of agents operating the system M^s. It is assumed that :

(1) each agent a_i $(i = 1, \ldots, n)$ knows all possible states of the system, i.e., the elements of W, and all possible direct transitions between states, i.e., the elements of R;
(2) each agent a_i $(i = 1, \ldots, n)$ knows the elements of S, Tr, and f;
(3) each agent a_i $(i = 1, \ldots, n)$ knows, for every $A \in \mathcal{A}$, not what will in fact happen, but what can happen if the action A is performed, i.e., for every $u \in W$ and every $A \in \mathcal{A}$, the agent a_i knows the elements of the set $\{w \in W : u\, A\, w\}$;

(Postulates (1)–(3) thus ascertain that the internal structure of the system $(W, R, \mathcal{A})$, as well as the situational envelope are completely transparent to the agents $a_1, \ldots, a_n$.)

(4) each agent a_i always knows the current situation of the system; in particular the agent knows the current state of the system;

(The postulates (1)–(4), thus, enable each agent a_i to state if a quite arbitrary action $A \in \mathcal{A}$ is performable (or totally performable) in the current situation.)

(5) if a_i is the agent of an action $A \in \mathcal{A}$ in a situation s and A is performable in s, then a_i knows what *will* happen, if he performs A in s.

The principle (5) thus says that the actions performed by the agent are completely subject to his free will. It is not assumed that to each atomic action A only one agent is assigned, where he is the same in all possible situations. The agents of A may vary from one situation to another. We state however that every action is carried out by at the most only one agent in a given situation (the agent who actually performs the action). Even if the agents are ideal, if the system is operated by more than one agent, they may not foresee the evolution of the system. The agent may not know *who* is going to perform an action in the next move, *what* action the other agent is going to perform and *how* the action will be performed. (The first case is obviously excluded in a game of chess but the other two are not.) It is conceivable that in some situation s there may be many agents who are allowed to perform certain actions. For example, agent a_1 could perform action A_1 or A_2 in s and agent a_2 could perform A_3, A_4 or A_5 in the same situation s. This may happen in some situational action systems. The relation Tr of direct transition between situations allows for many possible courses of events commencing with s. Whether agent a_1 or agent a_2 is going to make a move in s may be a random matter; the situation s and the relation Tr need not determine this accurately.

Free will, according to its lexical meaning, is the agent's power of choosing and guiding his actions (subject to limitation of the physical world, social environment and inherited characteristics). Free will enables the agent, in a particular situation in which he can act, both to choose an action he would like to perform (if there is more than one possible action he can perform in this situation) and to be completely free in his way of accomplishing it. In the framework of the formalism accepted here, the fact that an action A is performed according to the agent's free will means only that the moment the action is undertaken (a situation s), the system may pass to any of the states from the set $\{f(s') : s\, Tr\, s' \text{ and } f(s)\, A\, f(s')\}$. The free agent is not hindered from transferring the system to any of the states belonging to the above set (provided that the set is non-empty); his practical decision in this area is the result of a particular value, and so something that in the given situation the agent considers proper.

The notion of an agent should be widely understood; he can be a real man or a group of people (a collective agent); or a man cooperating with a machine or a robot. A detailed analysis of the concept of agent is not necessary for the present investigation. We want, however, this notion to be devoid of any anthropocentric connotation. (This view is not inconsistent with the above postulates (1)–(5) if the meaning of the phrase 'the agent knows' is properly understood.)

Realistic theories of action should be based on more restrictive postulates than those defining the epistemic and praxeological status of ideal agents. For instance, in the game of bridge the players never know the current distribution of cards—the postulate (4) is not true in this case. The agents may not know the members of W, as well as the pairs of states which are in the relation R. The agent of an action may not know all the possible performances of this action. The knowledge of S and Tr may also be fragmentary.

In Part II an approach to action theory is outlined which takes into account the above limitations and assumes that agents' actions are determined and guided by systems of norms. Norms determine what actions in given conditions are permitted and which are not. Agents' actions are then controlled by the norms. The relationship between this concept of action and deontic logic is also discussed there.

A theory of action is the meeting ground of many interests, and therefore many approaches. The formal approach attempts to answer the following question: how should actions be described in terms of set-theoretic entities? Any answer to this question requires, of course, some ontological presuppositions. In this book we propose, in a tentative spirit, a relatively shallow analysis of this subject. First, we assume that there exist states of affairs and processes. Furthermore, we claim that these simple ontological assumptions can be rendered adequately into the language of set theory. This leads to the definition of a discrete system. The set W of states is the mathematical counterpart of the totality of states of affairs, and R, the relation of direct transitions, represents possible processes (i.e., transitions) between states of affairs. Apart from these concepts the category of situations is singled out. It is an open problem whether situations can be faithfully represented by set theory. We do not discuss this issue here because we would have to begin with a general account of what a situation is. Instead, we aim at showing that in some simple cases the theory of sets provides a good framework for the concepts of action and situation.

2.6 Games as Action Systems

In this section, as an illustration of the theory expounded above, a class of idealized, infinite mathematical games will be presented. These games will be modeled as simple, situational action systems.

We recall that the symbols $\mathbb{N}$ and ω interchangeably denote the set of natural numbers with zero. In set theory the set of natural numbers is *defined* as the least non-empty limit ordinal. The elements of ω are also called *finite ordinals*. Thus $0 = \emptyset$ is the least natural number and $n + 1 = \{0, 1, \ldots, n\}$, for all $n \in \omega$, according to the standard set-theoretic definition of natural numbers.

If $n \in \omega$, then ${}^{n}A$ is the set of functions from n to A. Note that ${}^{0}A = \{\emptyset\}$. The empty function $\emptyset$ is also denoted by $\mathbf{0}$. The length of the sequence $\mathbf{0}$ is 0. $\boldsymbol{A} = (W, \Sigma, \Gamma, \delta, w_0, \boldsymbol{\alpha}, F)$.

We then define

$$^{<\omega}A := \bigcup_{n\in\omega} {}^{n}A$$

If $p \in {}^{n}A$, we also write $p = (a_0, a_1, \ldots, a_{n-1})$, or simply $p = a_0a_1 \ldots a_{n-1}$, where $a_k := p(k)$ for $k = 0, 1, \ldots, n-1$. $|p|$ is the length of p. (Formally, $|p|$ coincides with the domain of p.)

We consider the scheme of an infinite game between two players, denoted respectively by I (the first player) and II (the second player).

Let A be a fixed non-empty set, e.g., the set of natural numbers.

(i) The players alternately choose elements of A. Each choice is a *move* of the game; and each player before making each of his moves is privy to all the previous moves.
(ii) Infinitely many moves are to be performed;
(iii) Player I starts the game.

The resulting infinite sequence of elements of A

$$\boldsymbol{a} = a_0, a_1, \ldots, a_n, a_{n+1}, \ldots$$

is called a *play* of the game. The even numbered elements of $\boldsymbol{a}$, viz., $a_{2n}, n = 0, 1, \ldots$ are established by player I while the odd-numbered elements of $\boldsymbol{a}$, $a_{2n+1}, n = 0, 1, \ldots$ by player II.

Each set Z of infinite sequences of elements of A, $Z \subseteq {}^{\omega}A$ determines a game $G(Z)$. Z is called the *payoff* of the game $G(Z)$.

A play $\boldsymbol{a}$ of the game $G(Z)$ is won by player I if it is the case that $\boldsymbol{a} \in Z$. If $\boldsymbol{a} \notin Z$—the play $\boldsymbol{a}$ is won by II. Thus the game $G(Z)$ is decidable in the sense that for each play $\boldsymbol{a}$ either I or II wins $\boldsymbol{a}$ (there are no draws).

Every initial segment $\boldsymbol{a}\lceil n = a_0, a_1, \ldots, a_{n-1}$ of a possible play $\boldsymbol{a}$ is called a *partial play*. ${}^{<\omega}A$ is the set of all partial plays.

Let π_n be the projection of ${}^{\omega}A$ onto the nth axis, that is, $\pi_n(\boldsymbol{a})$ is the nth term of the sequence $\boldsymbol{a}$ for all $\boldsymbol{a} \in {}^{\omega}A$, $n \in \omega$.

Suppose that $\pi_n(Z)$ is a proper subset of A for some odd n. Then II wins the game irrespective of the first n moves $a_0, a_1, \ldots, a_{n-1}$ made by the players. In the nth step player II selects an arbitrary element $a_n \in A \setminus \pi_n(Z)$. Then for *any* play $\boldsymbol{a}$ being a continuation of the initial segment $a_0, a_1, \ldots, a_n$ it is the case that $\boldsymbol{a} \in Z$. (For if $\boldsymbol{a} \in Z$, then $\pi_m(\boldsymbol{a}) \in \pi_m(Z)$, for all $m \in \omega$, which is excluded.) Thus the play $\boldsymbol{a}$ is won by II. Intuitively, the winning strategy for II is any function S defined on all possible positions of even length such that $S(p) \in A \setminus \pi_n(Z)$ for any position $p = a_0, a_1, \ldots, a_{n-1}$ of length n with n defined as above.

We shall express the above remarks in a more formal setting as follows. We shall limit ourselves to deterministic games, i.e., the ones in for each player in any position there is only one option open to him according to which he continues the game, which means that the strategy available to him is a function.

Definition 2.6.1 Let A be a non-empty set.

1. The set $W := {}^{<\omega}A$ is called the set of *possible positions* or, in the terminology of action theory, it is called the set of *possible states*.
2. Any function $S : W \to A$ is called a *deterministic strategy* in A.
3. A play in A is any mapping $P : \omega \to A$. □

Lemma 2.6.2 *For any deterministic strategy S in A there exists a unique play P in A such that $P(n) = S(P\lceil n)$ for all $n \in \omega$.*

(Note that $P(0) = S(P\lceil 0) = S(P\lceil \emptyset) = S(\mathbf{0})$.)

Proof This is an application of the known Principle of Definability by Arithmetic Recursion. □

Putting $P(n) = a_n$ for all $n \in \omega$, the above lemma says that $a_0 = S(\mathbf{0})$ and $a_n = S(a_0, a_1, \ldots, a_{n-1})$ for all n. P is called the play *played according* to the strategy S.

From the perspective of action theory, each strategy S defines a binary relation R_S on the set $W = {}^{<\omega}A$ is of states—possible positions. R_S is the relation of direct transition between possible situations. Intuitively, for $u, w \in W$, the fact $u\, R_S\, w$ means that w results form u by an application of the strategy S to u and w is the position obtained from the sequence u by adjoining the element $S(u)$ at the end of u. Formally, $u\, R_S\, w \;\Leftrightarrow_{df}\; w = u,\ S(u)$.

Since the strategy S is deterministic, R_S is a function from W to W. It follows that (W, R_S) is a deterministic discrete system.

We are interested in a mathematical analysis of two-person games. To represent such games in terms of action systems, we define the operation of the merger of any two strategies.

The definition of the merger of two strategies takes into account the context-sensitive aspect of each play, represented by the situation whose component is not only the finite string of moves made by one player up to some phase of the play, but also the respective moves of his opponent. Formally, this situation is represented by the following mathematical construction. The players I and II have certain strategies

S_1 and S_2 at their disposal. The strategy S_1 defines the elementary action A_{I} of the player I on the set of states W. A_{I} is defined as above as the transition relations between positions imposed by the strategy S_1 on the set of positions. Thus, for any positions $p, q \in W$,

$$p\, A_{\mathrm{I}}\, q \quad \Leftrightarrow_{df} \quad q = p,\ S_1(p).$$

Analogously one defines the action of the other player: for $p, q \in W$,

$$p\, A_{\mathrm{II}}\, q \quad \Leftrightarrow_{df} \quad q = p,\ S_2(p).$$

It is clear that the relations A_{I} and A_{II} are both unary functions from W to W.

However, in two-person games the relation of direct transition between states is defined in a more involved way. According to the rules, each play is represented by a sequence

$$a_0, b_0, a_1, b_1, \ldots, a_n, b_n, a_{n+1}, \ldots$$

The moves made by I are influenced not only by his earlier choices $a_0, a_1, \ldots, a_n$ but also by the moves made by the second player, represented by the sequence $b_0, b_1, \ldots, b_n$. The choice of a_{n+1} is motivated not only by the moves $a_0, a_1, \ldots, a_n$ but also by $b_0, b_1, \ldots, b_n$. In other words, in the context-sensitive strategy available for I the consecutive moves performed by him are determined *not* only by $a_0, a_1, \ldots, a_n$ but by the sequence $a_0, b_0, a_1, b_1, \ldots, a_n, b_n$. Analogously, in the context-sensitive strategy of II the $(n+1)$-th move of II is determined *not* by $b_0, b_1, \ldots, b_n$, but by the sequence $a_0, b_0, a_1, b_1, \ldots, a_n, b_n, a_{n+1}$. This sequence determines the next move b_{n+1}.

To define properly the relation R of direct transition between states of W we first define the notion of the the merger of two strategies.

Definition 2.6.3 Let S_1 and S_2 be two strategies in A. The *merger* of S_1 and S_2 is the unique strategy $S_1 \oplus S_2$ in A such that for every $p \in {}^{<\omega}A$,

$$(S_1 \oplus S_2)(p) := \begin{cases} S_1(p) & \text{if } |p| \text{ is even} \\ S_2(p) & \text{if } |p| \text{ is odd.} \end{cases}$$ □

If $|p| = 0$, i.e., p is empty, $p = \mathbf{0}$, then

$$(S_1 \oplus S_2)(p) = S_1(\mathbf{0}),$$

and if $|p| = 1$, then

$$(S_1 \oplus S_2)(p) = S_2(p).$$

The above definition says that if $p = (a_0, b_0, a_1, b_1, \ldots, a_n, b_n)$, then

$$(S_1 \oplus S_2)(p) = S_1\big((a_0, b_0, a_1, b_1, \ldots, a_n, b_n)\big)$$

and if $p = (a_0, b_0, a_1, b_1, \ldots, a_n, b_n, a_{n+1})$, then

$$(S_1 \oplus S_2)(p) = S_2\big((a_0, b_0, a_1, b_1, \ldots, a_n, b_n, a_{n+1})\big).$$

In particular, $(S_1 \oplus S_2)(\mathbf{0}) = S_1(\mathbf{0}) = a_0$ and $(S_1 \oplus S_2)(\langle a_0\rangle) = S_2\big((a_0)\big) = b_0$.
Applying Lemma 2.6.2 to the strategy $S_1 \oplus S_2$ we get

Corollary 2.6.4 *For any strategies S_1, S_2 there exists a unique play P in A such that $P(n) = (S_1 \oplus S_2)(P\lceil n)$ for all $n \in \omega$.* □

It follows from the above corollary and the definition of $S_1 \oplus S_2$ that

$$P(2n) = S_1(P\lceil 2n) \quad \text{and} \quad P(2n+1) = S_2(P\lceil 2n+1)$$

for all $n \in \omega$.

The relation R of direct transition on W is defined as the transition relation determined by the merger $S_1 \oplus S_2$, that is, $R = R_{S_1 \oplus S_2}$. Thus, for any positions $p, q \in W$,

$$p\,R\,q \quad \Leftrightarrow_{df} \quad q = p,\ (S_1 \oplus S_2)(p).$$

We thus arrive at the definition of an elementary action system $\boldsymbol{M} = (W, R, \{A_{\mathrm{I}}, A_{\mathrm{II}}\})$ endowed with two atomic actions. $\boldsymbol{M}$ is called the action system *corresponding to the two-person game with strategies* S_1 and S_2.

Taking into account the typology of action systems adopted in Chap. 1, Definition 1.2.3, we have:

Proposition 2.6.5 *The system $\boldsymbol{M} = (W, R, \{A_{\mathrm{I}}, A_{\mathrm{II}}\})$ is deterministic and complete.*

Proof Immediate. □

The system $\boldsymbol{M}$ is not normal, because neither $A_{\mathrm{I}} \subseteq R$ nor $A_{\mathrm{II}} \subseteq R$. It may also happen that the intersection $A_{\mathrm{I}} \cap A_{\mathrm{II}}$ is non-empty. Therefore the system need not be separative. One may also easily check that the system is irreversible. This follows from the fact that $p\,R\,q$ implies that $|p| < |q|$, for any positions p and q.

The system $\boldsymbol{M}$ has another property: the set of states W is ordered by the relation $\leqslant$, where for $p, q \in W$, $p \leqslant q$ means that $p = q$ or p is a proper prefix of q. $\boldsymbol{M}$ is therefore an ordered action system. (Ordered systems are defined in Chap. 3.)

The system $\boldsymbol{M}$ is operated by two agents: I and II. Though each of them is able to foresee the course of states according to his own strategy, the situation changes when they play together. As a rule each player does not know his opponent's strategy and therefore is unable to predict consecutive positions that will be taken during the game. In other words, each of them knows the action he will perform, because each knows his own strategy. However, they individually may not know the relation of direct transition R because they do not necessarily know the merger $S_1 \oplus S_2$. Consequently, the players need not know the unique play P determined by $S_1 \oplus S_2$.

But during the course of the game, the consecutive positions $P(n)$ of the game are revealed to both of them. Summing up, according to the remarks from the preceding section, each of the agents I and II is an ideal one. However, the rules of the game exclude the possibility that each of them knows the strategy available to his opponent. In this sense they are *not* omniscient.

(The problem of agency in games is investigated by many thinkers—see e.g. van Benthem and Liu (2004). Various "playing with information" games, public announcements and, more generally, verbal actions are not analysed here.)

The above definition of a deterministic strategy is strong from the epistemic viewpoint. So as to continue the game, the strategy S refers to *all* earlier positions occurring in the course of the play determined by S. In other words, to make a successive move in a given position p, the agent *knows* the sequence p; in particular he knows all the prefixes of p, and therefore he knows all earlier positions in the play determined by S. Mathematically, this means that the strategy of an agent is a function defined on the set of *all* possible positions $W = {}^{<\omega}A$. In practice, the agent is not omniscient and he is able to remember at most two or three earlier moves. It is therefore tempting to broaden the definition of a deterministic strategy by allowing functions S which are defined not on ${}^{<\omega}A$ but merely on a subset of the form ${}^{n}A$ for some (not very large) n. The organization of the game is then different. For example, for $n = 2$, the play determined by such a new strategy $S : {}^{2}A \to A$ is defined by declaring in advance the first two moves a_0 and a_1. Then successive moves are determined by S; the function S takes into account merely the last two moves and determines the subsequent move. We may call such functions $S : {}^{n}A \to A$ *Fibonacci-style strategies*, by analogy with Fibonacci sequences.

Mathematically, such a broadening of the definition of a strategy does not change the scope of game theory because, as one can easily check, for every Fibonacci-like strategy S there is a "standard" strategy S' such that both S and S' determine the same play. However, this distinction between the standard and Fibonacci-style strategies may turn out to be relevant in various epistemic contexts in which agents with restricted memory resources compete.

The game is not fully encoded in the elementary action system $\boldsymbol{M}$. We must also take into account rules (i)–(iii) above that organize the game. They constitute the situational envelope of $\boldsymbol{M}$. More specifically, by a *possible situation* we shall understand any element of the set

$$S := W \times \{\mathrm{I}, \mathrm{II}\}.$$

Thus, S is a set of labeled situations with labels being the elements of $\{\mathrm{I}, \mathrm{II}\}$. The pair $(\mathbf{0}, \mathrm{I})$ is called the *initial situation*. (The symbol "S" also ranges over strategies, but such a double use of this letter should not lead to confusion.)

The relation of transition Tr between situation is defined in accordance with the above requirements (i)–(iii). Thus, if $s_1, s_2 \in S$ and $s_1 = (p, a)$, $s_2 = (q, b)$, then

$$s_1\, Tr\, s_2 \;\Leftrightarrow_{df}\; \text{either } p \text{ is a position of even length, } a = \text{I}, b = \text{II}, \text{ and } A_{\text{I}}(p, q)$$
$$\text{or } p \text{ is a position of odd length, } a = \text{II}, b = \text{I}, \text{ and } A_{\text{II}}(p, q).$$

Intuitively, if p is a position whose length is an even number, then in the situation (p, I) player I performs the action A_{I} and the situation (p, I) turns into (q, II), where $q = p, (S_1 \oplus S_2)(p)$ $(= p, S_1(p))$. Analogously, if p is a position whose length is an odd number, then in the situation (p, II) player II performs the action A_{II} and the situation (p, II) turns into (q, I), where $q = p, (S_1 \oplus S_2)(p)$ $(= p, S_2(p))$.

$f : S \to W$ is defined as the projection onto the first axis: if $s = (p, a)$, then $f(s) := p$, for all $s \in S$.

The relation R of direct transition between states of the above action system $\boldsymbol{M}$ is compatible with *Tr*; that is, for every pair $s_1, s_2 \in S$ of situations, if $s_1\, Tr\, s_2$ then $f(s_1)\, R\, f(s_2)$. Indeed, assume $s_1\, Tr\, s_2$. If $s_1 = (p, a)$, and the length of p is even, then $a = \text{I}$, $s_2 = (q, \text{II})$, where $q = p, (S_1 \oplus S_2)(p)$. Hence $p\, R\, q$, which means that $f(s_1)\, R\, f(s_2)$. The case when $s_1 = (p, a)$, and the length of p is odd, is analogously checked.

It follows from the above remarks that

$$\boldsymbol{M}^s = (W, R, \{A_{\text{I}}, A_{\text{II}}\}, S, Tr, f)$$

is a *situational action system*. $\boldsymbol{M}^s$ is the system *associated with* the above game. $\boldsymbol{M}^s$, together with the initial situation $(\mathbf{0}, \text{I})$ being distinguished, faithfully encodes all aspects of the game with one exception, however; the target of the game has not been revealed thus far. In other words, to have the game fully determined, the payoff set must be defined. Below we present some remarks on wining strategies.

Let Z be a subset of ${}^{\omega}A$ (i.e., Z is a set of infinite functions from ω to A). Let, as above, $S_1 \oplus S_2$ be the merge of strategies S_1 and S_2 and let $P = a_0, b_0, a_1, b_1, \ldots, a_n, b_n, a_{n+1}, \ldots$ be the play determined by $S_1 \oplus S_2$.

Player I *wins the play P for Z* if the infinite sequence P belongs to Z. Player II *wins the play P for Z* if P does *not* belong to Z.

S_1 is a *winning strategy* in Z (for player I) if for *every* strategy S_2 chosen by player II it is the case that the resulting play $P = a_0, b_0, a_1, b_1, \ldots, a_n, b_n, a_{n+1}, \ldots$ determined by $S_1 \oplus S_2$ belongs to Z.

S_2 is a *winning strategy* in Z (for the second player) if for every strategy S_1 chosen by player I it is the case that the play P determined by $S_1 \oplus S_2$ is *not* in Z.

Examples illustrating the above notions below are taken from Błaszczyk and Turek (2007), Sect. 17.3.

Example We assume $A = \omega$. Any two-person play consists in selection of a sequence $a_0, b_0, a_1, b_1, \ldots, a_n, b_n, \ldots$ of natural numbers.

Let $Z = {}^{\omega}\omega$. Then every strategy of I is a winning strategy for I, because for all n, each selection of numbers $a_0, a_1, \ldots, a_n$ by I guarantees that I wins every play (irrespective of the corresponding moves $b_0, b_1, \ldots, b_n$ made by II).

If $Z = \emptyset$, then every strategy of II is a winning strategy for II.

Let us consider a less trivial case: Z is a *countable* subset of ${}^{\omega}\omega$, i.e., Z is a countable set of infinite sequences z_n, $n \in \omega$. We define the following strategy for the second player II:

$$S_2(\mathbf{0}) := z_0(0) + 1,$$
$$S_2(c_0, c_1, \ldots, c_n) := z_{n+1}\big(2(n+1)\big) + 1,$$

for all $n \in \omega$ and all $(c_0, c_1, \ldots, c_n) \in {}^{n+1}\omega$.

S_2 is called the *diagonal* strategy. (Here $z_n = (z_n(0), z_n(1), \ldots)$ and $z_n\big(2(n+1)\big)$ is the value of z_n for $2(n+1)$).

Proposition 2.6.6 *The diagonal strategy is a winning strategy for the second player.*

Proof Let S_1 be an arbitrary strategy of player I. The play P defined by $S_1 \oplus S_2$ is represented by an infinite sequence

$$a_0, b_0, a_1, b_1, \ldots, a_n, b_n, a_{n+1}, b_{n+1}, \ldots$$

where $b_0 = z_0(0) + 1 = a_0 + 1$ and $b_n = z_n(2n) + 1$ for $n = 1, 2, \ldots$

It suffices to notice that $a_0, b_0, a_1, b_1, \ldots, a_n, b_n, a_{n+1}, \ldots$ is not in Z. Indeed, the above sequence can be written as

$$z_0(0)+1, b_0, z_1(2)+1, b_1, z_2(4)+1, b_2, z_3(6)+1, b_3, \ldots, z_n(2n)+1, b_n, \ldots \quad (*)$$

The above sequence differs from the sequences z_n, $n \in \omega$, because if $(*)$ were equal to some z_n, we would have that the element $z_n(2n)$ of z_n would be equal to the element of $(*)$ with index $2n$; i.e., it would be equal to $z_n(2n) + 1$, which is excluded. □

A set $Z \subseteq {}^{\omega}\omega$ is said to be *determined* if in the two-player game $G(Z)$ one of the players, I or II, has a winning strategy. According to the above proposition, every countable set $Z \subseteq {}^{\omega}\omega$ is determined. Determined sets, when taken collectively, do not show regular algebraic properties. For example, one cannot prove that they form a Boolean algebra. In 1976 Martin proved the following theorem:

Theorem *Assume the set theory ZF of Zermelo Fraenkel. Every Borel subset of $\mathbb{R}$ is determined.* □

The situation changes if one introduces the following set-theoretic axiom:

Axiom of Determinacy (AD) *Every subset of ${}^{\omega}\omega$ is determined.* □

It is easy to see that AD is equivalent to the following statement:

For every countably infinite set A, every subset of ${}^{\omega}A$ is determined. □

Other games, as e.g. Banach–Mazur games or parity games, can be also represented in a similar manner as situational action systems. (Parity games are played on a coloured directed graphs. These games are history-free determined. This means that if a player has a winning strategy then he has a winning strategy that depends only on the current position in the graph, and not on earlier positions.)

The games presented in this chapter are all qualified as games with perfect information, which means that they are played by ideal agents in the sense expounded in Sect. 2.5. Card games, where each player's cards are hidden from other players, are examples of games of imperfect information.

2.7 Cellular Automata

Cellular automata form a class of situational action systems. Here, we shall only outline the functioning of one-dimensional cellular automata, the latter being a particular case of the general notion. Suppose we are given an infinitely long tape (in both directions) divided into cells. From the mathematical viewpoint, the tape is identified with the set of integers Z. Cells are, therefore, identified with integers. Each cell has two immediate neighbors, viz. the left and the right adjacent cells. There are two possible states for each cell, labeled 0 and 1. The state of a given cell c, together with the states of its immediate neighbors, fully determines a *configuration* or *possible situation* of the cell c. There are $8 = 2^3$ possible situations of each cell. Possible configurations are written in the following order

$$1\underline{1}1,\quad 1\underline{1}0,\quad 1\underline{0}1,\quad 1\underline{0}0,\quad 0\underline{1}1,\quad 0\underline{1}0,\quad 0\underline{0}1,\quad 0\underline{0}0,$$

where the underlined digit indicates the current state of a particular cell c. The rules defining a cellular one-dimensional automaton specify possible transitions for each such situation. These rules are identified with functions assigning to each situation a state from 0, 1. There are, therefore, $256 = 2^8$ such rules. Each rule individually defines a separate cellular automaton and its action. For instance,

possible situations	$1\underline{1}1$,	$1\underline{1}0$,	$1\underline{0}1$,	$1\underline{0}0$,	$0\underline{1}1$,	$0\underline{1}0$,	$0\underline{0}1$,	$0\underline{0}0$,
new state for center cell	$\underline{0}$	$\underline{1}$	$\underline{1}$	$\underline{0}$	$\underline{1}$	$\underline{1}$	$\underline{1}$	$\underline{0}$

is a rule which assigns to each possible configuration of the center cell a new state of this cell.

Stephen Wolfram proposed a scheme, known as the *Wolfram scheme*, to assign to each rule a number from 0 to 255. This scheme has become standard (see e.g. Wolfram 2002).

As each rule is fully determined by the sequence forming new states for the center cell, this sequence is encoded first as the binary representation of an integer and then by its decimal representation. This number is taken to be the rule number of the automaton. For example, the above rule is encoded by 01101110_2 which is 110 in base 10. This is, therefore, Wolfram's rule 110.

In the above infinite model, each cell c has two immediate neighbors: $c-1$ (the left one) and $c+1$ (the right one). Each mapping $u : Z \to \{0, 1\}$ defines the *global state* of an automaton. (One may think of an automaton as a system of conjugate cells subject to a given transformation rule.) Each rule r determines transformations of global states of the automaton. Suppose that the following sequence of states (the second row) have been assigned to the displayed finite segment of the sequence of integers (cells):

$$\begin{array}{cccccccccccc} \ldots, & -3, & -2, & -1, & 0, & 1, & 2, & 3, & 4, & 5, & 6, & 7, \ldots \\ & 1, & 1, & 0, & 1, & 0, & 1, & 0, & 1, & 0, & 0, & 0, \ldots \end{array}$$

The rule 110 turns the above states assigned to $-2, -1, 0, 1, 2, 3, 4, 5, 6$ to the following sequence

$$\begin{array}{cccccccccccc} \ldots, & -3, & -2, & -1, & 0, & 1, & 2, & 3, & 4, & 5, & 6, & 7, \ldots \\ \ldots & & 1, & 1, & 1, & 1, & 1, & 1, & 1, & 0, & 0, & \ldots \end{array}$$

(The new states of the cells -3 and 7 are not displayed because the initial states assigned to the left neighbour of -3 and the right neighbour of 7 have not been disclosed.) Further iterations of the rule produce new global states.

One may also define an elementary cellular automaton with a finite number of cells, say $1, 2, \ldots, n$ by declaring that 1 is the right neighbour of n and n is the left neighbour of 1. (Of course, $n-1$ is the left neighbour of n and 2 is the right neighbour of 1.) It is also natural to take all the n complex roots of degree n of 1 as cells).

Each Wolfram rule r may be regarded as a certain context-dependent substitution. In each sequence of zeros and ones (of the type of integers) it replaces each consecutive element of the sequence by a new one by looking at the left and right immediate neighbours of this element in the sequence and then applying the adopted rule. Wolfram rules act as rules with limited memory. In order to make a successive substitutions in a 0-1 sequence, the rule merely assumes the knowledge of three consecutive elements of the sequence. One may complicate Wolfram rules in various ways such as by taking into account more distant neighbours.

From the algebraic viewpoint, each Wolfram rule may be treated as a ternary operation of the two-element (Boolean) algebra $\{0, 1\}$. In turn, each (infinite) elementary cellular automaton is regarded as an action of this algebra on the uncountable set $\{0, 1\}^Z$ in the way which was explained above.

It is worth noticing that the automaton determined by the rule 110 is capable of universal computation (Cook 2004) and is therefore equivalent to a universal Turing machine.

Suppose a Wolfram rule r has been selected. Each individual cell c may be then regarded as a simple two-state situational action system $\boldsymbol{M}_c = (W, R, \mathcal{A}, S, Tr, f)$ which we shall describe. Here $W = \{0, 1\}$ is the set of states. Possible situations s

are labeled states, $s = (u; a)$, where u is the state of s and a the label of s. Labels are ordered pairs of states, $a = (u_L, u_R)$, where u_L is a state of the left neighbour of c and u_R is a state of the right neighbour of c. Possible situations may therefore be identified with configurations of c. Thus, if e.g., $0\underline{1}1$ is a configuration of c, then $s = (1; 01)$ is the possible (labeled) situation corresponding to $0\underline{1}1$: 1 is the current state of c and 0 and 1 are the current states of the left and the right neighbour of c respectively. There are, of course, $4 = 2^2$ labels and 8 possible situations of $\boldsymbol{M}_c$. $Label = \{00, 01, 10, 11\}$ is the set of labels and S is the set of possible situations.

The work of $\boldsymbol{M}_c$ is not, however, organized in the same way as in the case of the situational action systems corresponding to iterative or pushdown algorithms.

Each label $a \in Label$ gives rise to an atomic deterministic action A_a on W. For suppose $u \in W$ and $a = (u_L, u_R)$. Then, (u_L, u, u_R) is a configuration. The rule r assigns to this configuration a unique state w. We then put: $A_a(u) := w$. (The correspondence $a \to A_a$, $a \in Label$, need not be one-to-one. It is dependent on the choice of a Wolfram rule.) We, therefore, have at most 4 atomic actions, all being unary functions). We put:

$$\boldsymbol{\mathcal{A}} := \{A_a : a \in Label\}.$$

Rule r determines transitions between states. As each state u is nested in some configuration (there are four such configurations for u), the rule r defines at the most four possible immediates successors of u. Thus,

$u\,R\,w$ if and only if for some configuration (u_L, u, u_R) it is the case that r assigns the state w to (u_L, u, u_R).

Equivalently,

$u\,R\,w$ if and only if $w = A_a(u)$ for some $a \in Label$.

It follows that $(W, R, \boldsymbol{\mathcal{A}})$ is a deterministic and normal action system. (But R need not be a function.)

The function f is a projection assigning to each possible situation $s = (u; a)$ the state u.

It remains to define the transition relation *Tr* between possible situations. How do situations evolve? Suppose $(u_L, \underline{u}, u_R)$ is a current configuration of the cell c, where $\underline{u}$ is the current state of c. Rule r merely unambiguously determines the next state of c. The clue is that the configuration being the successor of $(u_L, \underline{u}, u_R)$ depends not only on the current states u_L and u_R of the left and right neighbours of c, i.e. the current states of the cells $c - 1$ and $c - 1$, but it is also determined by the current states of farther neighbours, these being the cells $c - 2$ and $c + 2$. Thus, in order to predict at least on move in a trajectory of situations, one has to know a quintuple of states $(u_{LL}, u_L, \underline{u}, u_R, u_{RR})$ with u_{LL} being the current state of $c - 2$, the left neighbour of $c - 1$, and u_{RR} being the current state of $c + 2$, the right neighbour of $c+1$. Rule r then assigns to the configurations $(u_{LL}, u_L, \underline{u})$ and $(\underline{u}, u_R, u_{RR})$ certain states, say w_1, w_2, respectively, and it assigns to the configuration $(u_L, \underline{u}, u_R)$ a state

w. Taking this into account, we see that the configuration $(w_1, \underline{w}, w_2)$ is a successor of $(u_L, \underline{u}, u_R)$.

The above remarks enable one to define the relation *Tr* of direct transitions between possible situations. Note that generally *Tr* need *not* be a function because the configuration $(w_1, \underline{w}, w_2)$ depends not only on $(\underline{u}, u_R, u_{RR})$ but also on states u_{LL} and u_{RR}. On the other hand, any situation s has at the most four intermediate successors, according to *Tr*. Each such successor is determined by states of the cells $c - 2$ and $c + 2$.

The relation R of direct transition between states of the above action system $(W, R, \boldsymbol{\mathcal{A}})$ is compatible with *Tr*, i.e., for every pair $s_1, s_2 \in S$ of situations, it is the case that if $s_1 \, Tr \, s_2$ then $f(s_1) \, R \, f(s_2)$. It follows that $\boldsymbol{M}_c = (W, R, \boldsymbol{\mathcal{A}}, S, Tr, f)$ is a well-defined situational action system.

The action systems $\boldsymbol{M}_c$ are identical, for all cells c. The functioning of each system $\boldsymbol{M}_c$ is a kind of a local feedback among $\boldsymbol{M}_c$ and the surrounding systems $\boldsymbol{M}_{c-2}, \boldsymbol{M}_{c-1}$, $\boldsymbol{M}_{c+1}$ and $\boldsymbol{M}_{c+2}$, for all cells c. The structure of this feedback is determined by a definite Wolfram rule r.

The evolution of the system of automata $\boldsymbol{M}_c$, $c \in Z$, (Z—the set of integers) is initialized by declaring a *global initial state* (a *global configuration*) of the system. This is done by selecting a mapping $u : Z \to \{0, 1\}$. (For example, one may assume that u assigns the zero state to each cell.) The system is set in motion by the action of the Wolfram rule r for the cell $u(0)$ and its neighbours. Then the evolution of the system proceeds spontaneously.

The theory of one-dimensional cellular automata is a particular case of the more general theory of finitely many dimensional automata. We shall restrict our attention to the two-dimensional case because it gives a sufficient insight into the basic intuitions underlying the general definition. A standard way of depicting a two-dimensional cellular automaton is with an infinite sheet of graph paper and a set of rules for the cells to follow. Each square is called a 'cell' and each cell has two possible states, black and white. The 'neighbours' of a cell are the 8 squares touching it. Mathematically, each cell (or square) is identified with a pair of integers (a, b). Any pair of the form (c, d) with $c = a \pm 1$ or $d = b \pm 1$ is therefore a neighbour of (a, b). (All these pairs form the *Moore neighbourhood* of (a, b). One may also take a non-empty subset of the defined set of pairs (c, d) to form a narrower neighbourhood of (a, b); in a particular case one gets the *von Neumann neighbourhood* of (a, b) by taking the pairs (c, d) with $c = a \pm 1$ and $d = b \pm 1$.) For such a cell and its neighbours, there are 512 ($= 2^9$) possible patterns. For each of the 512 possible patterns, the rule table states whether the center cell will be black or white on the next time interval.

It is usually assumed that every cell in the universe starts in the same state, except for a finite number of cells in other states, often called a (global) *configuration*. More generally, it is sometimes assumed that the universe starts out covered with a periodic pattern and only a finite number of cells violate that pattern. The latter assumption is common in one-dimensional cellular automata.

2.8 A Bit of Physics

We have been concerned thus far with *discrete* situational action systems. A quite natural question which can be posed in this context is: what about *continuous* situational action systems? How should we understand them? We shall not dwell on this topic and discuss it thoroughly here. The paradigmatic examples of one-agent continuous situational action systems occupy a central place in classical mechanics and the behaviour of these systems is well described by appropriate differential equations. An analysis of the mathematical apparatus pertinent to classical mechanics might provide some clues which would enable one to build a theory of continuous action systems in a much wider conceptual setting than that provided by physics. We shall now briefly recall some basic facts, which will be very familiar to physicists. Suppose we are given a physical body consisting of finitely many particles whose individual sizes may be disregarded. We may therefore identify the body with a finite set of material points. Hamiltonian mechanics, being a reformulation of classical principles of Newtonian physics, was introduced in the 19th century by the Irish mathematician William Rowan Hamilton. The key role is played by the *Hamiltonian function* $\mathcal{H}$. The value of the Hamiltonian is the total energy of the system being described. For a closed system, it is the sum of the kinetic and potential energy of the system. The time evolution of the system is expressed by a system of differential equations known as the *Hamiltonian equations*. These equations are used to describe such systems as a pendulum or an oscillating spring in which energy changes from kinetic to potential and back again over time. Hamiltonians are also applied to model the energy of other more complex dynamic systems such as planetary orbits in celestial mechanics.

Each state w of the system is represented by a pair of vectors $w = (\mathbf{p}, \mathbf{q})$ from a vector space $\mathbf{V}$. To simplify matters, we shall assume that $\mathbf{V}$ is a finitely dimensional, real space, say $\mathbf{V} = \mathbb{R}^n$ for some positive n. (The number n is called the *degree of freedom* of the system.) $\mathbf{V}$ has therefore well-established topological properties. (Topology is introduced by any of the equivalent norms on $\mathbf{V}$.) The vector $\mathbf{q}$ represents the (generalized) coordinates of the particles forming the system, and $\mathbf{p}$ represents the (generalized) momenta of these particles (conjugate to the generalized coordinates). If $\mathbf{V}$ is endowed with a Cartesian coordinate system, then the vectors $\mathbf{p}$ are "ordinary" momenta. In the simplest case, when the systems consists of one particle only, not subject to external forces, one may assume that $\mathbf{q} = (x, y, z)$ is a vector in the three dimensional Euclidean space $\mathbb{R}^3$, being the coordinates of the particle, and the vector $\mathbf{p} = (p_x, p_y, p_y)$ representing the momentum of the particle.

$W := \mathbf{V} \times \mathbf{V}$ is the set of possible states of the system. W is called a *phase space*. We define an action system which is operated by one agent. We call him **Hamilton**. He performs one continuous action only. His action is identified with the Hamiltonian function $\mathcal{H}$. The Hamiltonian $\mathcal{H} = \mathcal{H}(\mathbf{p}, \mathbf{q}, t)$ is a (scalar valued) function and it specifies the domain T of values in which the time parameter t varies. The domain of $\mathcal{H}$ is a subset of $\mathbf{V} \times \mathbf{V} \times T$ but we assume that the Hamiltonian $\mathcal{H}$ makes sense for all $t \in T$. The set T is usually an open interval on the straight line $\mathbb{R}$ (or $\mathbb{R}$ itself).

By a *possible situation* we shall mean any pair $s = (w, t)$, where w is a state and t is real number representing time moments. In other words, possible situations are just members of the set $S := \mathbf{V} \times \mathbf{V} \times T \quad (= W \times T)$. We therefore identify possible situations with triples $s = (\mathbf{p}, \mathbf{q}, t)$, with $\mathbf{p}$ and $\mathbf{q}$ described as above. The members of T play the role of labels of possible situations. We define f to be the projection assigning to each possible situation $s = (w, t)$ the state w.

How does **Hamilton** act? Suppose we are given a quite arbitrary situation $s_0 = (\mathbf{p}_0, \mathbf{q}_0, t_0) \in S$ from the domain of $\mathcal{H}$. s_0 is called an *initial* situation. The action of **Hamilton** undertaken in the state $(\mathbf{p}_0, \mathbf{q}_0)$ labeled by t_0 carries over the system, for every moment t, from the state $(\mathbf{p}_0, \mathbf{q}_0)$ to a *uniquely* defined state $(\mathbf{p}, \mathbf{q})$ with label t. In other words, for every $t \in T$, **Hamilton** turns the initial situation s_0 into a uniquely defined situation $s = (\mathbf{p}, \mathbf{q}, t)$. We may therefore say that **Hamilton** assigns to each situation s_0 from its domain a *phase path* in the phase space, i.e., a continuous mapping from T to W. Moreover, this path passes through the state $(\mathbf{p}_0, \mathbf{q}_0)$ at t_0. It is now quite obvious how to define the transition relation Tr between possible situations. For any pair of situations $s_0 = (\mathbf{p}_0, \mathbf{q}_0, t_0)$, $s = (\mathbf{p}, \mathbf{q}, t) \in S$, it is the case that $s_0\ Tr\ s$ if and only if both situations s_0 and s belong to the domain of $\mathcal{H}$ and the phase path assigned to s_0 passes through the state $(\mathbf{p}, \mathbf{q})$ at t. In light of the above remarks, the relation Tr is reflexive on the domain of $\mathcal{H}$. How to compute this path? We shall discuss this issue below.

The action of **Hamilton** is a binary relation $A_{\mathcal{H}}$ defined on the set W. It is assumed that the transition relation R between states is identical with the relation $A_{\mathcal{H}}$. (**Hamilton** is omnipotent because he establishes the rules that govern mechanics.) Therefore the system we define is normal. In light of the above remarks, given a state $(\mathbf{p}_0, \mathbf{q}_0)$ and $t_0 \in T$, the set $f_R(\mathbf{p}_0, \mathbf{q}_0)$ contains in particular the range of the unique phase path assigned to the situation $s_0 = (\mathbf{p}_0, \mathbf{q}_0, t_0)$, provided that s_0 is in the domain of $\mathcal{H}$. We assume that for every state $(\mathbf{p}_0, \mathbf{q}_0)$, the set $f_R(\mathbf{p}_0, \mathbf{q}_0)$ is the union of co-domains of all possible phase paths assigned to $s_0 = (\mathbf{p}_0, \mathbf{q}_0, t_0)$, for all possible choices of $t_0 \in T$.

It follows from these definitions that the relation R is compatible with Tr; that is, for any situations $s_0 = (\mathbf{p}_0, \mathbf{q}_0, t_0)$, $s = (\mathbf{p}, \mathbf{q}, t)$, $s_0\ Tr\ s$ implies that $f(s_0)\ R\ f(s)$. Indeed, $s_0\ Tr\ s$ means that the phase path assigned to s_0 passes through the point $(\mathbf{p}, \mathbf{q})$ of the phase space at t. Hence $(\mathbf{p}, \mathbf{q})$ belongs to the range of this path. Consequently, $(\mathbf{p}, \mathbf{q}) \in f_R(\mathbf{p}_0, \mathbf{q}_0)$, i.e., $f(s_0)\ R\ f(s)$.

We have defined a one-agent situational action system

$$\boldsymbol{M}^s := (W, R, \{A_{\mathcal{H}}\}, S, Tr, f).$$

The class of all such systems $\boldsymbol{M}^s$ is called the *Hamiltonian class*.

How do we compute the phase paths involved in the above definitions? One assumes that:

(1) $\mathcal{H}$ possesses the continuous partial derivatives with respect to $\mathbf{p}, \mathbf{q}$ (in the sense of the vector space $\mathbf{V}$) and t.

(2) For every initial condition $s_0 = (\mathbf{p}_0, \mathbf{q}_0, t_0)$ belonging to the domain of $\mathcal{H}$, the phase path assigned to s_0 is a pair of continuous functions $(\mathbf{p}(t), \mathbf{q}(t))$, each function being a map from T to $\mathbf{V}$, such that $\mathbf{p}(t_0) = \mathbf{p}_0$ and $\mathbf{q}(t_0) = \mathbf{q}_0$. (In the Cartesian coordinate system, the vector function $\mathbf{q}(t)$ describes the trajectories of the particles in space while $\mathbf{p}(t)$ defines their momenta, and hence their velocities.) Moreover, by solving the Hamiltonian equations:

$$\frac{\mathrm{d}}{\mathrm{d}t}\mathbf{p}(t) = -\frac{\partial}{\partial \mathbf{q}}\mathcal{H}\big(\mathbf{q}(t), \mathbf{p}(t), t\big)$$

$$\frac{\mathrm{d}}{\mathrm{d}t}\mathbf{q}(t) = +\frac{\partial}{\partial \mathbf{p}}\mathcal{H}\big(\mathbf{q}(t), \mathbf{p}(t), t\big)$$

with the initial condition $\mathbf{p}(t_0) = \mathbf{p}_0$ and $\mathbf{q}(t_0) = \mathbf{q}_0$, one computes the values of $\mathbf{p}(t)$ and $\mathbf{q}(t)$ at any moment t belonging to T.

Chapter 3
Ordered Action Systems

Abstract Ordered action systems are systems whose sets of states are partially ordered. A typology of ordered systems depending on the properties of the component order relation is presented. A significant role of fixed points of the transition relation in ordered systems is emphasized. Numerous examples illustrate the theory.

For many years, methods based on the so-called semantics of fixed points have been extensively developed. This sort of semantics applies a range of results, conventionally called fixed-point theorems, to the definition or construction of abstract objects with given prescribed properties. The heart of the matter is the definition of abstract things as fixed points; and often as the least fixed point, of certain mappings of a partially ordered set into itself. In the simplest case, the defined object is approximated by certain constructions performed in a finite number of steps. By passing to the limit, one obtains the object with the required properties. This process takes place in ordered structures, in which the passage to the infinite is allowed and the existence of the limit is secured. The above fixed-point approach to semantics is, therefore, meaningful only for ordered structures which exhibit some completeness properties such as the existence of suprema for directed subsets or chains. The completeness of a poset guarantees that among its elements there always exists at least one for which the above infinite process terminates. The process of definability by means of fixed-points methods thus consists in picking out of the set of preexisted objects (which form a complete poset), the one with the desired properties.

In this chapter, we present some results about fixed points and show applications of these results in the theory of action. We are mainly concerned with fixed-point theorems for *relations* defined in posets and not only for mappings. (We note that traditionally the focus of mathematical literature is on fixed-point theorems for *functions* rather than relations.) We present here a bunch of such theorems and also show their applicability in the theory of action.

J. Czelakowski, *Freedom and Enforcement in Action*, Trends in Logic 42,
DOI 10.1007/978-94-017-9855-6_3

3.1 Partially Ordered Sets

A rudimentary knowledge of set theory and the theory of partially ordered sets is required in this section.

Let P be a set. A binary relation $\leqslant$ on P is an *order* (or *partial order*) on P if and only if $\leqslant$ satisfies the following conditions:

(i) $\leqslant$ is *reflexive*, i.e., $a \leqslant a$, for all $a \in P$;
(ii) $\leqslant$ is *transitive*, i.e., $a \leqslant b$ and $b \leqslant c$ implies $a \leqslant c$, for all $a, b, c \in P$;
(iii) $\leqslant$ is *antisymmetric*, i.e., $a \leqslant b$ and $b \leqslant a$ implies $a = b$, for all $a, b \in P$.

A *partially ordered set*, a *poset*, for short, is a set with an order defined on it.

Each order relation $\leqslant$ on P gives rise to a relation $<$ of *strict order*: $a < b$ in P if and only if $a \leqslant b$ and $a \neq b$.

Let $(P, \leqslant)$ be a poset and let X be a subset of P. Then, X inherits an order relation from P: given $x, y \in X$, $x \leqslant y$ in X if and only if $x \leqslant y$ in P. We then also say that the order on X is *induced* by the order from P.

In what follows we shall consider only *nonempty* posets $(P, \leqslant)$ (unless it is explicitly stated otherwise).

(1) An element $M \in X$ is called *maximal* in X whenever $M \leqslant x$ implies $M = x$, for every $x \in X$.
(2) An element $m \in X$ is called *minimal* in X whenever $x \leqslant m$ implies $m = x$, for every $x \in X$.
(3) An element $u \in P$ is called an *upper bound* of the set X if $x \leqslant u$, for every $x \in X$.
(4) An element $l \in P$ is called a *lower bound* of the set X if $l \leqslant x$, for every $x \in X$.
(5) An element $a \in P$ is called the *least upper bound* of the set X if a is an upper bound of X and $a \leqslant u$ for every upper bound u of X. If X has a least upper bound, this is called the *supremum* of X and is written as "sup(X)".
(6) An element $b \in P$ is called the *greatest lower bound* of the set X if b is a lower bound of X and $l \leqslant b$ for every lower bound l of X. If X has a greatest lower bound, this is called the *infimum* of X and is written as "inf(X)".

X may have more than one maximal element, or none at all. A similar situation holds for minimal elements.

Instead of 'sup(X)' and 'inf(X)', we shall often write $\vee X$' and $\wedge X$'; in particular, we write '$a \vee b$' and '$a \wedge b$' instead of 'sup($\{a, b\}$)' and 'inf($\{a, b\}$)'.

If the poset P itself has an upper bound u, then it is the only upper bound. u is then called the *greatest element* of P. The notion of the *least element* of P is defined in an analogous way.

A set $X \subseteq P$ is:

(a) an *upper directed* subset of P if for every pair $a, b \in X$ there exists an element $c \in X$ such that $a \leqslant c$ and $b \leqslant c$ (or, equivalently, if every finite nonempty subset of X has an upper bound which is an element of X);

(b) a *chain* in P if, for every pair $a, b \in X$, either $a \leqslant b$ or $b \leqslant a$ (that is, if any two elements of X are comparable);
(c) a *well-ordered subset* of P (or: a *well-ordered chain* in P) if X is a chain, in which every nonempty subset $Y \subseteq X$ has a minimal element (in Y). Equivalently, using a weak form of the Axiom of Choice, X is well ordered if and only if it is a chain and there is no strictly decreasing sequence $c_0 > c_1 > \cdots > c_n > \cdots$ of elements of X. Every well-ordered chain X is isomorphic with a unique ordinal, called the *type* of X.

A subset directed downwards is defined similarly; when nothing to the contrary is said, 'directed' will always mean 'directed upwards'. If the poset $(P, \leqslant)$ itself is a chain or directed, then it is simply called a chain or a directed poset.

Let $(P, \leqslant)$ be a poset and let X be a subset of P. The set X is *cofinal in* $(P, \leqslant)$ if for every $a \in P$ there exists $b \in X$ such that $a \leqslant b$. If X is cofinal in $(P, \leqslant)$, then $\sup(P)$ exists if and only if $\sup(X)$ exists. Furthermore, $\sup(P) = \sup(X)$.

Theorem 3.1.1 *Let $(P, \leqslant)$ be a poset. Every countable directed subset D of $(P, \leqslant)$ contains a well-ordered subset of type $\leqslant \omega$. In particular, every countably infinite chain contains a cofinite well-ordered subchain of type ω.* □

The above theorem is not true for uncountable directed subsets.

Theorem 3.1.2 (Zorn's Lemma) *If every nonempty chain in a poset P has an upper bound, then the set P contains a maximal element.* □

Zorn's Lemma (quantified over all posets) is an equivalent form of the Axiom of Choice (on the basis of Zermelo–Fraenkel set theory ZF).

Definition 3.1.3 Let $(P, \leqslant)$ be a poset.

(1) The poset $(P, \leqslant)$ is *directed-complete* if for every nonempty directed subset $D \subseteq P$, the supremum $\sup(D)$ exists in $(P, \leqslant)$.
(2) The poset $(P, \leqslant)$ is *chain complete* (or *inductive*) if for every nonempty chain $C \subseteq P$, the supremum $\sup(C)$ exists in $(P, \leqslant)$.
(3) The poset $(P, \leqslant)$ is *well-orderably complete* if for every nonempty well-ordered chain $C \subseteq P$, the supremum $\sup(C)$ exists in $(P, \leqslant)$. □

Note In the literature one also finds stronger versions of the above definitions in which one also assumes the existence of the supremum for the *empty* directed subset (the *empty* chain, respectively.) The supremum of the empty directed set is, of course, the least element in the given poset, denoted by $\mathbf{0}$. Definition 3.1.3 *do not* entail that the respective posets possess the zero element. However, when necessary, we shall separately assume the existence of the zero element in directed-complete or inductive posets. □

It is clear that every directed-complete poset is chain complete and every inductive poset is well-orderably complete. In the presence of the Axiom of Choice these three properties are known to be mutually equivalent.

The empty subset of a poset is well ordered. Hence, if $(P, \leqslant)$ is well-orderably complete and contains the least element $\mathbf{0}$, then the supremum of the empty subset exists and it is the least element in (P).

Let $(P, \leqslant)$ be a poset. A function $\pi : P \to P$ is *monotone*[1] if $a \leqslant b$ implies $\pi(a) \leqslant \pi(b)$ for every pair $a, b \in A$.

If π is monotone and C is a nonempty chain (a nonempty well-ordered chain) in $(P, \leqslant)$, then the image $\pi[C] := \{\pi(a) : a \in C\}$ is also a nonempty chain (a nonempty well-ordered chain). A similar implication holds for every nonempty directed set $D \subseteq P$.

3.2 Fixed-Point Theorems for Relations

Let $R \subseteq P \times P$ be a binary relation defined on a nonempty set P. An element $a^* \in P$ is called a *fixed point* of R if it is the case that $a^* \, R \, a^*$. (Fixed points of relations are also called *reflexive points*.) a^* is called a *strong fixed point* of R if it is a fixed point and, moreover, for every $b \in P$, if $a^* \, R \, b$, then $b = a^*$. (This means that a^* is the unique element b in P such that $a^* \, R \, b$, i.e., $\{b \in P : a^* \, R \, b\} = \{a^*\}$.)

Suppose that R is a function on P, that is, R satisfies the condition: for every $a \in P$ there exists a unique element $b \in P$ such that $a \, R \, b$ holds. Symbolically:

$$(\forall a \in P \; \exists! b \in P) \; a \, R \, b.$$

Then the notions of a fixed point and of a strong fixed point for R are equivalent.

We recall that a relation $R \subseteq P \times P$ is *total* (on P) if $Dom(R) = P$. Symbolically,

$$(\forall a \in P)(\exists b \in P) \; a \, R \, b. \tag{3.2.1}$$

Let $(P, \leqslant)$ be a poset. A relation $R \subseteq P \times P$ is called *inflationary* if it is total and included in $\leqslant$, i.e.,

$$(\forall a, b \in P) \; a \, R \, b \quad \text{implies} \quad a \leqslant b. \tag{3.2.2}$$

(The term 'inflationary relation' is borrowed from computer science—see Cai and Paige (1992), Desharnais and Möller (2005). In the literature the terms '∀-expansive relation' and '∃-expansive relation' are also used as names for inflationary and expansive relations, respectively.)

We note two simple but useful observations:

[1] The term 'isotone function' is also used.

Theorem 3.2.1 (The Fixed-Point Theorem for Inflationary Relations) *Let* $(P, \leqslant)$ *be a poset in which every nonempty chain has an upper bound. Let* $R \subseteq P \times P$ *be an inflationary relation. Then R has a fixed point* a^* *which additionally satisfies the following condition:*

$$\textit{for every } \; b \in P, \;\; \textit{if } \; a^* \, R \, b, \;\; \textit{then } \; b = a^*. \tag{3.2.3}$$

(This means that a^* *is the unique element b in P such that* $a^* \, R \, b$*, i.e.,* $\{b \in P : a^* \, R \, b\} = \{a^*\}$*.)*

Proof By Zorn's Lemma, applied to $(P, \leqslant)$, there exists at least one maximal element a^* (in the sense of $\leqslant$). We shall check that a^* is a fixed point for R. By totality, there exists $b \in P$ such that $a^* \, R \, b$. (3.2.2) implies that $a^* \leqslant b$. Since a^* is maximal, $a^* = b$. Hence, $a^* \, R \, a^*$. So a^* is a fixed point of R. Evidently, by maximality and (3.2.2), a^* also satisfies (3.2.3). □

Let $(P, \leqslant)$ be a poset. A relation $R \subseteq P \times P$ is called *expansive* if for every $a \in P$ there exists $b \in P$ such that $a \, R \, b$ and $a \leqslant b$; symbolically,

$$(\forall a \in P)(\exists b \in P) \; a \, R \, b \; \text{ and } \; a \leqslant b. \tag{3.2.4}$$

Every expansive relation is total and every inflationary relation is expansive. If R is a function from P to P, then the properties of being inflationary and expansive are equivalent for R.

Theorem 3.2.2 (The Fixed-Point Theorem for Expansive Relations) *Let* $(P, \leqslant)$ *be a poset in which every nonempty chain has an upper bound. Let* $R \subseteq P \times P$ *be an expansive relation. Then R has a fixed point* a^* *which additionally satisfies the condition:*

$$(\forall b \in P) \; a^* \, R \, b \textit{ and } \; a^* \leqslant b \; \textit{ implies } \; b = a^*. \tag{3.2.5}$$

Proof Define: $R_0 := R \cap \leqslant$. The relation R_0 is total and inflationary. By Theorem 3.2.1, R_0 has a fixed point a^* for which (3.2.3) holds. Consequently, $a^* \, R \, a^*$ and (3.2.5) readily follows. □

The proof of Theorem 3.2.1 employs the Axiom of Choice (in the form of Zorn's Lemma). But in fact, the set-theoretic status of Theorems 3.2.1 and 3.2.2 is the same—each of the above fixed-point theorems is equivalent to the Axiom of Choice.

Theorem 3.2.3 *On the basis of Zermelo–Fraenkel set theory* ZF, *the following conditions are equivalent:*

(a) The Axiom of Choice (AC).
(b) Theorem 3.2.1.
(c) Theorem 3.2.2.

Proof The implication (a) $\Rightarrow$ (b) directly follows from the proof of Theorem 3.2.1 because Zorn's Lemma is used here. The implication (b) $\Rightarrow$ (c) is present in the proof of Theorem 3.2.2. To prove (c) $\Rightarrow$ (a), we assume Theorem 3.2.2. We show that then Zorn's Lemma holds. Let $(P, \leqslant)$ be an arbitrary poset in which every nonempty chain has an upper bound. Let $R :=\leqslant$. The relation R is expansive. By Theorem 3.2.2, R has a fixed point a^* which satisfies (3.2.5). We show that a^* is a maximal element in $(P, \leqslant)$. Suppose $b \in P$ and $a^* \leqslant b$. Hence, $a^* R b$ and $a^* \leqslant b$, by the definition of R. Condition (3.2.4) then gives that $b = a^*$. This means that a^* is a maximal element in $(P, \leqslant)$. □

Example Let $P = [0, 1]$ be the closed unit interval of real numbers. The system $(P, \leqslant)$ with the usual ordering $\leqslant$ of real numbers satisfies the hypothesis of Theorems 3.2.1 and 3.2.2. If the relation R is taken to be equal to $\leqslant$ on P, then R is inflationary. Hence, it has a fixed point in $(P, \leqslant)$ which additionally satisfies (3.2.3). It is clear that 1 is the only such a fixed point of R. On the other hand, every element of P is a fixed point of R. □

Let $(P, \leqslant)$ be a poset. A function $\pi : P \to P$ is *expansive* if $a \leqslant \pi(a)$ for every $a \in P$.

Corollary 3.2.4 (Zermelo) *Let $(P, \leqslant)$ be a poset in which every nonempty chain has an upper bound. Let $\pi : P \to P$ be an expansive function. Then π has a fixed point, i.e., there exists $a^* \in P$ such that $\pi(a^*) = a^*$.*

Proof Following the set theory, each function is identified with its graph. Accordingly, let R be the graph of π. Thus, $a\, R\, b$ holds if and only if $b = \pi(a)$, for any $a, b \in P$. The relation R satisfies the assumptions of both Theorems 3.2.1 and 3.2.2. R is total because π is a function. R is inflationary because it is total and π is expansive. In virtue of Theorem 3.2.1, R has a fixed point a^*. In particular, $a^* R\, a^*$ holds, which means that $a^* = \pi(a^*)$. □

It is interesting to note that on the basis of Zermelo–Fraenkel set theory ZF, Corollary 3.2.4 is equivalent to the Axiom of Choice, as shown by Abian. However, if one strengthens the hypothesis of Corollary 3.2.4 and assumes that $(P, \leqslant)$ is an inductive poset with zero, the corollary modified in this way is provable in ZF and the Axiom of Choice is not needed here—see Moschovakis (1994), Chap. 7.

The above theorems prove the existence of rather big reflexive points—they are maximal elements in a given poset. The theorems we present further are concerned with the problem of finding smaller fixed points. Before passing to them we recall an axiom of set theory which is weaker than AC.

The Principle of Dependent Choices (DC) This principle asserts the following:

Let R be a total relation defined on a set A. Let $a_0 \in A$. Then there exists a function $f : \omega \to A$ such that $f(0) = a_0$ and $f(n)\, R\, f(n + 1)$ for all $n \in \omega$. □

Definition 3.2.5 Let $(P, \leqslant)$ be a poset. A binary relation $R \subseteq P \times P$ is called *monotone*[2] if it satisfies:

$$(\forall a, b, c \in P)(a \leqslant b \text{ and } a\,R\,c \text{ implies } (\exists d \in P)\; b\,R\,d \text{ and } c \leqslant d)$$

(see the figure below). □

If $\pi : P \to P$ is a function, then π is *monotone* if and only if the graph R of π is a monotone relation in the above sense.

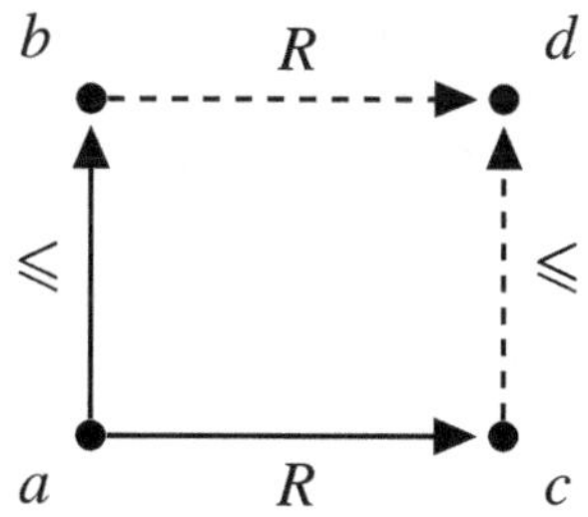

Fig. 3.1

A poset $(P, \leqslant)$ is called *σ-complete* if every chain in $(P, \leqslant)$ of type ω has a supremum.

Definition 3.2.6 Let $(P, \leqslant)$ be a σ-complete poset and let $R \subseteq P \times P$ be a binary relation.

1. R is called *σ-continuous* if R is monotone and it additionally satisfies the following condition:

$(\text{cont})_\sigma$ For every chain C in $(P, \leqslant)$ of type ω and every monotone function $f : C \to P$, if $a\,R\,f(a)$ for all $a \in C$, then $\sup(C)\,R\,\sup(f[C])$.

2. R is *σ-continuous in the stronger sense* if it is σ-continuous and satisfies:

$(*)_\sigma$ For every chain D in $(P, \leqslant)$ of type ω and for every element $a \in P$, if $a\,R\,d$ for all $d \in D$, then $a\,R\,\sup(D)$. □

$(\text{cont})_\sigma$ can be formulated equivalently as follows:

For every chain of elements of P of type ω, $a_0 < a_1 < \cdots < a_n < a_{n+1} < \cdots$ and every increasing sequence $b_0 \leqslant b_1 \leqslant \cdots \leqslant b_n \leqslant b_{n+1} \leqslant \cdots$ of elements of P, if $a_n R b_n$ for all n, then $\sup\{a_n : n \in \mathbb{N}\} R \sup\{b_n : n \in \mathbb{N}\}$.

It is not difficult to prove that R is σ-continuous in the stronger sense if and only if it is monotone and satisfies:

$(\text{cont})^*_\sigma$ For any two monotone functions $f : \mathbb{N} \to P$ and $g : \mathbb{N} \to P$, if $f(n)\,R\,g(n)$ for all $n \in N$, then $\sup(f[\mathbb{N}])\,R\,\sup(g[\mathbb{N}])$.

[2] The term 'isotone relation' is also used.

The antecedent of $(\text{count})_\sigma$ is not vacuously satisfied for certain relations:

Observation *Let $(P, \leqslant)$ be a poset. If a relation $R \subseteq P \times P$ is a monotone and total, then for every chain $a_0 < a_1 < \cdots a_n < a_{n+1} < \cdots$ in $(P, \leqslant)$ of type ω, there exists a chain $b_0 \leqslant b_1 \leqslant \cdots \leqslant b_n \leqslant b_{n+1} \leqslant \cdots$ in $(P, \leqslant)$ such that $a_n \, R \, b_n$ for all n.*

Indeed, let $a_0 < a_1 < \cdots a_n < a_{n+1} < \cdots$ be a countable chain of type ω. As R is total, there exists an element $b_0 \in P$ such that $a_0 \, R \, b_0$. As $a_0 \leqslant a_1$ and $a_0 \, R \, b_0$, the monotonicity of R implies the existence of an element $b_1 \in P$ such that $a_1 \, R \, b_1$ and $b_0 \leqslant b_1$. As $a_1 \leqslant a_2$ and $a_1 \, R \, b_1$, there exists an element $b_2 \in P$ such that $a_2 \, R \, b_2$ and $b_1 \leqslant b_2$, again by monotonicity. Going further, as $a_2 \leqslant a_3$ and $a_2 \, R \, b_2$, there exists an element $b_3 \in P$ such that $a_3 \, R \, b_3$ and $b_2 \leqslant b_3$. Continuing this procedure, we define the countable chain $b_0 \leqslant b_1 \leqslant \cdots \leqslant b_n \cdots b_{n+1} \leqslant \cdots$ in $(P, \leqslant)$ such that $a_n \, R \, b_n$ for all n. (The Principle of Dependent Choices is used here.) □

We recall that if R is a binary relation R on a set P and $a \in P$, then $\delta_R(a) := \{b \in P : a \, R \, b\}$. The set $\delta_R(a)$ is called the *R-image* of the element a.

Given a poset $(P, \leqslant)$, a binary relation R on P and $a \in P$, we define:

$$\uparrow a := \{b \in : a \leqslant b\},$$
$$\downarrow a := \{b \in p : b \leqslant a\}$$

The basic fact concerning fixed points of σ-continuous relations is provided by the following theorem:

Theorem 3.2.7 *Let $(P, \leqslant)$ be a σ-complete poset with zero $\mathbf{0}$. Every σ-continuous relation $R \subseteq P \times P$ such that the set $\delta_R(\mathbf{0})$ is nonempty has a fixed point a^*. Furthermore, a^* can be assumed to have the following property: for every $y \in P$, if $\delta_R(y) \subseteq \downarrow y$, then $a^* \leqslant y$.*

Proof Since R is σ-continuous, R is a monotone relation in $(P, \leqslant)$. We define:

$$Q := \{x \in P : \delta_R(x) \cap \uparrow x \neq \emptyset \ \text{ and } \ (\forall y \in P)(\delta_R(y) \subseteq \downarrow y \ \text{ implies } \ x \leqslant y)\}.$$

Evidently, $\mathbf{0} \in Q$. Hence, the set Q is nonempty.

Lemma 1 *$R\lceil Q$, the restriction of R to Q, is expansive in the poset $(Q, \leqslant)$.*

Proof of the lemma. Let $a \in Q$. As $\delta_R(a) \cap \uparrow, a \neq \emptyset$, there exists $b \in P$ such that

$$a \, R \, b \quad \text{and} \quad a \leqslant b. \tag{3.2.6}$$

We prove that $b \in Q$. This will show that $R\lceil Q$ is expansive. By (3.2.6) and the monotonicity of R in $(P, \leqslant)$ there exists $c \in P$ such that $b \, R \, c$ and $b \leqslant c$ (see the figure below).

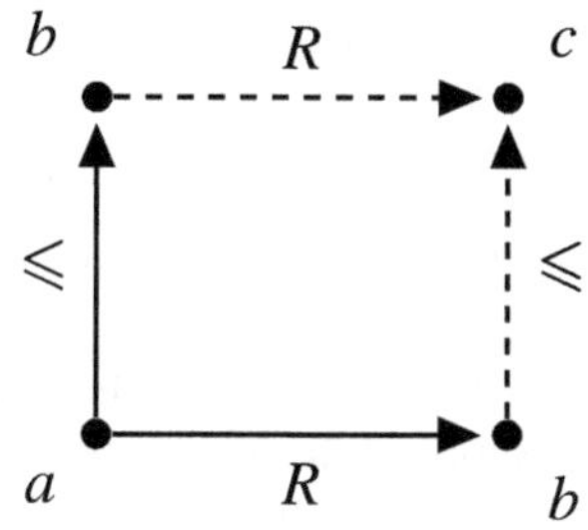

Fig. 3.2

This means that $c \in \delta_R(b) \cap \uparrow b$. Hence,

$$\delta_R(b) \cap \uparrow b \neq \emptyset. \tag{3.2.7}$$

Now let y be an arbitrary element of P such that $\delta_R(y) \subseteq \downarrow y$. We show $b \leqslant y$. As $a \in Q$ and $\delta_R(y) \subseteq \downarrow y$, we have that $a \leqslant y$, by the definition of Q. As $a\, R\, b$, the monotonicity of R implies the existence of an element $z \in P$ such that $y\, R\, z$ and $b \leqslant z$ (see the diagram below). As $z \in \delta_R(y)$ and $\delta_R(y) \subseteq \downarrow y$, we get that $z \leqslant y$. Consequently, $b \leqslant z \leqslant y$ which gives that $b \leqslant y$. This proves that $b \in Q$. □

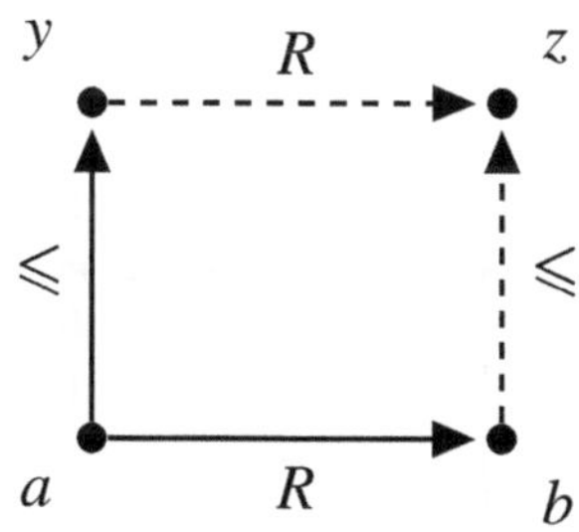

Fig. 3.3

If C is a chain in $(P, \leqslant)$ of type ω and $f : C \to P$ is a monotone function, then $f[C]$ is a chain of type $\leqslant \omega$. Hence, $\sup(f[C])$ exists in $(P, \leqslant)$. Evidently, every chain in $(Q, \leqslant)$ is also a chain in $(P, \leqslant)$. Consequently, by the σ-continuity of R:

(a) *For every chain C in $(Q, \leqslant)$ of type ω and for every monotone function $f : C \to P$ such that $x\, R f(x)$ for all $x \in C$, it is the case that* $\sup(C)\ R\ \sup(f[C])$.

Lemma 2 *Let C be a chain in $(Q, \leqslant)$ of type ω. Suppose that there exists a monotone function $f : C \to Q$ such that $x\, R f(x)$ and $x \leqslant f(x)$ for all $x \in C$. Then* $\sup(C)$ *belongs to Q.*

Proof of the lemma. Let $M := \sup(C)$. We show $M \in Q$. By (a) we have that $M\ R\ \sup(f[C])$. Furthermore, as $x \leqslant f(x)$ for all $x \in C$, it follows that $M = \sup(C) \leqslant \sup(f[C])$. This shows that $\sup(f[C]) \in \delta_R(M) \cap \uparrow M$. Hence,

$$\delta_R(M) \cap \uparrow M \neq \emptyset. \tag{3.2.8}$$

Now let $y \in P$ be an element such that $\delta_R(y) \subseteq \downarrow y$. We claim that $M \leqslant y$. As $C \subseteq Q$, we have that $x \leqslant y$, for all $x \in C$, by the second conjunct of the definition of Q. It follows that $M = \sup(C) \leqslant y$. This and (3.2.8) prove that $M \in Q$. □

We pass to the proof of the theorem. We inductively define a strictly increasing sequence $a_0 < a_1 < \ldots < a_n < a_{n+1} < \ldots$ of elements of Q. (The Principle of Dependent Choices is applied here.) The type of the sequence is $\leqslant \omega$.

We define: $a_0 := \mathbf{0}$. Let us suppose the elements $a_0 < a_1 < \ldots < a_n$ have been defined. As $a_n \in Q$ and, by Lemma 1, the relation $R\lceil Q$ is expansive, there exists an element $b \in Q$ such that $a_n \leqslant b$ and $a_n \, R \, b$. If $b = a_n$, the defining procedure terminates. In this case $a^* := a_n$ is already a fixed point of R. As a^* belongs to Q, the second statement of the thesis of the theorem evidently holds for a^*. If $b \neq a_n$, we put: $a_{n+1} := b$. Clearly, $a_n < a_{n+1}$.

It remains to consider the case when the sequence $a_0 < a_1 < \ldots < a_n < a_{n+1} < \ldots$ has type ω. In this case, we put $C := \{a_n : n \in \mathbb{N}\}$ and define $f : C \to P$ by $f(a_n) := a_{n+1}$ for all $n \in \mathbb{N}$. f is well-defined and monotone. As $C \subseteq Q$, $a \, R f(a)$ and $a \leqslant f(a)$ for all $a \in C$, the supremum $\sup(C)$ belongs to Q, by Lemma 2. Furthermore, $\sup(C) \; R \; \sup(f[C])$, by the σ-continuity of R. But, evidently, $\sup(C) = \sup(f[C])$ because $a_0 = \mathbf{0}$ (see the figure below). Putting $a^* := \sup(C)$, we thus see that $a^* \, R \, a^*$. So a^* is a fixed point of R. Since $a^* \in Q$, it follows that for every $y \in P$, if $\delta_R(y) \subseteq \downarrow y$, then $a^* \leqslant y$. □

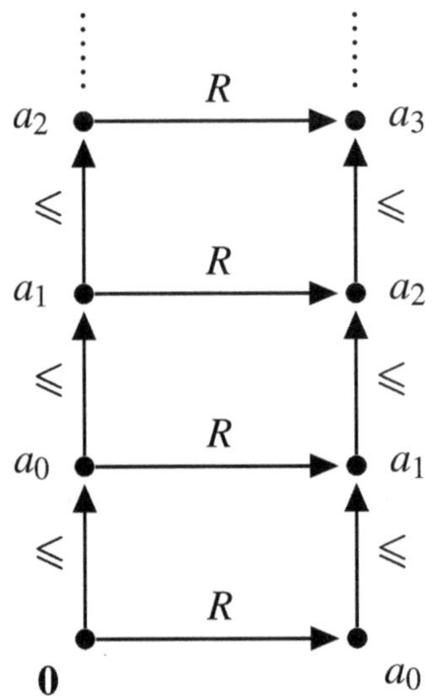

Fig. 3.4

Notes (1). The hypothesis of monotonicity of R in Theorem 3.2.7 is essential and cannot be dropped. Let $P = [0, 1]$ be the closed unit interval of real numbers. The system $(P, \leqslant)$ with the usual ordering $\leqslant$ of real numbers is a σ-complete poset with zero. The relation $R \subseteq P \times P$ is defined as follows: $a \, R \, b$ if and only if $(\exists n \in \mathbb{N}, n \geqslant 1) \; |a - b| = (1/2)^n$. It is easy to see that:

(a) R is symmetric and total,
(b) R is not monotone,
(c) R does not possess a fixed point.

As to (b), for the numbers $1/2$ and 1, we have $1/2 \, R \, 1$ and $1/2 < 1$. But there does not exist a number d in $[0, 1]$ such that $1 \, R \, d$ and $1 \leqslant d$.

(2). The assumption that $(P, \leqslant)$ possesses zero is essential in Theorem 3.2.7. For let $P = \{1, 2, a, b\}$, where $1 < 2$ and $a < b$. Let $R := \{(1, a), (a, 1), (2, b), (b, 2)\}$. The relation R is symmetric and monotone on P. R is also σ-continuous. But R does not possess a fixed point. The poset $(P, \leqslant)$ is σ-complete but it is lacking the zero element.

(3). On the basis of Zermelo–Fraenkel set theory ZF, Theorem 3.2.7 is equivalent to the Principle of Dependent Choices (DC). □

Let $(P, \leqslant)$ be a σ-complete poset. A function $\pi : P \to P$ is called *σ-continuous* if it is monotone and $\pi(\sup(C) = \sup(\pi[C])$ for every chain C in $(P, \leqslant)$ of type ω.

Since for any monotone function $\pi : P \to P$, the image $\pi[C]$ of any chain C in $(P, \leqslant)$ is a chain as well, we see that, in view of the σ-completeness of $(P, \leqslant)$, the above definition thus postulates the equality of the two suprema and not their existence.

A function $\pi : P \to P$ is σ-continuous in the above sense if and only if the graph R of π is σ-continuous as a binary relation.

Corollary 3.2.8 *Let $(P, \leqslant)$ be a σ-complete poset with zero $\mathbf{0}$. Every σ-continuous function $\pi : P \to P$ has a least fixed point a^*, i.e., $\pi(a^*) = a^*$ and $(\forall b \in P)$ $(\pi(b) \leqslant b$ implies $a^* \leqslant b)$.*

Proof We work with the graph R of π and proceed as the proof of Theorem 3.2.7. Since π is σ-continuous, the graph R is a σ-continuous relation and $\delta_R(\mathbf{0}) = \{\pi(\mathbf{0})\}$. By Theorem 3.2.7 there is a fixed point a^* of R such that for every $b \in P$, $\delta_R(b) \subseteq \downarrow b$ implies that $a^* \leqslant b$. But the last condition simply says that for every $b \in P$, $\pi(b) \leqslant b$ implies $a^* \leqslant b$. The chain $a_0 \leqslant a_1 \leqslant \cdots \leqslant a_n \leqslant a_{n+1} \leqslant \cdots$, defined as in the proof of Theorem 3.2.7, has the following properties: $a_0 = \mathbf{0}$, $a_{n+1} = \pi(a_n)$, for all n. Furthermore, $a^* = \sup(\{a_n : n \in \mathbb{N}\})$. □

The above chain $a_0 \leqslant a_1 \leqslant \cdots \leqslant a_n \leqslant a_{n+1} \leqslant \cdots$ is known as *Kleene's approximation sequence* (Kleene (1952). It can also be found in Tarski's (1955) classic paper. Many properties of such objects are compiled in Chap. 4 of Davey and Priestley (2002).

Corollary 3.2.8 is provable in ZF—see Moschovakis (1994 pp. 108–109).

3.3 Ordered Action Systems: Applications of Fixed-Point Theorems

We now show how to apply one of the above fixed-point theorems for relations—Theorem 3.2.7—to the proof of the downward Löwenheim–Skolem–Tarski Theorem for countable languages. This theorem belongs to model theory. It states, roughly, that every infinite model $\boldsymbol{A}$ has an elementary submodel of any intermediate power between the cardinality of the language and the cardinality of $\boldsymbol{A}$. It is illuminating to see the proof of this theorem in the context of elementary action systems. We apply the standard model-theoretic notation. The notions we use, such as a submodel, an elementary submodel, etc. are not defined here. Readers unfamiliar with them are advised to consult any of the standard textbooks on the theory of models. A *language* is a set that is the union of threes sets: a set of relational symbols (predicates), a set of function symbols, and a set of constant symbols. (Constant symbols are often viewed as nullary function symbols.) If L is a language, then $For(L)$ denotes the set of first-order formulas of L.

Theorem 3.3.1 (Downward Löwenheim–Skolem–Tarski Theorem) *Let L be a countable language and let α and β be cardinal numbers such that $|For(L)| \leqslant \beta \leqslant \alpha$. Let $\boldsymbol{A}$ be a model for L of cardinality α. Then $\boldsymbol{A}$ has an elementary submodel of cardinality β. In fact, for any subset $X_0 \subseteq \boldsymbol{A}$ of power $\leqslant \beta$, the model $\boldsymbol{A}$ has an elementary submodel of power β which contains X_0 as a subset of its universe.*

Proof Let P be the family consisting of all subsets $X \subseteq A$ such that $|X| = \beta$ and $X_0 \subseteq X$. Evidently, P is nonempty because $|A| \geqslant \beta$. We have:

(A) *The family P, ordered by inclusion, is σ-complete.*

Indeed, for any chain $\boldsymbol{C}$ of subsets of A (of type ω) such that $|X| = \beta$ for all $X \in \boldsymbol{C}$, the union $\bigcup \boldsymbol{C}$ has cardinality β. Moreover, the set $\mathbf{0} := X_0$ is the least element of P.

We define the following binary relation on the poset $(P, \subseteq)$: for $X, Y \in P$,

$X R Y$ if and only if $X \subseteq Y$ and for every formula $\Phi(x, x_1, \ldots, x_n) \in For(L)$ and any sequence $a_1, \ldots, a_n \in X$ (of length n) such that $\boldsymbol{A} \models (\exists x)\ \Phi[a_1, \ldots, a_n]$ there exists $b \in Y$ such that $\boldsymbol{A} \models \Phi[b, a_1, \ldots, a_n]$.

$X R Y$ thus says that the set Y includes X and, furthermore, for each formula $\Phi(x, x_1, \ldots, x_n)$ and any n-tuple $a_1, \ldots, a_n \in X$ satisfying $(\exists x)\Phi$ in $\boldsymbol{A}$, the set Y contains at least one $b \in A$ such that the $(n+1)$-tuple $b, a_1, \ldots a_n$ satisfies $\Phi(x, x_1, \ldots, x_n)$ in $\boldsymbol{A}$. The set Y may also contain some other elements of A, but the cardinality of Y does not exceed β.

As $|For(L)| \leqslant \beta$, the crucial observation is that

(B) *The relation R is total, i.e., for any $X \in P$ there exists $Y \in P$ such that $X R Y$.*

Furthermore, the definition of R gives that for any $X, Y, X_1, X_2, Y \in P$:

(C) *If $X R Y$ then $X \subseteq Y$.*

(D) $X_1 \subseteq X_2$ *and* $X_2 RY$ *implies* $X_1 RY$.

It follows from (B) and (C) that R is inflationary. But, more interestingly,

(E) *The relation* R is *σ-continuous (in the stronger sense).*

We first check that R is monotone in $(P, \subseteq)$. We assume $X, Y, Z \in P$ so that $X \subseteq Y$ and XRZ. Evidently, $Y \cup Z$ belongs to P. By (B) and (C) there exists a set W such that $Y \cup ZRW$ and $Y \cup Z \subseteq W$. As $Y \cup ZRW$, (D) gives that YRW. As $Z \subseteq W$, monotonicity follows.

Suppose we are given two nonempty chains of elements of P,

$$Y_0 \subseteq Y_1 \subseteq \ldots \subseteq Y_n \subseteq Y_{n+1} \subseteq \ldots \quad \text{and} \quad Z_0 \subseteq Z_1 \subseteq \ldots \subseteq Z_n \subseteq Z_{n+1} \subseteq \ldots$$

of types $\leqslant \omega$ such that $Y_n \, RZ_n$ for all n. The sets $\bigcup_{n\in\mathbb{N}} Y_n$ and $\bigcup_{n\in\mathbb{N}} Z_n$ belong to P. We claim that $\bigcup_{n\in\mathbb{N}} Y_n \; R \; \bigcup_{n\in\mathbb{N}} Z_n$.

The fact that $Y_n \, RZ_n$ for all n implies that $Y_n \subseteq Z_n$ for all n. Hence $\bigcup_{n\in\mathbb{N}} Y_n \subseteq \bigcup_{n\in\mathbb{N}} Z_n$. Let $Y := \bigcup_{n\in\mathbb{N}} Y_n$ and $Z := \bigcup_{n\in\mathbb{N}} Z_n$. Let $\Phi(x, x_1, \ldots, x_k)$ be a formula in $For(L)$ and let $a_1, \ldots, a_k$ be a sequence of elements of Y (of length k) such that $\boldsymbol{A} \models (\exists x)\, \Phi[a_1, \ldots, a_k]$.

There exists $n \in \mathbb{N}$ such that $a_1, \ldots, a_k \in Y_n$. Since $Y_n \, RZ_n$, there exists $b \in Z_n$ such that $\boldsymbol{A} \models \Phi[b, a_1, \ldots, a_n]$. Consequently, there exists $b \in Z$ such that $\boldsymbol{A} \models \Phi[b, a_1, \Phi, a_n]$. As $\Phi(x, x_1, \ldots, x_k)$ is an arbitrary formula in $For(L)$ and $a_1, \ldots, a_k$ are arbitrary elements of Y, this proves that $Y : R : Z$. So R is σ-continuous.

As R is total, then $\delta_R(\mathbf{0}) \neq \emptyset$. The system $(P, \subseteq, R)$ thus satisfies the assumptions of Theorem 3.2.7. It follows that the relation R has a fixed point in $(P, \subseteq)$, say B. It is then easy to verify that B is the universe of an elementary submodel $\boldsymbol{B}$ of $\boldsymbol{A}$. As B belongs to P, the cardinality of $\boldsymbol{B}$ is equal to β. □

The poset $(P, \subseteq)$, defined as in the proof of Theorem 3.3.1, is furnished with an infinite family of atomic actions, each action being associated with a formula of L. Indeed, for each formula $\Phi(x, x_1, \ldots, x_k)$ with at least one free variable x, we define the binary relation $A_\Phi \subseteq P \times P$ as follows: for $X, Y \in P$,

> $X \, A_\Phi \, Y$ iff $X \subseteq Y$ and for every sequence $a_1, \ldots, a_n \in X$ (of length n) such that $\boldsymbol{A} \models (\exists x)\, \Phi[a_1, \ldots, a_n]$ there exists $b \in Y$ such that $\boldsymbol{A} \models \Phi[b, a_1, \ldots, a_n]$.

A_Φ is called the *action of the formula* Φ.

It follows from the definition of the relation R that

$$R = \bigcap \{A_\Phi : \Phi \in For(L)\}.$$

(B) and (F) imply that for any $X \in P$, each action A_Φ is $\exists$-performable at X but it need not be $\forall$-performable. But (F) also implies that each realizable performance of whichever action A_Φ is also a realizable performance of each of the remaining actions. Thus any agent that successfully performs one action also performs the other actions. Each fixed point of R is uniformly a fixed point of all the actions A_Φ, $\Phi \in For(L)$, i.e., X^* is a fixed point of R if and only if it is a fixed point of each relation A_Φ, $\Phi \in For(L)$.

As the ordered action system $(P, \subseteq, R, \{A_\Phi : \Phi \in For(L)\})$ is not normal, we may define another transition relation between states such that the resulting action system is normal. The fact that the system is normal ensures that every possible performance of an arbitrary action is realizable. We put:

$$R' := \bigcup \{A_\Phi : \Phi \in For(L)\}.$$

The system

$$(P, \subseteq, R', \{A_\Phi : \Phi \in For(L)\})$$

is normal, i.e., every atomic action A_Φ is $\forall$-performable in every state $X \in P$.

R' has many fixed points. In fact, the set of fixed points of R' is the union of the set of fixed points of the actions A_Φ, $\Phi \in For(L)$. But not all fixed points of R' qualify as solutions of the problem posed in Theorem 3.3.1, i.e., they need not even be submodels of $\boldsymbol{A}$. The right solutions are provided by states $X \in P$ which are uniformly fixed points of all the actions A_Φ, $\Phi \in For(L)$.

The above remarks on ordered action systems are encapsulated by the following general definitions.

Definition 3.3.2 An *ordered elementary action system* is a quadruple

$$\boldsymbol{M} = (W, \leqslant, R, \boldsymbol{\mathcal{A}}), \tag{3.3.1}$$

where the reduct $(W, \leqslant)$ is a poset and the reduct $(W, R, \boldsymbol{\mathcal{A}})$ is an elementary action system. □

Thus, in every ordered action system, the set W of states is assumed to be ordered. In practice, the system (3.3.1) is subject to further constraints. For example, it is often assumed that

> $(W, \leqslant)$ is a tree-like structure or it is σ-complete and R satisfies some continuity or expansivity conditions with respect to $\leqslant$.

In some action systems, e.g., in stit semantics (see Chap. 5), the relation R itself is assumed to be an order and is identified with $\leqslant$. Everybody knows the frustrating experience of performing a string of actions commencing with some state u, only to end up back in state u. In this case a loop of positive length k of transitions between states may occur: $u \, R \, w_1 \, R \ldots R \, w_k \, R \, u$. Any effective action system excludes loops, because actions involving them are futile and systems containing them are poorly organized.

A binary relation R on a nonempty set W is *loop-free* if for any *positive* integer $k \geqslant 1$ there does not exist a sequence of states $u, w_1, \ldots, w_k$ such that $u\, R\, w_1\, R \ldots R\, w_k\, R\, u$. The fact that R is loop-free does not preclude the existence of reflexive points of R, i.e., elements u such that $u\, R\, u$.

We recall that R^* is the Kleene closure of R, that is, R^* is the least reflexive and transitive relation including R. The following observation is immediate:

Proposition 3.3.3 *R is loop-free if and only if R^* is a partial order.*

Proof For any $u, w \in W$, $u\, R^*\, w$ if and only if $u = w$ or for some $k \geqslant 1$ there exists a sequence of states $w_1, \ldots, w_k$ such that $u\, R\, w_1\, R \ldots R\, w_k\, R\, w$.

($\Rightarrow$). If R is loopless, then one directly verifies that R^* is antisymmetric. As R^* is reflexive and transitive, it follows that R^* is a partial order.

($\Leftarrow$). As any partial order is loop-free, any subrelation of a partial order is loop-free as well. Hence R, being a subrelation of the order R^*, is loop-free. □

The main goal that underlines introducing ordered action systems on stage is the need to make room for some infinite procedures in action. We call these procedures *approximations*. Mathematics admits such procedures. The process of computing consecutive decimal approximations of an irrational number is an example of such an (potentially) infinite procedure. We write 'potentially' here because in practice this process halts at some stage. This is caused by both the limitations of the computer capacities and finite time resources. But, theoretically, this procedure can be made as long as one wishes. In this case, the members of W represent different approximations of a given irrational number. We may therefore identify each element of W with a pair consisting of a rational number r together with a measure of accuracy of the approximation provided by r. (It is assumed that such a measure is available.)

The states of (3.3.1), that is, the members of W, represent various possible phases of the performance of a certain task by the agents. It may be the process of building a house, sewing a dress, and so on. Intuitively, the order relation in (3.3.1) makes it possible to compare degrees of completion. The fact that $u \leqslant v$ means that at the state v the process is more advanced than at u. For example, it may mean that the state of the house built represented by v is more complete than that represented by u.

Obviously, various courses of actions are possible. We may have the situation that $u \leqslant v_1$ and $u \leqslant v_2$, where the states v_1 and v_2 are incomparable. This means that if the system is at the state u (i.e., u is a phase of realizing the task), several further courses of action, leading to more advanced stages of the performance represented either by v_1 or by v_2, are conceivable. The fact that v_1 and v_2 are incomparable means that further strings of actions undertaken at u may ramify. It may happen that there exists a state w such that $v_1 \leqslant w$ and $v_2 \leqslant w$, which means that the ramified strings of actions converge. But, generally, the task may not be unambiguously determined and several options for completing the action may be admissible. (For example, it may happen that the architect has designed various alternative versions of the house.) In this situation such a state w may not exist.

Definition 3.3.4 Let $\boldsymbol{M} = (W, \leqslant, R, \boldsymbol{\mathcal{A}})$ be an ordered action system.

(1) $\boldsymbol{M}$ is said to be *σ-complete* (*inductive*) if the poset $(W, \leqslant)$ is σ-complete (inductive, respectively).
(2) An element $u^* \in W$ is called a *fixed point* of $\boldsymbol{M}$ if it is a fixed point of R, i.e., $u^* \, R \, u^*$.
(3) An element $u^* \in W$ is called a *maximal fixed point* of $\boldsymbol{M}$ if it is a fixed point of $\boldsymbol{M}$ and, furthermore, for every $w \in W$, if $u^* \, R \, w$, then $u^* \leqslant w$. □

The justification of (2) of the above definition is as follows. The action system $\boldsymbol{M}$ is 'activated' in the zero state of $\mathbf{0}$. The task is to lead the system to a state u^*, this being a fixed point of the relation R and the final stage of the action of $\boldsymbol{M}$. The state u^* is reached by making a possibly transfinite sequence of atomic actions from $\boldsymbol{\mathcal{A}}$, which results in a transfinite monotone sequence of states of the space W; the state u^* is a final element of this sequence or its supremum.

The above definition can be strengthened in various directions by, e.g., accepting that each fixed point of $\boldsymbol{\mathcal{A}}$ is at the same time a fixed point of some relation (action) $A \in \boldsymbol{\mathcal{A}}$ or that it is a fixed point of all atomic actions of $\boldsymbol{\mathcal{A}}$.

The action system defined as in the proof of Theorem 3.3.1 is σ-complete with zero, while the fixed points investigated there are actually fixed points of this system. Since $R = \bigcap\{A_\Phi : \Phi \in For(L)\}$, any fixed point u^* of $\boldsymbol{M}$ is a fixed point of all actions A_Φ, i.e., $u^* \, A_\Phi \, u^*$ for all Φ. $\boldsymbol{M}$ has yet another property: for every pair u, w of states, if $u \, R \, v$, then $u \leqslant v$. The same implication holds for every action A of this system: if $u \, A \, v$, then $u \leqslant v$, for all states u, w. This observation gives rise to the following definition:

Definition 3.3.5 An ordered action system $\boldsymbol{M} = (W, \leqslant, R, \boldsymbol{\mathcal{A}})$ is called *constructive* if the relation R and the union $\bigcup \boldsymbol{\mathcal{A}}$ are both subsets of $\leqslant$.

A system $M = (W, \leqslant, R, \boldsymbol{\mathcal{A}})$ is *destructive* if R and the union $\bigcup \boldsymbol{\mathcal{A}}$ are both subsets of the dual order $\geqslant$. □

Destructive systems are thus constructive *a rebouir*—they are constructive in the sense of the dual order $\geqslant$.

The system $(P, \subseteq, R, \{A_\Phi : \Phi \in For(L)\})$ is obviously constructive.

(Constructive action systems should not be confused with the concept of constructive mathematics. The term 'constructive mathematics' has a well-established meaning which is not captured by the above definition.)

If an elementary action system $\boldsymbol{M} = (W, \leqslant, R, \boldsymbol{\mathcal{A}})$ is constructive, then every fixed point of $\boldsymbol{M}$ (if there is any) is a strong fixed point. Indeed, assume that a^* is a fixed point of $\boldsymbol{M}$. Hence, $a^* \, R \, a^*$. We assume furthermore that $a^* \, R \, w$ for some state w. As R is a subset of $\leqslant$, we have that $a^* \leqslant w$.

Theorem 3.3.6 *Let $\boldsymbol{M} = (W, \leqslant, R, \boldsymbol{\mathcal{A}})$ be a σ-complete action system with zero $\mathbf{0}$ such that $R \subseteq \bigcup \boldsymbol{\mathcal{A}}$. If R is σ-continuous and $\delta_R(\mathbf{0}) \neq \emptyset$, then $\boldsymbol{M}$ has a fixed point.*

Proof By Theorem 3.2.7, R has a fixed point. □

In ordered action systems $\boldsymbol{M} = (W, \leqslant, R, \boldsymbol{\mathcal{A}})$, tasks for $\boldsymbol{M}$ may take the form $(\{\mathbf{0}\}, \Psi)$, where $\mathbf{0}$ is the least element of $(W, \leqslant)$ and Ψ is a set of fixed points of $\boldsymbol{M}$. The process of carrying the system $\boldsymbol{M}$ from the state $\mathbf{0}$ to a Ψ-state may not be finitary since it usually involves various infinite limit passages. These passages are strictly linked with the process of approximating fixed points of R by performing possibly infinite strings of consecutive actions of the system. In the simplest case, a fixed point can be reached in ω steps.

The following notion is convenient in the context of *ordered* action systems.

Definition 3.3.7 Let $\boldsymbol{M} = (W, \leqslant, R, \boldsymbol{\mathcal{A}})$ be a σ-complete elementary action system. The ω*-reach* of $\boldsymbol{M}$ is the binary relation $\boldsymbol{Re}^{\omega}_{\boldsymbol{M}}$ on W defined as follows. For $u, v \in W$,

> $\boldsymbol{Re}^{\omega}_{\boldsymbol{M}}(u, v)$ if and only if *either* $u = w$ and $u\ R\ w$ *or* there exists a chain of states $u_0 \leqslant u_1 \leqslant \cdots \leqslant u_n \leqslant u_{n+1} \leqslant \ldots$ of type $\leqslant \omega$ and a sequence of atomic actions $A_0, A_1, \ldots, A_n, A_{n+1}, \ldots$ such that $u = u_0$, $v = \sup(\{u_n : n = 0, 1, \ldots\})$ and $u_n\ A_n, R\ u_{n+1}$, for all n. □

$\boldsymbol{Re}^{\omega}_{\boldsymbol{M}}(u, v)$ thus states that v is achieved from the state u by means of performing a possibly countably infinite string of actions A_n, $n = 0, 1, \ldots$ so that the outcomes of the actions form an increasing chain of states. The state v is the supremum of the resulting sequence of states. (It is not assumed here that the state v is a fixed point of R and hence of $\boldsymbol{M}$.)

The above chain $u_0 \ldots u_1 \leqslant \cdots \leqslant u_n \leqslant u_{n+1} \leqslant \cdots$ has the property that $u_n R u_{n+1}$, for all n. If the chain is infinite, it need not hold that $u_m\ R\ \sup(\{u_n : n \in \mathbb{N}\})$ for all (and even for some) m. This means that, generally, it is not possible to reach the state $v = \sup(\{u_n : n \in \mathbb{N}\})$ from the states u_m by means of finitary procedures, i.e., by accomplishing a finite string of actions from $\boldsymbol{\mathcal{A}}$. This situation occurs in the system $(P, \subseteq, R, \{A_\Phi : \Phi \in For(L)\})$, where, generally, it is not possible to produce an elementary submodel from the set $\mathbf{0}$ in a finite number of steps by applying the actions from $\{A_\Phi : \Phi \in For(L)\}$.

In the system $(P, \subseteq, R, \{A\Phi : \Phi \in For(L)\})$, every pair $(\mathbf{0}, X)$ such that X is the universe of an elementary submodel of power β of the given model $\boldsymbol{A}$, belongs to the reach $\boldsymbol{Re}^{\omega}_{\boldsymbol{M}}$.

3.4 Further Fixed-Point Theorems: Ordered Situational Action Systems

The *back and forth method* is particularly useful in many branches of algebra and model theory. It dates back to the proof of Cantor's famous theorem stating that any two countable linear dense orders without endpoints are isomorphic. The back and forth argument has proved convenient in many branches of model theory, especially in the theory of saturated models. In order to find a plausible abstract formulation of the above construction, we shall discuss some refinements of the fixed-point theorems presented in Sect. 3.2.

The notions we present in this section are modifications of the concepts we have discussed in previous sections. These modifications are twofold in character. First, some of the concepts defined in Sect. 3.2 are restricted to certain subsets of posets. Second, a certain weaker counterpart of the notion of expansivity is formulated here: conditional expansivity. We also define a certain restricted version of the σ-continuity of a relation.

Let $(P, \leqslant)$ be a poset. A function $\pi : P \to P$ is called *quasi-expansive* (or *conditionally expansive*) if it satisfies the following quasi-identity:

$$(\forall x)\big(x \leqslant \pi(x) \to \pi(x) \leqslant \pi(\pi(x))\big).$$

Clearly, every expansive function is quasi-expansive. But also every monotone function is quasi-expansive. Thus, the notion of a conditionally expansive function is a common generalization of the above two types of functions associated with posets.[3]

Given an element $a \in P$, we define: $\pi^0(a) := a$ and $\pi^{n+1}(a) := \pi(\pi^n(a))$, for all $n \geqslant 0$.

If $(P, \leqslant)$ has zero $\mathbf{0}$ and $\pi : P \to P$ is conditionally expansive, then the set $\{\pi^n(\mathbf{0}) : n \in \mathbb{N}\}$ forms an increasing chain of type $\leqslant \omega$.

We are concerned with the following counterpart of the notion of quasi-expansivity for relations. A binary relation R on a poset $(P, \leqslant)$ is called *quasi-expansive* (or *conditionally expansive*) whenever:

$$(\forall a, b \in P)(a \leqslant b \ \wedge \ a\,R\,b \ \to \ (\exists c \in P)\ b\,R\,c \ \wedge \ b \leqslant c)$$

(see the diagram below).

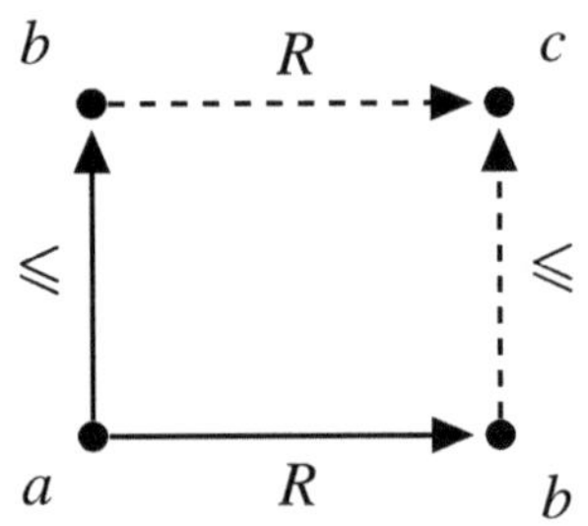

Fig. 3.5

Every monotone relation, defined as in Sect. 3.2, is trivially quasi-expansive. Furthermore, every expansive relation is quasi-expansive. It is also evident that if R is

[3]David Makinson considered functions with this property in the context of his research on modal logics.

the graph of a function $\pi : P \to P$, then the relation R is quasi-expansive in the above sense if and only if the function π is quasi-expansive.

Let $(P, \leqslant)$ be a σ-complete poset. A relation $R \subseteq P \times P$ is said to be *conditionally σ-continuous* (or *quasi-σ-continuous*) if the following conditions are met:

(1) R is quasi-expansive,
(2) For every chain C in $(P, \leqslant)$ of type $\leqslant \omega$ and for every monotone and expansive function $f : C \to P$, if $a R f(a)$ for all $a \in C$, then $\sup(C)\ R\ \sup(f[C])$.

Due to the monotonicity of $f : C \to P$, (2) implies that the set $f[C]$ is a countable chain of type $\leqslant \omega$ and therefore $\sup(f[C])$ exists. It is also clear that every σ-continuous relation is conditionally σ-continuous.

A function $\pi : P \to P$ is *conditionally σ-continuous* (or *quasi-σ-continuous*) if its graph has this property.

Theorem 3.4.1 *Let $(P, \leqslant)$ be a σ-complete poset. A function $\pi : P \to P$ is conditionally σ-continuous if and only if it is quasi-expansive and for every chain $a_0 \leqslant a_1 \leqslant \cdots \leqslant a_n \leqslant a_{n+1} \leqslant \cdots$ of elements of P of type $\leqslant \omega$, if $a_n \leqslant \pi(a_n)$ and $\pi(a_n) \leqslant \pi(a_{n+1})$ for all $n \in \mathbb{N}$, then $\pi\big(\sup(\{a_n : n \in \mathbb{N}\})\big) = \sup\big(\{\pi(a_n) : n \in \mathbb{N}\}\big)$.*

The proof is easy and is omitted. □

It follows from these remarks that if $\pi : P \to P$ is conditionally σ-continuous, where $(P, \leqslant)$ is a σ-complete poset with zero $\mathbf{0}$, then $\{\pi^n(\mathbf{0}) : n \in \mathbb{N}\}$ is an increasing chain and, moreover,

$$\pi\big(\sup(\{\pi^n(\mathbf{0}) : n \in \mathbb{N}\})\big) = \sup\big(\{\pi(\pi^n(\mathbf{0})) : n \in \mathbb{N}\}\big).$$

Consequently, the element $a^* := \sup(\{\pi^n(\mathbf{0}) : n \in \mathbb{N}\})$ is a fixed point of π.

We shall now modify a little the above definitions by relating the above properties to certain selected subsets of posets.

Let $(P, \leqslant)$ be a σ-complete poset and P_0 a subset of P. Furthermore, let R be a binary relation on P. We say that R is *conditionally σ-continuous relative to P_0* whenever:

(1) R is conditionally expansive on P_0, i.e., for every pair $a, b \in P_0$ such that $a \leqslant b$ and $a R b$ there exists $c \in P_0$ such that $b R c$ and $b \leqslant c$.
(2) For every countable chain $C \subseteq P_0$ of type ω and every monotone and expansive function $f : C \to P_0$, if $a R f(a)$ for all $a \in C$, then $\sup(C)\ R\ \sup(f[C])$.

(Note: $\sup(C)$ and $\sup(f[C])$ may not belong to P_0.)

A closer look at the proof of Theorem 3.2.7 reveals that its conclusion can be reached under weaker assumptions:

Theorem 3.4.2 *Let $(P, \leqslant)$ be a σ-complete poset with zero $\mathbf{0}$ and let P_0 be a subset of P. Suppose that a relation $R \subseteq P \times P$ is conditionally σ-continuous relative to P_0. If $\mathbf{0} \in P_0$ and the set $P_0 \cap \delta_R(\mathbf{0})$ is nonempty, then R has a fixed point in P.*

Proof We argue as in the proof of Theorem 3.2.7. Let a_0 be an arbitrary element of $P_0 \cap \delta_R(\mathbf{0})$. As the elements $\mathbf{0}$ and a_0 belong to P_0 and $\mathbf{0}\, R\, a_0$ and $\mathbf{0} \leqslant a_0$, there exists, by the conditional monotonicity of R on P_0, an element $a_1 \in P_0$ such that $a_0\, R\, a_1$ and $a_0 \leqslant a_1$. Taking the pair (a_0, a_1) and applying the conditional monotonicity of R on P_0, we see that there exists an element $a_2 \in P$ such that $a_1\, R\, a_2$ and $a_1 \leqslant a_2$. Taking in turn the pair (a_1, a_2), we find an element $a_3 \in P_0$ such that $a_2\, R\, a_3$ and $a_2 \leqslant a_3$, again by monotonicity. Continuing this pattern of argument, we define a countable chain: $\mathbf{0} \leqslant a_0 \leqslant a_1 \leqslant \cdots \leqslant a_n \leqslant a_{n+1} \leqslant \cdots$ of elements of P_0 such that $\mathbf{0} R a_0 R a_1 R \ldots R a_n R a_{n+1} R \ldots$ (see the figure following the proof of Theorem 3.2.7).

It follows from the conditional σ-continuity of R relative to P_0 that

$$\sup\{\mathbf{0}, a_0, a_1, \ldots, a_n, \ldots\}\ R\ \sup\{a_0, a_1, \ldots, a_n, \ldots\}.$$

Putting $a^* := \sup\{a_0, a_1, \ldots, a_n, \ldots\}$, we see that $a^*\, R\, a^*$. So a^* is a fixed point of R. □

We discuss further modifications of the above definitions.

Let $(P, \leqslant)$ be a poset and let P_0 be a subset of P. Furthermore, let R_1 and R_2 be *two* binary relations on P. We say that R_1 and R_2 are *adjoint on* P_0 if the following two conditions hold:

$$(\forall a_1, b_1 \in P_0)\big[a_1 \leqslant b_1 \ \wedge\ a_1\, R_1\, b_1 \ \rightarrow\ (\exists c_1 \in P_0)\ b_1\, R_2\, c_1 \ \wedge\ b_1 \leqslant c_1\big] \quad (3.4.1)$$

$$(\forall a_2, b_2 \in P_0)\big[a_2 \leqslant b_2 \ \wedge\ a_2\, R_2\, b_2 \ \rightarrow\ (\exists c_2 \in P_0)\ b_2\, R_1\, c_2 \ \wedge\ b_2 \leqslant c_2\big] \quad (3.4.2)$$

(see the figures below).

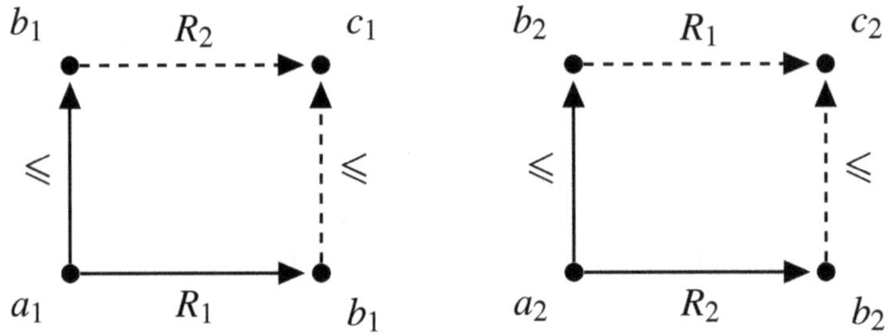

Fig. 3.6

The above two conditions can be neatly defined as one property on the direct products $P \times P$. Define the direct power $(P \times P, \leqslant)$ of the poset $(P, \leqslant)$ with the usual coordinate-wise definition of the order on $P \times P$. Thus,

$$(a_1, a_2) \leqslant (b_1, b_2) \quad \text{if and only if} \quad a_1 \leqslant b_1 \;\text{ and }\; a_2 \leqslant b_2 \;\text{ in }\; (P, \leqslant),$$

for all pairs $(a_1, a_2), (b_1, b_2) \in P \times P$. Furthermore, if R_1 and R_2 are binary relations on P, then $R_1 \times R_2$ is the direct product of R_1 and R_2. Thus, for all $(a_1, a_2), (b_1, b_2) \in P \times P$,

$$(a_1, a_2)\, R_1 \times R_2\, (b_1, b_2) \quad \text{if and only if} \quad a_1\, R_1\, b_1 \;\text{ and }\; a_2\, R_2\, b_2.$$

Properties (3.4.1) and (3.4.2) are expressed as a single property of the poset $(P \times P, \leqslant)$:

(A) For every two pairs $(a_1, a_2), (b_1, b_2) \in P_0 \times P_0$, if $(a_1, a_2) \leqslant (b_1, b_2)$ and $(a_1, a_2)\, R_1 \times R_2\, (b_1, b_2)$, then there exists a pair $(c_1, c_2) \in P_0 \times P_0$ such that $(b_1, b_2) \leqslant (c_1, c_2)$ and $(b_1, b_2)\, R_2 \times R_1\, (c_1, c_2)$.

Notice that in the antecedent of (A) the relation $R_1 \times R_2$ occurs. This relation is replaced in the consequent by $R_2 \times R_1$. Thus, (A) does *not* express the property of quasi-expansivity of the relation $R_1 \times R_2$ on $P_0 \times P_0$.

The relations R_1 and R_2 are also called the *forth* and the *back relations on* P_0, respectively. The pair (R_1, R_2) is also called the *back and forth pair* (of relations) relative to P_0.

Let $(P, \leqslant)$ be a σ-complete poset and let P_0 be a subset of P. A pair (R_1, R_2) of binary relations on P is said to be *σ-continuously adjoint relative to* P_0 if R_1 and R_2 are adjoint on P_0 and, furthermore, for every chain $C = \{a_n : n \in \mathbb{N}\}$ in P_0 of type $\leqslant \omega$ and for every monotone and expansive function $f : C \to P_0$ such that

$$a_{2n}\, R_1 a_{f(2n)} \;\text{ and }\; a_{2n+1}\, R_2\, a_{f(2n+1)}, \quad \text{for all } n \in \mathbb{N},$$

it is the case that

$$\sup(C)\, R_1\, \sup(f[C]) \;\text{ and }\; \sup(C)\, R_2\, \sup(f[C]).$$

(We note that $\sup(C)$ and $\sup(f[C])$ need not belong to P_0.)

An element $a^* \in P$ is called a *fixed point* of the pair (R_1, R_2) if a^* is a fixed point of both relations R_1 and R_2, i.e., $a^*\, R_1\, a^*$ and $a^*\, R_2\, a^*$ hold.

Theorem 3.4.3 *Let $(P, \leqslant)$ be a σ-complete poset with zero $\mathbf{0}$ and let P_0 be a subset of P. Suppose that a pair (R_1, R_2) of binary relations on P is σ-continuously adjoint relative to P_0. If $\mathbf{0} \in P_0$ and the set $P_0 \cap \delta_{R_1}(\mathbf{0})$ is nonempty, then the pair (R_1, R_2) has a fixed point in P.*

Proof We suitably modify the proof of Theorem 3.4.2. We define a countable chain C (of type $\leqslant \omega$) $a_0 \leqslant a_1 \leqslant \cdots \leqslant a_n \leqslant a_{n+1} \leqslant \cdots$ of elements of P_0. We put $a_0 := \mathbf{0}$. Let a_1 be an arbitrary element of $P_0 \cap \delta_{R_1}(\mathbf{0})$. As $a_0, a_1 \in P_0$, $a_0 \leqslant a_1$ and $a_0\ R_1\ a_1$, there exists, by (3.4.1), an element $a_2 \in P_0$ such that $a_1 \leqslant a_2$ and $a_1\ R_2\ a_2$. Taking then the pair a_1, a_2 and applying (3.4.2), we see that there exists an element $a_3 \in P_0$ such that $a_2 \leqslant a_3$ and $a_2\ R_1\ a_3$. Then, applying (3.4.1) to the pair a_2, a_3, we find an element $a_4 \in P_0$ such that $a_3 \leqslant a_4$ and $a_3\ R_2\ a_4$ (see the figure below). Continuing, we define an increasing chain $C = \{a_n : n \in \mathbb{N}\}$ in P_0 such that $a_0\ R_1\ a_1\ R_2\ a_2\ R_1\ a_3\ R_2\ a_4 \ldots a_{2n}\ R_1\ a_{2n+1}\ R_2\ a_{2n+2} \ldots$ The function $f : C \to C$ defined by $f(a_n) := a_{n+1}$, for all $n \in \mathbb{N}$, is expansive and monotone. Furthermore,

$$a_{2n}\ R_1 f(a_{2n}) \text{ and } a_{2n+1}\ R_2 f(a_{2n+1}), \quad \text{for all } n \in \mathbb{N}.$$

As the pair (R_1, R_2) is σ-continuously adjoint relative to P_0, we have that $\sup(C)\ R_1\ \sup(f[C])$ and $\sup(C)\ R_2\ \sup(f[C])$. Let $a^* := \sup(C)$. Then $a^* = \sup(f[C])$. It follows that $a^*\ R_1\ a^*$ and $a^*\ R_2\ a^*$. This concludes the proof of the theorem. □

As a somewhat trivial application of Theorem 3.4.3 we give a simple proof of the mentioned theorem of Cantor:

Theorem 3.4.4 *Every two countable linear and dense orders without end points are isomorphic.*

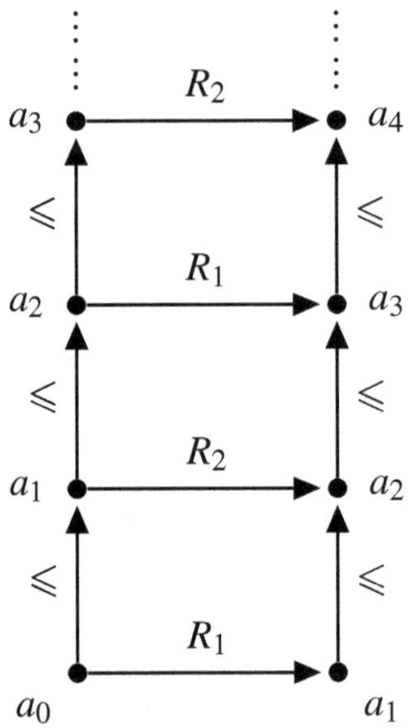

Fig. 3.7

Proof Let $(X_1, \leqslant_1)$ and $(X_2, \leqslant_2)$ be two such orders. A *partial isomorphism* (from $(X_1, \leqslant_1)$ to $(X_2, \leqslant_2)$) is any partial function $f : X_1 \to X_2$ such that f is injective on its domain $Dom(f)$ and, furthermore, for any elements $x, y \in Dom(f)$, $x \leqslant_1 y$ if and only if $f(x) \leqslant_2 f(y)$. A partial isomorphism $f : X_1 \to X_2$ is *finite* if its domain $Dom(f)$ is a finite set. $\mathbf{0}$ denotes the empty partial isomorphism. A (partial) isomorphism f is *total* if $Dom(f) = X_1$ and the co-domain $CDom(f)$ is equal to X_2.

Let P be the set of all partial isomorphisms from $(X_1, \leqslant_1)$ to $(X_2, \leqslant_2)$. P is partially ordered by the inclusion relation $\subseteq$ between partial isomorphisms. (Each partial isomorphism is a subset of the product $X_1 \times X_2$.) The poset $(P, \subseteq)$ is σ-complete because the union of any ω-chain of partial isomorphisms is a partial isomorphism. Moreover, the empty isomorphism $\mathbf{0}$ is the least element in $(P, \subseteq)$.

We define two relations R_1 and R_2 on P. As X_1 and X_2 are countably infinite, we can write $X_1 = \{a_n : n \in \mathbb{N}\}$ and $X_2 = \{b_n : n \in \mathbb{N}\}$. Given partial isomorphisms f and g, we put:

$f\ R_1\ g$ if and only if *either* f is a total isomorphism and $g = f$ *or* f is a finite isomorphism and $g = f \cup \{(a_m, b_n)\}$, where

(1) m is the smallest i such that $a_i \notin Dom(f)$,
(2) n is the smallest j such that $b_j \notin CDom(f)$ and $f \cup \{(a_m, b_j)\}$ is a partial isomorphism.

(Note that the choice of n depends on the definition of m.)

$f\ R_2\ g$ if and only if *either* f is a total isomorphism and $g = f$ *or* f is a finite isomorphism and $g = f \cup \{(a_m, b_n)\}$, where

(3) n is the smallest j such that $b_j \notin CDom(f)$,
(4) m is the smallest i such that $a_i \notin Dom(f)$ and $f \cup \{(a_i, b_n)\}$ is a partial isomorphism.

(The choice of m depends on the definition of n.)

Let $P_0 \subseteq P$ be the set of all finite isomorphisms. Using the fact that the orders $(X_1, \leqslant_1)$ and $(X_2, \leqslant_2)$ are linear, dense, and without endpoints, it is easy to verify that (R_1, R_2) is a back and forth pair of relations relative to P_0. But, more interestingly, the pair (R_1, R_2) is also σ-continuously adjoint relative to P_0. The set $P_0 \cap R_1[\mathbf{0}]$ is nonempty. Applying Theorem 3.4.3, we obtain that the pair (R_1, R_2) has a fixed point in $(P, \subseteq)$, say f^*. It follows from the definition of R_1 and R_2 that f^* is a total isomorphism between $(X_1, \leqslant_1)$ and $(X_2, \leqslant_2)$. □

The above relations R_1 and R_2 are partial functions with the same domain, viz. the set of finite partial isomorphisms plus the set of total isomorphisms. E.g., partial isomorphisms f such that either $Dom(f) = X_1$ and $CDom(f) \neq X_2$ or $Dom(f) \neq X_1$ and $CDom(f) = X_2$ do not belong to the common domain of R_1 and R_2. Furthermore, the existence of a total isomorphism between $(X_1, \leqslant_1)$ and $(X_2, \leqslant_2)$ is already implicit in the fact that the pair (R_1, R_2) is σ-continuously adjoint relative to P_0, as can be checked. Thus, the above proof does not give a new insight into the original proof by Cantor. The significance of the above proof is in the fact that it is uniformly formulated in the general and abstract framework provided by Theorem 3.4.3.

The above example gives rise to the construction of a certain simple situational action system. The set of states of the system coincides with the set P of partial isomorphisms between $(X_1, \leqslant_1)$ and $(X_2, \leqslant_2)$. The system is equipped with two elementary actions: R_1 and R_2, the forth and the back action, respectively. By way of analogy with game of chess, it is assumed that the system has two agents: BACK and FORTH. The agent FORTH performs the action R_1 while BACK performs R_2. The agents are cooperative—they obey the rule that their actions are performed alternately. Furthermore, they are successful in action—at the limit they are able to define a required total isomorphism between the posets $(X_1, \leqslant_1)$ and $(X_2, \leqslant_2)$.

A *possible situation* is a pair $s = (u, \alpha)$, where u is a partial isomorphism and α is an agent, i.e., $\alpha \in \{\text{BACK}, \text{FORTH}\}$. u is the (unique) state of the system corresponding to the situation s. This state is marked as $f(s)$. The situation s is read: 'The agent α performs his action in the state u'. The situation $(\mathbf{0}, \text{FORTH})$ is called *initial*. (We recall that $\mathbf{0}$ is the empty partial isomorphism.)

Let S be the set of possible situations. The *transition relation* Tr between situations is defined as follows. For any situations s and t,

$$s\, Tr\, t \quad \text{if and only if } \textit{either } s = (u, \text{FORTH}),\, t = (v, \text{BACK}) \text{ and } u\, R_1\, v$$
$$\textit{or } s = (u, \text{BACK}),\, t = (v, \text{FORTH}) \text{ and } u\, R_2\, v.$$

Thus, if $s\, Tr\, t$, then either $f(s)\, R_1 f(t)$ or $f(s)\, R_1 f(t)$. The union $R_1 \cup R_2$ is called the *transition relation between states*. Note that Tr is a (partial) function.

$\boldsymbol{M}^s := (P, R_1 \cup R_2, \{R_1, R_2\}, S, Tr, f)$ is a situational action system. $\boldsymbol{M}^s$ is called the *system of the back and forth actions*. $\boldsymbol{M}^s$ is ordered in the sense that its set of states P is ordered by $\leqslant$.

The extended structure $(P, \leqslant, R_1 \cup R_2, \{R_1, R_2\}, S, Tr, f)$ is an example of an ordered situational action system. (See the remarks placed at the end of this section.)

The back and forth method has a wider range of applications than the proof of the above theorem of Cantor. It was used by Fraïssé in his characterization of elementary equivalence of models. The method was formulated as a game by Ehrenfeucht. The players BACK and FORTH often go by different names: Spoiler and Duplicator, (H)eloise and Abelard, ∀dam and ∃va; and are often denoted by ∃ and ∀.

We recall that an elementary action system $\boldsymbol{M} = (W, R, \boldsymbol{\mathcal{A}})$ is *constructive* if the relation R and each of the actions $A \in \boldsymbol{\mathcal{A}}$ are subsets of $\leqslant$. If $\boldsymbol{M}$ is constructive, the relation R and the actions $A \in \boldsymbol{\mathcal{A}}$ need not be expansive and nondecreasing because they are not assumed to be total. By the same reason, the above relations need not be expansive. In the above example, the reduct $\boldsymbol{M} := (P, R_1 \cup R_2, \{R_1, R_2\})$ of $\boldsymbol{M}^s$ is a constructive, σ-complete elementary action system.

A relation $\ll$, defined on a set S, is called a *quasi-order* if $\ll$ is reflexive and transitive. If $\ll$ is a quasi-order, then the binary relation ρ defined on S is as follows:

$$a\, \rho\, b \quad \text{if and only if} \quad a \ll b \text{ and } b \ll a, \tag{3.4.3}$$

$(a, b \in S)$ is an equivalence relation on S. Let $[a]_\rho$ stands for the equivalence class of the element a relative to $\rho : [a]_\rho := \{x \in S : a \,\rho\, x\}$. The quotient set

$$S/\rho := \{[a]_\rho : a \in S\}$$

is partially ordered by the relation $\leqslant$, where

$$[a]_\rho \leqslant [b]_\rho \quad \text{if and only if} \quad a \ll b$$

$(a, b \in S)$.

Definition 3.4.5 Let $\boldsymbol{M}^s = (W, R, \boldsymbol{\mathcal{A}}, S, Tr, f)$ be a situational action system. The system $\boldsymbol{M}^s$ is *ordered* if its reduct $\boldsymbol{M} := (W, R, \boldsymbol{\mathcal{A}})$ is an ordered action system, i.e., the set W is endowed with a partial order $\leqslant$. □

In what follows ordered situational action systems are marked as

$$\boldsymbol{M}^s = (W, \leqslant, R, \boldsymbol{\mathcal{A}}, S, Tr, f), \tag{3.4.4}$$

thus explicitly indicating the order relation $\leqslant$ on the set of states W.

Given an ordered situational action system (3.4.4), we see that the partial order $\leqslant$ can be lifted to a quasi-order on the set S of possible situations. For any situations $a, b \in S$, we define:

$$a \ll b \quad \text{if and only if} \quad f(a) \leqslant f(b), \tag{3.4.5}$$

i.e., the state $f(a)$ corresponding to a is equal or less than the state $f(b)$ corresponding to b. $\ll$ is a quasi-order on S.

The following fact is immediate:

Proposition 3.4.6 *Let $\boldsymbol{M}^s = (W, \leqslant, R, \boldsymbol{\mathcal{A}}, S, Tr, f)$ be an ordered situational system. Let the quasi-order $\ll$ and the equivalence relation ρ on the set S of situations be defined as in (3.4.5) and (3.4.3). Then, for any $a, b \in S$:*

(8) $[a]_\rho = [b]_\rho$ *if and only if* $f(a) = f(b)$.
(9) $[a]_\rho \leqslant [b]_\rho$ *if and only if* $f(a) \leqslant f(b)$.
(10) The function f^*, which to each equivalence class $[a]_\rho$ assigns the state $f(a)$, is well-defined. Furthermore, f^* is an isomorphism between the posets $(S/\rho, \leqslant)$ and $(W, \leqslant)$. □

An ordered situational action system $\boldsymbol{M}^s = (W, \leqslant, R, \boldsymbol{\mathcal{A}}, S, Tr, f)$ is *σ-complete* (*inductive*) if its reduct $(W, \leqslant)$ is a σ-complete poset (inductive poset, respectively).

In the context of ordered situational systems, it is useful to have another infinite notion of a reach, more closely related to the properties of the set S of situations of the system (cf. Definition 3.3.7).

Let $\boldsymbol{M}^s = (W, \leqslant, R, \boldsymbol{\mathcal{A}}, S, Tr, f)$ be a σ-complete situational system. By the *situational ω-reach of the system $\boldsymbol{M}^s$* we mean a binary relation $\boldsymbol{SRe}^\omega$ on the set S defined as follows. For any situations $a, b \in S$,

$\boldsymbol{SRe}^{\omega}(a, b)$ if and only if *either* $a = b$ and $a\ Tr\ b$ *or* there exists a possibly infinite sequence $a_0, a_1, \ldots, a_n, a_{n+1}, \ldots$ of situations of S and a sequence $A_0, A_1, \ldots, A_n, A_{n+1}, \ldots$ of actions of $\boldsymbol{\mathcal{A}}$ such that

(i) $a_n\ Tr\ a_{n+1}$ and $f(a_n) \leqslant f(a_{n+1})$, for all n,
(ii) $a_n\ A_n\ a_{n+1}$, for all n,
(iii) $a_0 = a$ and $f(b) = \sup(\{f(a_n) : n = 0, 1, \ldots\})$.

Note that (i) can be formulated as:

(i)* $a_n \ll a_{n+1}$, for all n.

Since $a_n\ Tr\ a_{n+1}$ implies that $f(a_n)\ R\,f(a_{n+1})$, for all n, it is easy to see that if $\boldsymbol{SRe}^{\omega}(a, b)$, then $Re^{\omega}_{\boldsymbol{M}}\big(f(a), f(b)\big)$ in the sense of Definition 3.3.7, but not conversely. We recall that $\boldsymbol{Re}^{\omega}_{\boldsymbol{M}}$ is the ω-reach of the elementary system $\boldsymbol{M} = (W, \leqslant, R, \boldsymbol{\mathcal{A}})$.

For example, in the situational system $\boldsymbol{M}^s = (P, R_1 \cup R_2, \{R_1, R_2\}, S, Tr, f)$ of the back and forth actions, defined as above, there exists a unique total isomorphism g between the posets $(X_1, \leqslant_1)$ and $(X_2, \leqslant_2)$ such that $\boldsymbol{SRe}^{\omega}\big((\mathbf{0}, \text{FORTH}), (g, \text{FORTH})\big)$ and $\boldsymbol{SRe}^{\omega}\big((\mathbf{0}, \text{FORTH}), (g, \text{BACK})\big)$.

Part II
Freedom and Enforcement in Action

Chapter 4
Action and Deontology

Abstract This chapter is concerned with the deontology of actions. According to the presented approach, actions and not propositions are deontologically loaded. Norms direct actions and define the circumstances in which actions are permitted, prohibited, or mandated. Norms are therefore viewed as deontological rules of conduct. The definitions of permission, prohibition, and obligatoriness of an action are formulated in terms of the relation of transition of an action system. A typology of atomic norms is presented. To each atomic norm a proposition is associated and called the normative proposition corresponding to this norm. A logical system, the basic deontic logic, is defined and an adequate semantics based on action systems is supplied. The basic deontic logic validates the closure principle. (The closure principle states that if an action is not forbidden, it is permitted.) A system which annulls this principle is also presented. In this context the problem of consistency of norms is examined. The problem of justice is discussed. The key role plays here is the notion of a righteous family of norms.

On Norms

The whole life of man progresses among norms; they are components of our everyday existence, which we are hardly conscious of. Norms govern our actions and behaviors. Every man is biologically conditioned and socially involved. We belong to different communities and groups, be they social, professional, denominational, family, or political, which results in the imposition of various systems of norms on our lives and activities from the ones that determine what is and what is not allowed to the ones that specify what is demanded and what is prohibited. Norms are of different strength and rank. They can be as simple as God's commandments and as complex as regulations of the Civil Law. They relate to a variety of spheres and aspects of life, including the biological dimension. Norms are hierarchized. Norms can be general or detailed, concrete or abstract. We recognize norms and meta-norms: norms concerning norms. Norms shape a person's life from the cradle to the grave. The addressees of particular norms may be either all people or some members of a given community. The fact of being subject to norms does not necessarily mean that we are under the yoke. Although we have examples of communities kept in fetters by systems of totalitarian

J. Czelakowski, *Freedom and Enforcement in Action*, Trends in Logic 42,
DOI 10.1007/978-94-017-9855-6_4

norms (or, more frequently, communities in which those in power violate the letter of law in a brutal way), we also have examples of communities in which lives can be happily led. What is more, man can enjoy the benefits which flow from his freewill. Paradoxically, a suitable system of norms can basically *broaden* the sphere of man's freedom in society. 'Good' and respected systems of norms protect a community from chaos and anarchy, and in consequence multiply the effectiveness of actions undertaken both by individual members of a community and smaller subgroups (business people, office workers, medical doctors, politicians, etc.). Looking at the question from the contemporary language-related perspective, 'good' systems of norms (and not just legal ones) are the initial conditions which serve to secure stability of social life of both individuals and groups in a society. Such systems can create the space for actions that lead to the positive development and prosperity of the community as a whole.

The above general (and rather banal) statements introduce the proper subject of this chapter: what are 'good' sets of norms? We shall try to answer the question from the precise perspective of formal logic. We shall have to pay a price of our choice, though: the simplification of a complex matter and the exclusion of some of the most vital aspects of the problem. Still, a simplified perspective will allow us, I believe, to get to the essential heart of the matter.

A person's behavior and actions have roots and motivations. For example, the feeling of hunger or thirst forces man to take actions intended to satisfy these basic needs. His actions are determined by biological constraints. Further actions aimed at meeting biological needs are channeled through norms which determine what can be done. If he lives in a developed society, he will go to a restaurant to eat or to a supermarket to buy food to bring home to prepare a meal with. Ordinarily, a person will not kill another human being in order to satisfy their hunger, be they a member of the same or another community. (A community of cannibals may seem to provide an obvious exception but in many recorded examples of such groups, human flesh is not eaten to satisfy hunger but for other reasons, such as to absorb the spirit of the deceased.) Societies also permit people to kill animals for food but here too both what they can kill and how they can kill it are regulated by secular and religious laws.

If one lives in a developed country and falls ill, it is norms that determine what one needs to do. One calls on the doctor or, if it is more serious, one calls for an ambulance (though whether one does either may also depend on whether one has health insurance in a society that requires it). Those who are well-off can bypass bureaucracy and head off to the private clinic of their choice. At the extreme, one can take the ultimate measure and commit suicide.

We treat behaviors as actions. If someone's *conduct* was morally good, then they *did* something that was in compliance with the moral code accepted by them, although it may perhaps have violated other norms such as legal ones. Certain conducts are biologically or physiologically conditioned, e.g., through pain, the feeling of hunger, or lack of sleep. Nevertheless, behavior always consists in carrying out a certain activity. When we are in pain, we typically behave in ways to ease or eliminate it, e.g., through taking a prescribed drug or a palliative. When we feel sleepy, we normally go to bed.

The term *norm* itself acquires, in this text, a broader meaning than in jurisprudence. We invest it here with a different sense. It covers moral norms, legal norms, norms of language, norms of social coexistence which define models of behavior in various communities, norms of professions, norms of production which are binding in business, norms of conventional conducts or etiquette, and principles that regulate the use of all kinds of appliances, from cars to coffeemakers. Grice's Maxims may be regarded as norms of effective communication as a form of purposeful activity. In the framework presented herewith, norms play a role similar to the inference rules in logic, yet this analogy is rather loose. To put it in a nutshell, *norms are rules of actions* including conduct. Norms direct actions and define the circumstances in which actions are permitted, prohibited, or mandated. To each norm a certain proposition is associated; it is called the *normative proposition* corresponding to the norm. In the simplest case of atomic norms, it takes the form of an implication of two propositions. Norms, however, are not reducible to propositions; they are objects of another order.

One may separate a special category of norms; we call them *praxeological norms*. (The criteria of separation are not clear-cut however.) Praxeological norms are exemplified by the principles that serve producing a car or a TV set. They concern technological and organizational aspects of producing a commodity. Praxeological norms are concerned with efficient ways and means of attaining the planned goals. Are such norms deontologically loaded? The norms that govern manufacturing should not violate law regulations, e.g., the regulations concerning air or water pollution or child labor. That is obvious. But manufacturing is more concerned with keeping up technological regimentation and lowering production costs than with the principles of Civil Code. Praxeological norms are in this sense deontologically loaded—the manufacturer performs actions that are in accordance with the adopted technology regime.

Norms delineate the direction of scientific research, not to say they condition the scientific method. Proving mathematical theorems and defining notions take place according to principles of logic. Rules of reasoning can be perceived as norms determining the features of a well-conducted proof. Only results obtained in the way of deduction, thus in accordance with principles of logic, attain the status of a mathematical result, inasmuch as they are original. All other methods, such as illumination or persuasion, are not accepted. Here, the famous *dictum* of Jan Łukasiewicz comes to mind: logic is the morality of thought and language.

There are strict principles which are binding in the natural sciences; that is, norms relating to acknowledgment of scientific discoveries of the empirical nature, as well as to the acknowledgment of scientific theories of a varying range. Empirical facts confirmed repeatedly by independent research teams are considered irrefutable. There exist accepted procedures for acknowledging or rejecting hypotheses and empirical theories. There exist also criteria by which to determine the reliability of research apparatus and technical devices built on the basis of an accepted theory. A good example here is the laser range finder—a precise and tested tool popularly applied in the construction industry and in land surveying. This tool would not have been possible if it were not for quantum mechanics.

On Deontic Logic

Two fundamentally different approaches to deontological problems should be mentioned. (In order to facilitate the analysis we shall confine ourselves to propositional deontic systems.)

The dominant type of deontic logic singles out a list of deontic operators such as:

O 'It is obligatory'
P 'It is permitted'
F 'It is forbidden'

and treats them as monadic logical connectives that assign propositions to propositions. Thus, the deontic operators are viewed as proposition-forming functors defined on *propositions*, i.e., sets of states of affairs. This approach, initiated in the 1950s, following the pioneering work of von Wright, was subsequently enriched and developed in various directions, e.g., by incorporating temporal or quantificational indexing explicitly into formalism or by introducing dyadic deontic operators in order to do justice to ordinary normative discourse. Gabbay et al. (2014/2015) provide a comprehensive account of deontic logic.

Here, we discuss a somewhat different approach to the above issues. It conceives of the deontic operators as proposition-forming functors defined on *actions*. This approach also originates from the papers of G.H. von Wright (and M. Fisher). Actions (or deeds) bear deontic values (being obligatory, forbidden, permitted). We also want to delineate a certain research program and to indicate how it could possibly be implemented. The crucial point consists, of course, in a clear explanation of how actions are related to deontic operators. We also discuss the notion of the so-called *atomic norm* and its relationship to the above interpretation of the deontic operators. (The term *basic norm*, in German: *Grundnorm*, can be also used. But here the term *norm* receives a wider meaning than in jurisprudence, referring not only to legal norms only but also moral norms, linguistic norms (in German: *Sprachliche Normen*), social norms that provide patterns of behavior in social communities, etc.)

Trypuz and Kulicki (2014) write: "Deontic considerations are usually conducted in one of the following contexts: (i) general norms expressed in: legal documents, regulations, or implicitly present in the society in the form of moral or social rules; (ii) specific norms, i.e., duties of particular agents in particular situations. Norms of the two kinds refer to obligatory, permitted, and forbidden actions or states. (…) In deontic logic the two types of norms are not usually present together in *formal* systems. They are often regarded in deontic literature as linguistic variants of the same normative reality." Trypuz and Kulicki call the norms of the two types *a-norms* (for action norms) and *s-norms* (for state norms), respectively. Both kinds can be present in the same normative systems. For example, they remark after Atienza and Manero (1998) that both kinds of norms are present in the Spanish constitution.

In this book the crucial definition of obligatoriness of an action is entirely formulated in terms of this action and of the relation of direct transition in a given elementary action system. A similar remark applies to the remaining deontic operators. These definitions give rise to a deontic logical system, denoted by ***DL***, which is adequately characterized by natural and intuitive axioms. ***DL*** is decidable. The logical validity of deontic formulas can be determined by an algorithm which is a straightforward extension of the familiar zero-one method of classical propositional logic.

However, there are certain philosophical problems that should be discussed. One of these is the question whether norms can be said to bear a truth value. According to our standpoint, to each elementary norm there corresponds a unique proposition which is called the *normative proposition* corresponding to the norm. The key issue is to provide criteria for the truth and falsehood of these propositions. We believe the solution we propose sheds some light on the problem how one can meaningfully apply truth functional connectives to norms, as is (tacitly) assumed in deontic logic. The sentences which are built up from action variables by applying deontic operators comply with the principle of bivalence: they are either true or false. The conception presented here is thus based on the following three assumptions:

1. *Actions* (and not state of affairs) are permitted, forbidden, or obligatory.
2. Norms are certain rules for actions.
3. To each atomic norm an elementary proposition is assigned which is called a 'normative proposition' in the terminology of von Wright.

Norms, however, are not reducible to propositions. In our approach norms play a role somewhat similar to that of rules of inference in logic—norms guide actions and determine circumstances under which some actions are permitted, forbidden, or obligatory. Action theory investigates, describes, and classifies all forms of human activity from a universal and unifying deontic perspective. This perspective is constituted by the collections of norms. Whether it be proving a mathematical theorem, building a house, making a movie or writing a novel, each of these highly diverse actions is conjoined with a specific set of norms (rules) that one should obey so as to achieve the desired goal. In other words, these sets of norms, being a specific envelope of action, form the basis for action. Every movie director, writer, or architect knows the rules of his/her profession, that is, he/she knows certain well-established collections of norms, specific to this profession, that he/she should obey while producing a movie, writing a book with an intriguing plot, or designing a new house.

The notion of an elementary action system plays a crucial role in the approach presented below.

4.1 Atomic Norms

In his *Norm and Action*, Georg Henrik von Wright defines norms as entities that govern actions. This viewpoint is adopted also in this monograph. He divides norms into rules, prescriptions, customs, directives, moral norms and ideals, and presents a deeper analysis of these categories of norms. For example, prescriptions are subsequently divided into commands, permissions, and prohibitions. In his book he also investigates the problem of the truth value of norms. He argues that rules and prescriptions do not carry truth values and are therefore outside the category of truth. In this book we follow this path. There is, however, a difference in the apprehension of norms; in this book the central role is played by elementary action systems as primary units. Norms in the semantic stylization are strictly conjoined with action systems but the category of atomic norms is secondary. Action systems go first and norms are associated with action systems. The issue whether a norm is a rule, a prescription, etc., depends on the category (or rather a class) of distinguished action systems—the character of a given class of action systems determines the respective category of associated norms. It should be added that within the class of elementary action systems the discussion on categories of norms in the von Wright style is very limited because elementary action systems are devoid of many aspects which are necessary for a deeper discussion of norms. For instance, the category of an agent of actions lies outside the vocabulary of elementary systems. The much wider notion of a *situational* action system provides the right framework for a discussion of norms because this notion incorporates necessary ingredients thereby making such a discussion possible.

Von Wright discerns two groups of norms, *viz.* norms concerning *action* (acts and forbearances) and norms concerning *activity*. (Forbearances are not discussed in this book—they are linked rather with situational action systems, not elementary ones.) Both types of norms are common and important. 'Close the door' orders an act to be done. 'Smoking allowed' permits an activity. 'If the dog barks, don't run' prohibits an activity. The theory presented here does not differentiate actions from activities. We claim that within the adopted formalism the latter are reducible to the former. We accept the standpoint that from the formal viewpoint each activity is a class of actions, obtained by distinguishing a single action in each action system from a *class* of such systems. On the other hand, an action as a one-time act is a concrete action belonging to a set of actions (though not necessarily atomic) of an individually defined and clearly named system of action. An instance of such an action is the stabbing to death of Julius Cesar.

In this section we confine ourselves only to discussing a particular case of norms—the notion of an atomic norm. The notion is relativised to a definite elementary action system $\boldsymbol{M} = (W, R, \boldsymbol{\mathcal{A}})$. From the semantic viewpoint, these simplest norms will be identified with figures of the form $(\Phi, A, +)$, $(\Phi, A, -)$ or $(\Phi, A, !)$, where A is an atomic action of $\boldsymbol{\mathcal{A}}$ and Φ is an elementary proposition, i.e., $\Phi \subseteq W$.

The commonly understood norms do not form a homogeneous set. The family of all norms has a rich hierarchical structure—there can be less or more general norms, as well as less or more important norms. Norms are usually accompanied by a certain preferential structure. This structure determines the order and principles according to which the norm is applied.

A reasonable theory of norms presupposes a definite ontology of the world. (In Chapter II the role of situations in the theory of action is underlined.) On the ground level the category of *individuals* is distinguished. Each individual is equipped with a certain repertoire of atomic actions he is able to perform. We may thus speak about the norms guiding the actions of individuals. The next level is occupied by *collectives* (or *groups*) of individuals. There are also specific norms which the members of collectives are expected to obey. Any religious community is an example of such a collective.

In philosophy one also distinguishes the category of *individual concepts* such as the president of the United States or the king of France. The words ‘notion’ and ‘concept’ are treated here as synonyms. Without entering into details, we assume that notion is an abstract or general idea inferred or derived from specific instances. For example, the concept of a horse is not identical with the animals in the world grouped by this concept—or the reference class or extension.

One may also distinguish *collective notions* (or *universal concepts*). The latter category mainly comprises such objects as the notion of a horse, bachelor, worker, etc., i.e., the notions labeled by common nouns. (In a simplified picture, by a name we shall understand any proper or common noun in the sense of traditional grammar.) We put an emphasis on the fact that according to our standpoint every concept is viewed as an abstract entity reducible neither to the name of this notion nor to the extension of the concept. (By analogy, when colored bricks lie on the floor in the kindergarten, they form a set, which is an abstract entity, although the elements of this set are material objects.) To each collective notion (and, hence, to the name associated with such a notion) is thus assigned the totality (collective) of all individuals being denotations of this name and called the extension of this notion. In fact, these are just collective notions that serve distinguishing definite collectives, as e.g. the collective of teachers.)

There are many theories of concept (the classical theory, prototype theory, etc.). We shall not tackle here the general issues pertaining the concept of concept, because they are outside the scope of our theory.

An individual may belong to many collectives. This is often a source of the conflict of norms, especially when the relations between the groups to which an individual belongs are characterized by hostility. For example, a particular individual may belong to a definite religious community and at the same time be a member of another, say professional, group which is hostile or indifferent to any religion. There can also be conflicts between norms guiding the actions of an individual and the norms regulating the actions of a collective to which the individual belongs.

Norms define the ways and circumstances in which actions are performed, e.g., by specifying the place, time, and order of their performance, or by forbidding their use for the second time. For example, in a chess game the instruction that *allows* one to perform a small castling is an example of a norm—it permits that a certain action be performed, i.e., moving the king and a castle in a particular situation on the chessboard. But we have also another norm that allows one to castle only once. It *forbids* one to perform a castling for a second time. The terms ‘use of a norm’ or ‘application of a norm’ are given in the intuitive sense. In view of their role in an explication of the notion of a norm, it seems that a theory of action should attach definite meanings to these terms.

The circumstances in which an atomic norm is used are entirely determined by states of the elementary action system under consideration. The atomic norms are neutral with respect to the factors constituting the situational envelope of the system; these factors (and the agents of actions in particular) are simply not taken into account by any atomic norm.

The assumption that *actions* (and not state of affairs) are permitted, forbidden or obligatory, departs, to a significant extent, from the traditional format of deontic logic according to which deontic operators act as unary logical connectives assigning propositions to propositions. In our approach the deontic operators

O 'It is obligatory'
P 'It is permitted'
F 'It is forbidden'

are proposition-forming functors defined on actions. Each of the above operators applied to an action A defines a certain proposition. Thus, $\boldsymbol{O}A$, $\boldsymbol{P}A$, and $\boldsymbol{F}A$ are subsets of W, for any atomic action A.

At first, we shall present a few remarks concerning the semantic aspects of norms. (Questions pertaining to their syntax will be discussed later.) The starting point here is the notion of an elementary action system as an appropriate semantic unit. Here are formal definitions.

Let $\boldsymbol{M} = (W, R, \boldsymbol{\mathcal{A}})$ be an elementary action system. A *positive* (or a *permissive*) *atomic norm* for the system $\boldsymbol{M}$ is any triple

$$(\Phi, A, +), \tag{4.1.1}$$

where $A \in \boldsymbol{\mathcal{A}}$ and Φ is an elementary proposition about $\boldsymbol{M}$, i.e., $\Phi \subseteq W$.

(4.1.1) is read:

$(1)^{*}_{+}$ 'It is the case that Φ, therefore A is permitted to perform,'

or in short: 'Φ, therefore A is permitted.' Φ is called the *hypothesis* of (4.1.1).

To the norm (4.1.1) an elementary proposition is assigned,

$$\Phi \to \boldsymbol{P}A, \tag{4.1.2}$$

where $\to$ represents the implication operation on the power set $\wp(W)$, i.e., $\Phi \to \Psi := \neg\Phi \cup \Psi$ for all $\Phi, \Psi \subseteq W$. (4.1.2) is called the *normative proposition corresponding to the norm* (4.1.1). $\Phi \to \boldsymbol{P}A$ is a subset of W. It will be defined in the next section.

A *negative* (or a *prohibitive*) atomic norm for $\boldsymbol{M}$ is any triple

$$(\Phi, A, -), \tag{4.1.3}$$

where Φ and A are described as above. (4.1.3) is read:

$(1)^{*}_{-}$ 'It is the case that Φ, therefore A is forbidden to perform,'

or, in short: 'Φ, therefore A is forbidden.'

To the norm (4.1.3) an elementary proposition is assigned, viz.

$$\Phi \to \boldsymbol{F}A, \tag{4.1.4}$$

defined in Sect. 4.2 below, and called the *normative proposition corresponding to the norm* (4.1.3).

The circumstances in which a positive or negative atomic norm can be used depend only on one factor: the set of states of the system encapsulated by the proposition Φ. The moment the proposition Φ is ascertained to be true, a positive norm allows the performance of a certain action from $\mathcal{A}$, while a negative norm forbids it.

The norm (4.1.1) does not decide the issues concerning the performability of the action A in circumstances in which Φ, the antecedent of the proposition (4.1.2), is true. It may happen that in each state $u \in \Phi$ the action A is not performable. In this case the executive powers of the norm (4.1.1) are nil.

We shall now discuss some limit cases of atomic norms. They are singled out on the following short list:

$$(W, A, +) \tag{4.1.5}$$

$$(W, A, -) \tag{4.1.6}$$

$$(\emptyset, A, +) \tag{4.1.7}$$

$$(\emptyset, A, -). \tag{4.1.8}$$

Proposition W (the full proposition) is satisfied in every state of the system; in particular W is satisfied before any performing of an arbitrary atomic action A. Thus, the norm

$$(W, A, +), \tag{4.1.5}$$

can be read as follows:

$(2)^*_+$ 'A is permitted to perform', shortly 'A is permitted.'

The norms of the form

$$(W, A, -), \tag{4.1.6}$$

are dual to (4.1.5). The norm (4.1.6) has the following obvious meaning:

$(2)^*_-$ 'A is forbidden to perform', shortly 'A is forbidden.'

The empty proposition $\emptyset$ is never satisfied; in particular it is not satisfied before any performance of an arbitrary action A. The triple

$$(\emptyset, A, +), \tag{4.1.7}$$

then can be read as follows:

$(3)^*_+$ 'A is not permitted to perform', shortly 'A is not permitted.'

Obviously, the prohibition $(3)^*_+$ refers to any state of the system. If one accepts the meta-norm saying that what is not permitted is forbidden (see the discussion of the Closure Principle below), we see that the positive norm (4.1.7) has the same meaning as the negative norm (4.1.6). We may write down this equivalence as follows: (4.1.7) $\equiv$ (4.1.6).

The norms of the form (4.1.5) with the full condition W are called *categorical positive norms*. In turn, the negative norms (4.1.6) with the full condition W are called *categorical negative norms*. They forbid the performance of some actions irrespective of the state of the system. Some of the ten commandments, as, e.g., 'Thou shalt not kill,' 'Thou shalt not steal' assume the form of categorical negative norms.

The prohibitive norm

$$(\emptyset, A, -), \tag{4.1.8}$$

say that the action A is (never) forbidden. If one accepts the Closure Principle, (4.1.8) is equivalent to the permissive norm (4.1.5), i.e., (4.1.8) $\equiv$ (4.1.5), as the set of states of the system in which performing the action A is prohibited is empty. We, thus, see that, in view of the above equivalences, the categorical negative atomic norms could be equally well defined as permissive norms of the form (4.1.7).

The general idea behind the above semantic definitions of norms is this: norms are viewed as deontically loaded rules of conduct. Norms are guides, enabling one to perform (or to refrain from performing) actions and to build action plans. Elementary norms do not specify the order in which actions are to be performed (these factors belong to the situational envelope of a given action system). Once the norm $(\Phi, A, +)$ is accepted, performing the action A is permitted whenever the action system is in a Φ-state. Analogously, once the norm $(\Phi, A, -)$ is accepted, performing the action A is canceled whenever the action system is in a Φ-state. In the next section, obligatory norms are discussed. Their role in action is highlighted.

4.2 Norms and Their Semantics

Obligatory norms need a discussion of their own. They have greater executive power than permissive norms. From the formal point of view, the simplest obligatory norms, atomic obligatory norms, are identified with the triples

$$(\Phi, A, !), \tag{4.2.1}$$

where Φ is an elementary proposition and A is an action belonging to $\boldsymbol{\mathcal{A}}$. (Here, $\boldsymbol{M} = (W, R, \boldsymbol{\mathcal{A}})$ is assumed to be a fixed elementary action system.) The norm (4.2.1) is read:

$(1)_!$ 'It is the case that Φ, therefore A is obligatory,'

or in short: 'Φ, therefore A is obligatory.'

To the norm (4.2.1) an elementary proposition is assigned, viz.

$$\Phi \rightarrow \boldsymbol{O}A, \tag{4.2.2}$$

called the *normative proposition* corresponding to the norm (4.2.1). It will be defined later.

While driving a car, we reach a crossroads where there is a road sign bearing a white rightward arrow against the blue background. In this situation, we have to turn right and drive on, following the direction indicated by the arrow in the road sign.

Obligatory norms can also have the form of a branched norm. For example, the road sign bearing a two-cleft arrow is associated with the norm ordering the driver to turn at the nearest crossroads behind the road sign to one of the streets that follows the direction of the arrow, e.g., to the left or to the right but not straight ahead.

The action ordered by an obligatory norm is naturally also permitted. What makes obligatory norms differ from permissive ones? Undoubtedly, it is a certain categorical character that forces the agent to perform in a given state (or situation) the action indicated by an obligatory norm, i.e., the action A in (4.2.1); other norms, even those involving the proposition Φ, are insignificant in Φ. Obviously, one can imagine a situation where the agent would feel the pressure of many obligatory norms at the same time. In this situation, say s, he/she would have to perform a number of actions simultaneously. It is usually impossible. It may well be that there is a contradiction in a system of obligatory norms. Suppose we are given two obligatory norms

$$(\Phi, A, !), \quad (\Phi, B, !) \tag{4.2.3}$$

with the same antecedent Φ. Let us assume, moreover, that the sets of possible effects of A and B in the states of Φ are disjoint, i.e., $\delta_A(u) \cap \delta_B(w) = \emptyset$ for all $u, w \in \Phi$. The system of norms (4.2.3) is obviously contradictory in the intuitive sense—the two actions A and B cannot be performed in any state $u \in \Phi$. (We abstract here from the situational envelope of the action system; thus, we cannot say that one action can be performed before another, or that they are performed by different agents. These components are not included into the notion of a state and consequently cannot be articulated by norms of the form (4.2.1).)

In what way, then, do obligatory norms come before permissive ones? In trying to answer this question, we will comment on two approaches to the problem of obligatory norms.

The fact that a given action is ordered can be seen in a wider context. Agents are usually involved in various action plans. In the more formal language we have been using, one can say that an action plan is, in the simplest case, a certain finite sequence of actions the agents intend to perform so that starting from a certain state at the outset, they can attain a certain intended state of the system. Without a loss of generality we may identify action plans with algorithms. The adoption by the agents of a certain action plan gives obligatory strength to the actions included in it. In implementing a given action plan, the agents are obliged to perform the actions included in the plan in a definite order. Acceptance of the plan sanctions the priority of actions that make the plan. Other actions, though permitted, are regarded then as irrelevant or contingent. Therefore, one can say that an action is obligatory with respect to the given action plan. However, it is not obligation in itself—the actions included in the plan are to be performed in a definite order. So, if an action A is to be performed as the third or the fifth, its obligation is just associated with this fact; the action A is obligatory on account of the fact it is to be performed as the third or fifth and not, for example, the seventh. The obligatory norms under which the above situation falls will not be, however, atomic norms, i.e., they will not be of the form $(\Phi, A, !)$, where Φ is an elementary proposition. The fact is that the order in which the actions involved in an action plan are to be performed is indescribable by sentences, whose meanings are elementary propositions. The above interpretation of the obligatoriness of an action, i.e., an occurrence of the action in the accepted action plan, leads to norms of a richer structure than the norms of the form $(\Phi, A, !)$. Nevertheless, the Yiddish proverb 'Man plans and God laughs' adequately reflects the significance of the plans we make and the actions that they render obligatory.

Yet another interpretation now appears here; let us call it the *inner-system interpretation*, referring to the relation R of direct transition in the system $(W, R, \mathcal{A})$. Our point of view can be described as follows. To each atomic action $A \in \mathcal{A}$ an agent is assigned, who is the performer of this action. The same agent may also perform other atomic actions. The correspondence between actions and their agents need not be functional. It is conceivable that the same action A may be performed, in different states, by different agents. However, for the sake of simplicity, we assume that there is only *one* agent of all the actions of $\mathcal{A}$. In other words, the system is operated by only one agent. This assumption enables us to restrict considerations to the formalism of elementary action systems. Let $u \in W$ and suppose $\mathcal{B} \subseteq \mathcal{A}$ is a (nonempty) family of atomic actions that are performable in u in the sense of Definition 1.4.1. Thus, for every $A \in \mathcal{B}$, there exists a state w such that $u\, R, A\, w$. What does it mean that the agent has the freedom in the state u to perform any action from the family $\mathcal{B}$? One should distinguish here two situations which we shall briefly describe.

CASE 1. The agent *has* to perform one of the actions from $\mathcal{B}$ in the state u, but it is only up to him/her which action from $\mathcal{B}$ to choose.

CASE 2. The agent *does not have* to undertake in the state u any actions from $\mathcal{B}$.

If Case 1 holds, we say that the agent is in the situation of enforcement to perform a certain action from $\mathcal{B}$ *and*—at the same time—has a free choice of each action from the family $\mathcal{B}$ which he/she will perform. (It is obvious that if $\mathcal{B}$ consists only of one action, free choice is illusory as the agent is forced to perform only one action.)

The relation R of direct transition, and only this relation, imposes a form of enforcement upon the agent irrespective of who he/she is. Hence, speaking about enforcement in the context of a given elementary action system $(W, R, \mathcal{A})$, we should provide the word with an additional letter R and talk about *R-enforcement* as it is entirely determined by the relation R.

If several relations of transition are defined on W, different *forms* of enforcement and obligations are imposed on the agent. For example, we can distinguish economic, administrative, or legislative enforcement. This would give rise to a hierarchy of the agent's obligations being in compliance with the force that different relations R would create. The relations expressing physical necessity (e.g., laws of gravity) would have the strongest obligatory power, whereas obligations imposed by certain conventions the weakest. Such an approach makes the mathematical formalism of the action theory more involved for it leads to action systems of the form

$$(W, R_1, \ldots, R_n, \mathcal{A}) \tag{4.2.4}$$

with a complicated network of different relations $R_1, \ldots, R_n$ of direct transitions. The simplifications we have made enable us to consider, instead of the system (4.2.4), the family of systems $(W, R_i, \mathcal{A})$, $i = 1, \ldots, n$, and so the systems that fall under Definition 1.2.1.

For a given state u, Case 1 is mathematically expressed as follows:

$$(\forall A \in \mathcal{B})(\delta_A(u) \cap \delta_R(u) \neq \emptyset) \ \ \& \\ (\forall w \in W)\big(w \in \delta_R(u) \Rightarrow (\exists A \in \mathcal{B})(w \in \delta_A(u))\big). \tag{4.2.5}$$

The first conjunct of (4.2.5) ascertains that each of the actions $A \in \mathcal{B}$ is performable in u, whereas the second conjunct, which can be briefly written as follows

$$\delta_R(u) \subseteq \bigcup\{\delta_A(u) : A \in \mathcal{B}\}, \tag{4.2.6}$$

says that *every* direct transition uRw from u to an arbitrary state w is accomplished by *some* action of $\mathcal{B}$.

Suppose now that A is a single atomic action. When does the agent find himself in the situation of being forced to perform the action A? Such a situation is a particular instance of Case 1, i.e., when $\mathcal{B} = \{A\}$. Condition (4.2.5) then takes the form $\delta_A(u) \cap \delta_R(u) \neq \emptyset$ & $\delta_R(u) \subseteq \delta_A(u)$ or, equivalently,

$$\emptyset \neq \delta_R(u) \subseteq \delta_A(u). \tag{4.2.7}$$

(4.2.7) thus states that u is not terminal (which means that $\emptyset \neq \delta_R(u)$) and every transition uRw originated in u is accomplished by A. In other words, every transition from the state u into any state admissible by R can be executed solely through performing the enforced action, i.e., the action A; hence, there is no possibility of 'bypassing' this action. The assumption of nonterminality of the state u is a necessary condition ensuring the nontriviality of all possible enforcements pertinent to this state.

The agent is *not* R-obliged to perform the action A in u if and only if (4.2.7) does not hold. The negation of (4.2.7) is equivalent to the disjunction:

$$\delta_R(u) = \emptyset \quad \text{or} \quad \delta_R(u) \setminus \delta_A(u) \neq \emptyset. \tag{4.2.8}$$

The first disjunct says that u is a terminal state, while the second states that there is a transition from u which is not accomplished by A.

Case 2 is equivalent to the fact that the agent is not enforced to perform any action that belongs to $\mathcal{B}$ in the state u. Thus, Case 2 is equivalent to

$$(\forall A \in \mathcal{B})(\delta_R(u) = \emptyset \quad \text{or} \quad \delta_R(u) \setminus \delta_A(u) \neq \emptyset). \tag{4.2.9}$$

In particular, if $u \in \delta_R(u)$ and, for every $A \in \boldsymbol{\mathcal{B}}$, $u \notin \delta_A(u)$, the agent does not have to perform any action of $\boldsymbol{\mathcal{B}}$ in the state u.

The above remarks give rise to a typology of atomic actions in the system $(W, R, \boldsymbol{\mathcal{A}})$.

Definition 4.2.1 A is called an *R-obligatory* atomic action in the state u if and only if $\emptyset \neq \delta_R(u) \subseteq \delta_A(u)$ (i.e., (4.2.7) holds). Equivalently, we shall say: 'It is obligatory to perform A in u,' when R is clear from context. □

Definition 4.2.2

(i) A is an *R-forbidden* atomic action in u if and only if $\delta_R(u) \cap \delta_A(u) = \emptyset$ (i.e., A is not R-performable in u). Equivalently, we shall say: 'It is forbidden to perform A in u.'

(ii) A is an *R-permitted* action in u if and only if A is not R-forbidden in u, i.e., when $\delta_R(u) \cap \delta_A(u) \neq \emptyset$. We then say: 'It is permitted to perform A in u.' □

According to the above conception, if an action A is obligatory (in a state u), then it is also permitted in u, but the converse need not hold.

For reasons of economy, when the main concern is the effectiveness of the system (especially when the system itself is to be a well-planned and working production system), the 'futile' states of the system are rejected. These are states in which some actions A are permitted but are not obligatory. In such a system, in each state u which belongs to the domain of A, the action will be obligatory.

The fact that the action A is obligatory is represented by the elementary proposition $\boldsymbol{O}A$. Thus, we assume:

(4.2.10) The proposition $\boldsymbol{O}A$ is true in u (i.e., $u \in \boldsymbol{O}A$) if and only if $\emptyset \neq \delta_R(u) \subseteq \delta_A(u)$.

The proposition $\Phi \rightarrow \boldsymbol{O}A$ ascertains that it is obligatory to perform the action A whenever the system is in any state belonging to Φ. Hence,

$(4.2.10)_\Phi$ $\Phi \rightarrow \boldsymbol{O}A$ is true in u if and only if $u \in \Phi$ implies that $u \in \boldsymbol{O}A$ (i.e., $u \notin \Phi$ or $\emptyset \neq \delta_R(u) \subseteq \delta_A(u)$.)

The arrow $\rightarrow$ in the above proposition has thus the same meaning as material implication.

We can also establish truth conditions for normative propositions corresponding to atomic permissive or prohibitive norms, according to Definition 4.2.2.

The fact that the action A is permitted is represented by the proposition $\boldsymbol{P}A$. Thus:

(4.2.11) The proposition $\boldsymbol{P}A$ is true in u (i.e., $u \in \boldsymbol{P}A$) if and only if $\delta_R(u) \cap \delta_A(u) \neq \emptyset$ (i.e., A is performable in u).

This is the weak interpretation of the permissibility of an action—A is permitted in u if at least one performance of A in u is admissible by the relation R. Then, we have:

$(4.2.11)_\Phi$ $\Phi \rightarrow \boldsymbol{P}A$ is true in u if and only if $u \in \Phi$ implies that $u \in \boldsymbol{P}A$.

Analogously,

(4.2.12) The proposition $\boldsymbol{F}A$ is true in u (i.e., $u \in \boldsymbol{F}A$) if and only if $\delta_R(u) \cap \delta_A(u) = \emptyset$ (i.e., A is unperformable in u).

It follows from the above definitions that both $\boldsymbol{O}A$ and $\boldsymbol{P}A$ are false in any terminal state u (that is, when $\delta_R(u) = \emptyset$). Hence, in view of (4.2.12), $\boldsymbol{F}A$ is true in any terminal state, as expected. We also have:

$(4.2.12)_\Phi$ $\Phi \to \boldsymbol{F}A$ is true in u if and only if $u \in \Phi$ implies $u \in \boldsymbol{F}A$.

If $A \subseteq R$, conditions (4.2.10), (4.2.11) and (4.2.12) reduce, respectively, to the following ones:

(4.2.10)* $u \in \boldsymbol{O}A$ if and only if $\emptyset \neq \delta_A(u) = \delta_R(u)$
(4.2.11)* $u \in \boldsymbol{P}A$ if and only if $\delta_A(u) \neq \emptyset$
(4.2.12)* $u \in \boldsymbol{F}A$ if and only if $\delta_A(u) = \emptyset$.

Note One may also distinguish a strong form of permission of a given action A, which we write as $\boldsymbol{P}_S A$ and read: 'The action A is strongly permitted.' $\boldsymbol{P}_S A$ is the elementary proposition defined as follows:

$$\boldsymbol{P}_S A \text{ is true in } u \text{ (i.e., } u \in \boldsymbol{P}_S A\text{) if and only if } \emptyset \neq \delta_A(u) \subseteq \delta_R(u).$$

The proposition $\boldsymbol{P}_S A$ is equivalent to the total performability of A: $u \in \boldsymbol{P}_S A$ if and only if A is totally performable in u. $\boldsymbol{P}_S A$ implies $\boldsymbol{P}A$, i.e., $\boldsymbol{P}_S A \subseteq \boldsymbol{P}A$ for every action A.

The stipulation that $\delta_A(u)$ be nonempty in the definition of $\boldsymbol{P}_S A$ is made so as to exclude the states u in which A is vacuously permitted, i.e., the states u in which $f_A(u) = \emptyset$.

The distinction between the above two forms of permission, that is, between the propositions $\boldsymbol{P}A$ and $\boldsymbol{P}_S A$, is relevant only in the case when the action A is not deterministic.

In a similar way one may define strong forms of the prohibition and obligatoriness of a given action A. We shall say that A is *strongly forbidden* in u, in symbols: $u \in \boldsymbol{F}_S A$, if and only if either $\delta_A(u) \cap \delta_R(u)$ is empty or at least one possible performance of A in u is not in R. Thus,

$$u \in \boldsymbol{F}_S A \text{ if and only if } u \in \boldsymbol{F}A \text{ or } (\exists w \in W) A(u, w) \,\&\, \neg R(u, w).$$

A is *strongly obligatory* in u, symbolically: $u \in \boldsymbol{O}_S A$, if and only if $\emptyset \neq \delta_A(u) = \delta_R(u)$. □

Permissive legal regulations as a rule take the form of a weak permission $\boldsymbol{P}A$, where A is a certain action considered by the law. For instance, the norms which regulate the showing of pornographic movies have just such a form. Films of this type can only be screened in certain places: in special cinemas but not in churches or schools. (We obviously disregard here the situational context or the fact that this action is not atomic.) The legal norm which allows capital punishment can be seen in a similar light. Inasmuch the legislator generally approves of this form of dealing out justice, the specific regulations clearly define and regulate the ways the punishment is executed. Not all forms of legal execution (in our jargon—not all possible performances of this action) can be approved by the law. For example, it may be possible to apply the electric chair, but the use of the guillotine may be forbidden.

Remark 4.2.3 In situational contexts, a definite action is linked with its agents. The latter may vary depending on the situation. (All nontrivial forms of agency are expressed in situational action systems.) Thus, instead of saying that the given action A is obligatory in u, one says 'It is obligatory for the given agent to perform A in u.' This leads to the concept of deontic functors defining *agents*' duties and freedoms. This concept differs from the one presented in this section, where we only speak about the obligatoriness of *actions* as such. In consequence, one may distinguish four interesting situations depending on the state u of the system:

$\boldsymbol{O}_a^+ A$ The agent a is obliged to perform the action A;
$\boldsymbol{O}_a^- A$ The agent a is obliged *not* to perform the action A;
$\neg\boldsymbol{O}_a^+ A$ The agent a is *not* obliged to perform the action A;
$\neg\boldsymbol{O}_a^- A$ The agent a is *not* obliged *not* to perform A.

If the system is operated by only one agent a, the first and the third utterances $\boldsymbol{O}_a^+ A$ and $\neg\boldsymbol{O}_a^+ A$ are reducible to elementary norms. The crucial point is thus to find a plausible semantics for the second of the above utterances. One of the possible solutions is to assume that the agent a must refrain from doing A (i.e., he/she is obliged not to perform the action A) in u if and only if A is totally unperformable in u. Such a solution reduces $\boldsymbol{O}_a^- A$ to an elementary proposition:

$$u \in \boldsymbol{O}_a^- A \text{ if and only if } \delta_R(u) \cap \delta_A(u) = \emptyset.$$

In the light of the definition (4.2.12), the proposition $\boldsymbol{O}_a^- A$ would be equivalent to $\boldsymbol{F}A$. □

Linking the deontic functors $\boldsymbol{P}$, $\boldsymbol{O}$, and $\boldsymbol{F}$ with the relation R of direct transition may seem to be a controversial assumption. The identification of permission with the performability of an action (more strictly: with R-performability) particularly raises doubts, because the notion of the performability of an action has clear connotations related rather to the technical side of an action: an action is performable if there exists a way, device, or machine that ensures achievement of the goal. (The goal is conceived as one of many possible effects of the action.) Since the relation of direct transition admits a large number of interpretations, each of them implies one specific understanding of the propositions $\boldsymbol{P}A$, $\boldsymbol{O}A$ or $\boldsymbol{F}A$, for any action A. If R is understood to be a familiar collection of action regulations (e.g., traffic regulations or the rules of a one person game), the above conception of deontic functors agrees with our intuitive understanding of them. On the other hand, when the relation R expresses physical permission, and so specifies the range of admissible transformations of the system whose behavior complies with definite physical laws, the meaning of the functor $\boldsymbol{P}$ is closer to the meaning of the word 'possible' than 'permitted,' and $\boldsymbol{O}$ becomes almost synonymous with 'necessary'.

Note One may consider more complex semantic norms. For example, they may be tree-like structures. Here is a tree with eleven vertices.

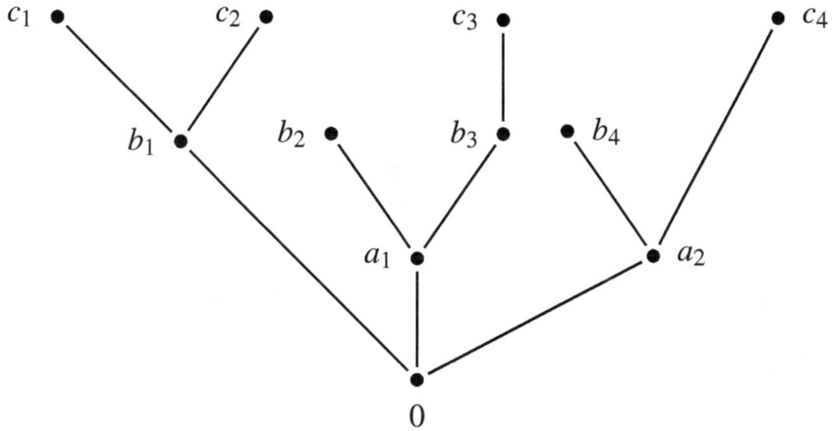

Fig. 4.1

A poset $(W, \leq, \mathbf{0})$ with zero $\mathbf{0}$ is a *tree* if for any $w \in W$, the set $\downarrow w = \{u \in W : u \leq w\}$ is well ordered. Pragmatic norms take the form of labeled trees. Formally, each tree-like positive norm is a pair $(\boldsymbol{T}, +)$, where $\boldsymbol{T}$ is a finite labeled tree with root $\mathbf{0}$ in which to each edge ab, where b is a child of a, an atomic action A_{ba} (from child b to its parent a) is assigned and to each vertex a a certain subset Φ_a of W is assigned (Φ_a is thus the label of a). These norms are discussed in Sect. 6.3 in the context of non-monotonic reasonings (but the discussion is limited there only to certain (totally) permitted norms). Tree-like norms are well suited for action systems operated by many agents.

For example, if the tree has three vertices and in the labeled form $\boldsymbol{T}$ it is depicted as

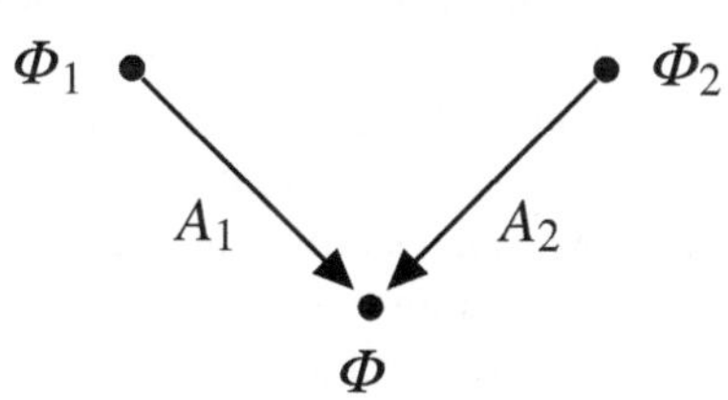

Fig. 4.2

then the normative proposition assigned to the norm $(\boldsymbol{T}, +)$ consists of pairs of states (u_1, u_2) such that whenever $u_1 \in \Phi_1$ and $u_2 \in \Phi_2$ then there are states $w_1, w_2 \in \Phi$ such that (u_1, w_1) is a realizable performance of A_1 and (u_2, w_2) is a realizable performance of A_2. The above proposition is therefore a subset of the product $W \times W$ and not a subset of W. If $\Phi = W$, the above condition implies that the product of the normative propositions corresponding to the atomic norms $(\Phi_1, A_1, +)$ and $(\Phi_2, A_2, +)$ contained in the normative proposition corresponding to the norm $(\boldsymbol{T}, +)$. (The arrows reverse the natural order of the tree.)

If the tree has two vertices and in the labeled form $\boldsymbol{T}$ it is depicted as

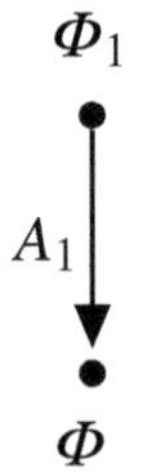

Fig. 4.3

then the normative proposition assigned to the norm $(\boldsymbol{T}, +)$ consists of states u_1 such that whenever $u_1 \in \Phi_1$ then there is a state $w_1 \in \Phi$ such that (u_1, w_1) is a realizable performance of A_1. If $\Phi = W$, the norm $(\boldsymbol{T}, +)$ is identified with the atomic norm $(\Phi_1, A_1, +)$. Thus positive atomic norms are limit cases of tree-like positive norms. □

The above semantics for atomic norms makes it possible to construct a simple logical deontic system.

Example 4.2.4 The model we outline below represents (in a simplistic way) the road traffic in the suburbs of a big American city. The street topography is represented by a planar graph with a regular tartan pattern. The nodes represent crossroads and the arrows between edges represent the direction of the traffic between the crossroads. A single arrow $a \to b$ indicates that there is only one-way traffic from a to b, while the double arrow $a \leftrightarrow b$ indicates that in the street segment between a and b the traffic runs in both directions. (Thus, $a \leftrightarrow b$ is an abbreviation for $a \to b$ and $b \to a$.) In the graph under consideration the above-mentioned arrows are either horizontal or vertical.

The options for the driver to maneuver at the crossroads depend, obviously, on which side he/she approaches the crossroads and on the road sign set up at the entrance to the crossroads. Each driver has only three actions in his/her repertoire. At each crossroads he/she may turn left $(\leftarrow D)$, turn right $(\to D)$, or drive straight on $(\uparrow D)$. There are no U-turns. Note, however, a simple but relevant fact: the qualification of the directions in the graph above (straight, left, right) depends, of course, not only at which crossroads the car finds itself, but also on the direction from which the car has entered the crossroads. Thus, let us assume that the state of the car and (the state of its driver) at crossroads b is determined not only by the very fact itself that his/her car is at b, but also by the direction which the car has reached b from.

The graph models the system of crossroads in the city from the viewpoint of an external observer (e.g., in a plane). Still, the directions (straight on, left, right) for the car are defined locally with reference to the concrete position of the car approaching the crossroads.

A state is therefore a pair (a, b) indicating that the driver finds himself/herself at crossroads b which he/she reached from the nearest crossroads a, in compliance with the traffic regulations. The crossroads a and b are therefore adjacent. We shall then identify possible states with pairs (a, b) of adjacent crossroads joined with an arrow (horizontal or vertical) commencing at a passing toward b in compliance with the graph. Let W be the set of states.

Each of the actions $\leftarrow D$, $\rightarrow D$, $\uparrow D$ is therefore identified with the set of pairs of states (and hence a set of pairs of pairs of adjacent crossings. More specifically,

$\leftarrow D := \{((a, b), (c, d)) : b = c$ and the arrow leading from c to d points leftwards for the driver moving toward $b = c$ from the direction $a\}$.

E.g., the following diagrams provide instances of $\leftarrow D$

$$\begin{array}{c} d \\ \uparrow \\ a \rightarrow b\ (= c) \end{array}$$

or

$$\begin{array}{c} a \\ \uparrow \\ (c =)\ b \rightarrow d \quad \text{etc.} \end{array}$$

The other actions are defined similarly.

The transition relation R between the states is defined as follows. Suppose $u_1 = (a_1, b_1)$, and $u_2 = (a_2, b_2)$ are states. Then, $u_1\ R\ u_2$ if and only if $a_2 = b_1$. In other words, $u_1\ R\ u_2$ if and only if the car approaching the crossroads b_1 directly from that of a_1 (in compliance with the traffic regulations), can then drive directly from the crossroads b_1 ($=a_2$) to that of b_2. The resulting triple $(W, R, \mathcal{A})$ with $\mathcal{A} = \{\leftarrow D, \rightarrow D, \uparrow D\}$ is therefore an elementary action system.

In this system, the actions of $\mathcal{A}$ are deontologically evaluated in particular states. E.g., for the action $\uparrow D$ and a state $u_1 = (a_1, b_1)$ we have:

> $\uparrow D$ is *obligatory* in $u_1 = (a_1, b_1)$ if and only if for the car approaching the crossroads b_1 from a_1 there is only one arrow in the graph commencing at b_1 pointing out the direction that enables continuation of the ride: straight on.

$$\begin{array}{ccccc} & & a_2 & & \\ & & \downarrow & & \\ a_1 & \rightarrow & b_1 & \rightarrow & c_1 \\ & & \uparrow & & \\ & & a_3 & & \end{array}$$

In the above figure, the car running from a_1 to b_1 has only one path of continuation of its ride: straight on from b_1 to c_1.

In turn, in the following part of the graph

$$\begin{array}{ccccc} & & a_2 & & \\ & & \downarrow & & \\ a_1 & \rightarrow & b_1 & \rightarrow & c_1 \\ & & \downarrow & & \\ & & a_3 & & \end{array}$$

the action $\uparrow D$ is no longer obligatory in the state $u_1 = (a_1, b_1)$ because the car may also turn right from b_1 to a_3. Similar stories hold for the remaining actions. □

4.3 The Logic of Atomic Norms

The formal language we wish to define contains the following syntactic categories:

Sentential variables:	$p_0, p_1, \ldots$
Action variables:	$\alpha_0, \alpha_1, \ldots$
Boolean connectives:	$\wedge, \vee, \neg$
Deontic symbols:	$\boldsymbol{P}, \boldsymbol{F}, \boldsymbol{O}$
Auxiliary symbols:	(,).

From these disjoint sets the set *Sent* of sentences is generated in the following way:

(i) every sentential variable p is a sentence
(ii) for every action variable α, $\boldsymbol{F}\alpha$, $\boldsymbol{P}\alpha$ and $\boldsymbol{O}\alpha$ are sentences
(iii) if ϕ and ψ are sentences then so are $\phi \wedge \psi$, $\phi \vee \psi$ and $\neg\phi$
(iv) nothing else is a sentence.

(In clauses (ii) and (iii) we adopt the usual convention of suppressing the outermost parentheses in sentences.) There are no special action-forming functors in the vocabulary of *Sent*.

Formulas of the shape $\phi \to \psi$ and $\phi \leftrightarrow \psi$ are defined in the standard way as abbreviations for $\neg\phi \vee \psi$ and $(\phi \to \psi) \wedge (\phi \to \psi)$, respectively.

The triples of the form

$$(\phi, \alpha, +), \quad (\phi, \alpha, -), \quad (\phi, \alpha, !) \tag{4.3.1}$$

are regarded as permissive, prohibitive, and obligatory norms in the *syntactic* form, respectively, where $\phi \in$ *Sent* and α is an action variable. They are respectively read: 'It is the case that ϕ, therefore α is permitted,' 'It is the case that ϕ, therefore α is forbidden,' 'It is the case that ϕ, therefore α is obligatory.' In turn, the sentences of the form

$$\phi \to \boldsymbol{P}\alpha, \quad \phi \to \boldsymbol{F}\alpha, \quad \phi \to \boldsymbol{O}\alpha, \tag{4.3.2}$$

are called the *sentences corresponding* to the norms $(\phi, \alpha, +)$, $(\phi, \alpha, -)$, $(\phi, \alpha, !)$, respectively. While norms (4.3.1) themselves do not carry a truth value, the sentences (4.3.2) are logically meaningful in the sense that logical values, viz. truth or falsity, are assigned to them. (4.3.2) are called *normative sentences* or *norm sentences*.

A *model* for *Sent* is a structure

$$M = (W, R, V_0, V_1), \tag{4.3.3}$$

where

(v) (W, R) is a discrete system,
(vi) V_0 is a *valuation* that assigns to each sentential variable p a subset $V_0(p)$ of W,
(vii) V_1 is an operator which assigns to each action variable α a binary relation $V_1(\alpha)$ on W.

If (4.3.3) is a model, the triple $(W, R, \{V_1(\alpha_n) : n \in \mathbb{N}\})$ is an elementary action system.

The property 'ϕ is true (holds) at u in M', symbolized $M \models_u \phi$, is defined by induction on the number of symbols of the formula ϕ as follows (the prefix M may be dropped if it is clear which model is intended):

$M \models_u p$	if and only if	$u \in V_0(p)$
$M \models_u \boldsymbol{P}\alpha$	if and only if	$(\exists w \in W)(uRw \ \& \ u\, V_1(\alpha)\, w)$
$M \models_u \boldsymbol{F}\alpha$	if and only if	$\neg(\exists w \in W)(uRw \ \& \ u\, V_1(\alpha)\, w)$
$M \models_u \boldsymbol{O}\alpha$	if and only if	$(\exists w \in W)(uRw) \ \& \ (\forall w \in W)(uRw \Rightarrow u\, V_1(\alpha)\, w)$
$M \models_u \phi \wedge \psi$	if and only if	$M \models_u \phi$ and $M \models_u \psi$
$M \models_u \phi \wedge \psi$	if and only if	$M \models_u \phi$ or $M \models_u \psi$
$M \models_u \phi \wedge \psi$	if and only if	not $M \models_u \phi$ or $M \models_u \psi$
$M \models_u \neg\phi$	if and only if	not $M \models_u \phi$.

A set $\Gamma \subseteq Sent$ is true at u in M, symbolically: $M \models_u \Gamma$, if and only if every sentence Φ of Γ is true at u in M.

The above clauses enable us to extend the valuation V_0 onto the set *Sent* by putting:

$$u \in V_0(\phi) \quad \text{if and only if} \quad M \models_u \phi,$$

for all u and ϕ.

It follows that $V_0(\boldsymbol{O}\alpha) = \boldsymbol{O}(V_1(\alpha))$, $V_0(\boldsymbol{P}\alpha) = \boldsymbol{P}(V_1(\alpha))$ and $V_0(\boldsymbol{F}\alpha) = \boldsymbol{F}(V_1(\alpha))$, where the propositions on the right-hand sides are defined in accordance with formulas (4.2.10)–(4.2.12) of Sect. 4.2.

We say that ϕ is *true* in M, denoted $M \models \phi$, if ϕ is true at every $u \in W$. ϕ is *tautological* if and only if it is true in all models M.

A model $M = (W, R, V_0, V_1)$ is *normalized* if and only if $R = \bigcup\{V_1(\alpha_n) : n \in \mathbb{N}\}$.

Proposition 4.3.1 *For every model $M = (W, R, V_0, V_1)$ there exists a normalized model $M' = (W, R', V_0', V_1')$ with the same set of states such that for every sentence $\phi \in Sent$ and any $u \in W$,*

$$M \models_u \phi \quad \textit{if and only if} \quad M' \models_u \phi.$$

Proof Define $R' := \bigcup\{R \cap V_1(\alpha_n) : n \in \mathbb{N}\}$, $V_0' := V_0$ and $V_1'(\alpha) = R \cap V_1(\alpha)$ for every α. Then, for any sentential variable p and any action variable α,

$$
\begin{array}{lll}
M \models_u p & \text{if and only if} & M' \models_u p, \\
M \models_u \boldsymbol{P}\alpha & \text{if and only if} & M' \models_u \boldsymbol{P}\alpha, \\
M \models_u \boldsymbol{F}\alpha & \text{if and only if} & M' \models_u \boldsymbol{F}\alpha, \\
M \models_u \boldsymbol{O}\alpha & \text{if and only if} & M' \models_u \boldsymbol{O}\alpha.
\end{array}
$$

From the above facts the thesis follows. □

We thus see that a sentence ϕ is tautological if and only if it is true in all normalized models.

So far we have introduced semantic concepts of validity. Now we shall introduce syntactic counterparts by presenting an axiomatic system. *Modus Ponens* is the only primitive rule of inference. We shall need a number of axioms. First, we take a set of axioms adequate for classical sentential logic. Second, we want special axioms for the deontic operators.

The *basic deontic* logic is the smallest set $\boldsymbol{DL}$ of sentences of *Sent* that satisfies:

(i) $\boldsymbol{DL}$ contains all instances of the schemata

$$\phi \to (\psi \to \phi) \tag{4.3.4}$$

$$(\phi \to (\psi \to \xi)) \to ((\phi \to \psi) \to (\phi \to \xi)) \tag{4.3.5}$$

$$\neg\neg\phi \to \phi; \tag{4.3.6}$$

(ii) $\boldsymbol{DL}$ contains all instances of

$$\boldsymbol{O}\alpha \to \boldsymbol{P}\alpha \tag{4.3.7}$$

$$\boldsymbol{F}\alpha \leftrightarrow \neg\boldsymbol{P}\alpha; \tag{4.3.8}$$

(iii) $\boldsymbol{DL}$ is closed under Detachment, i.e.,

$$\phi, \phi \to \psi \in \boldsymbol{DL} \quad \text{only if} \quad \psi \in \boldsymbol{DL}.$$

As it is well known, (i) and (iii) provide an adequate basis for classical propositional logic ($\boldsymbol{CL}$), so that every instance in *Sent* of a classical tautology belongs to $\boldsymbol{DL}$. (4.3.7) represents the old Roman maxim *Lex neminem cogit ad impossibila.* (4.3.7) excludes the states when the obligation of an action cannot be fulfilled; that is, the action is not realizable (*Obligatio impossibilium*). (4.3.8) (in fact, in its logically equivalent form $\boldsymbol{P}\alpha \leftrightarrow \neg\boldsymbol{F}\alpha$) is often called the Closure Principle (see Sect. 4.4).

If Γ is a subset of *Sent*, we say that ψ is *deducible from* Γ in $\boldsymbol{DL}$, in symbols $\Gamma \vdash \phi$, if there exists a finite sequence $\phi_1, \ldots, \phi_n$ of sentences such that $\phi_n = \phi$, and for all $i \leq n$, either $\phi_i \in \Gamma$ or ϕ_i is one of the axioms (4.3.4–4.3.8), or there are $j, k < i$ such that ϕ_k is $\phi_j \to \phi_i$ (so that ϕ_i is deducible from ϕ_j and ϕ_k by Detachment). The string $\phi_1, \ldots, _n$ is called a *Γ-proof* of ϕ. As a special case of this relation, we put $\vdash \phi$ if and only if $\emptyset \vdash \phi$. Thus, $\vdash \phi$ if and only if $\phi \in \boldsymbol{DL}$.

Every sentence of ***DL*** is tautological. This follows from the fact that in any model M: (i) all instances of the axioms (4.3.4)–(4.3.8) are true in M, and (ii) if ϕ and $\phi \to \psi$ are true in M, then ψ is true as well. Thus, the logic ***DL*** is sound with respect to the above relational semantics.

Theorem 4.3.2 (The Completeness Theorem) *For every sentence* ϕ, $\phi \in \boldsymbol{DL}$ *if and only if* ϕ *is tautological.*

The proof of the above theorem employs the well-known Lindenbaum techniques. In order to apply them to ***DL***, we shall make a list of the properties of ***DL*** which will intervene in the proof of the Completeness Theorem.

The presence of classical logic in ***DL*** suffices to establish the Deduction Theorem for ***DL***:

$$\Gamma \cup \{\phi\} \vdash \psi \quad \text{if and only if} \quad \Gamma \vdash \phi \to \psi. \tag{4.3.9}$$

(4.3.9) is used to prove, still only using ***CL***, the following equivalence:

$$\Gamma \vdash \psi \quad \text{if and only if} \quad \vdash \phi_1 \wedge \cdots \wedge \phi_n \to \psi \ \text{ for some } \phi_1, \ldots, \phi_n \in \Gamma. \tag{4.3.10}$$

A set $\Gamma \subseteq Sent$ is *consistent* if $\Gamma \nvdash \phi \wedge \neg\phi$.

Lemma 4.3.3 *Suppose* $M \models_u \Gamma$ *for some model* M *and a state* u. *Then the set* Γ *is consistent.*

Proof Suppose by way of contradiction that $\Gamma \vdash \phi \wedge \neg\phi$. Then, by (4.3.10), $\phi_1 \wedge \cdots \wedge \phi_n \to \phi \wedge \neg\phi$ is tautological for some $\phi_1, \ldots, \phi_n \in \Gamma$. As $M \models_u \Gamma$, we get that $M \models_u \phi \wedge \neg\phi$, which is impossible. □

A set $\Gamma \subseteq Sent$ is defined to be *maximal* if Γ is consistent, and for any $\phi \in Sent$, either $\phi \in \Gamma$ or $\neg\phi \in \Gamma$. (This is equivalent to requiring that Γ need not be a subset of any other consistent set.)

Lemma 4.3.4 *Suppose* ∇ *is maximal.*

$$\nabla \vdash \phi \ \textit{ implies } \ \phi \in \nabla. \tag{4.3.11}$$

$$\textit{If } \phi \notin \nabla, \ \textit{ then } \ \nabla \cup \{\phi\} \ \textit{ is not consistent.} \tag{4.3.12}$$

$$\boldsymbol{DL} \in \nabla. \tag{4.3.13}$$

$$\textit{For any sentence } \phi, \ \textit{exactly one of } \phi \textit{ and } \neg\psi \textit{ belongs to } \nabla,$$
$$\textit{i.e.,} \neg\psi \in \nabla \quad \textit{if and only if} \quad \psi \notin \nabla. \tag{4.3.14}$$

$$\phi \to \psi \in \nabla \quad \textit{if and only if} \quad (\phi \in \nabla \ \textit{ implies } \ \psi \in \nabla). \tag{4.3.15}$$

$$\phi \wedge \psi \in \nabla \quad \textit{if and only if} \quad \phi \in \nabla \ \textit{ and } \ \psi \in \nabla. \tag{4.3.16}$$

$$\phi \vee \psi \in \nabla \quad \textit{if and only if} \quad \phi \in \nabla \ \textit{ or } \ \psi \in \nabla. \tag{4.3.17}$$

$$\phi \leftrightarrow \psi \in \nabla \quad \textit{if and only if} \quad (\phi \in \nabla \ \textit{ if and only if } \ \psi \in \nabla). \tag{4.3.18}$$

The proof is easy and is omitted. □

The next result is a variant of the well-known Lindenbaum's Lemma.

Lemma 4.3.5

(i) *Every consistent set of sentences is contained in a maximal set.*
(ii) $\Gamma \vdash \phi$ *if and only if* ϕ *belongs to every maximal extension of* Γ.
(iii) $\vdash \phi$ *if and only if* ϕ *belongs to every maximal set.*

We omit the proof. □

Lemma 4.3.6 *The set* $\{\boldsymbol{O}\alpha_n : n \in \omega\}$ *is consistent.*

Proof Construct a model $M = (W, R, V_0, V_1)$ for *Sent* such that $M \models_u \{\boldsymbol{O}\alpha_n : n \in \mathbb{N}\}$ for some state $u \in W$. □

Corollary 4.3.7 *For every maximal set* ∇ *and every action variable* α *there exists a maximal set* ∇' *such that* $\boldsymbol{O}\alpha \in \nabla'$ *and* $\{\boldsymbol{O}\beta : \boldsymbol{O}\beta \in \nabla\} \subseteq \nabla'$.

To the logic $\boldsymbol{DL}$ a certain elementary action system

$$\boldsymbol{M}(\boldsymbol{DL}) = (W(\boldsymbol{DL}), R, \boldsymbol{\mathcal{A}})$$

is assigned and called the *canonical action system* (for $\boldsymbol{DL}$). $W(\boldsymbol{DL})$ is the set of all maximal sets in *Sent*. The relation R of direct transition is defined by the condition:

$$\nabla R \nabla' \quad \text{if and only if} \quad \{\boldsymbol{O}\alpha : \boldsymbol{O}\alpha \in \nabla\} \subseteq \nabla'. \tag{4.3.19}$$

For each action variable α the atomic action $A(\alpha)$ on $W(\boldsymbol{DL})$ is defined as follows:

$$\nabla A(\alpha) \nabla' \quad \text{if and only if} \quad \nabla R \nabla' \ \& \ \boldsymbol{P}\alpha \in \nabla \ \& \ \boldsymbol{O}\alpha \in \nabla'. \tag{4.3.20}$$

We put $\boldsymbol{\mathcal{A}} := \{A(\alpha_n) : n \in \mathbb{N}\}$.

Definitions (4.3.19) and (4.3.20) immediately imply that $A(\alpha) \subseteq R$ for every action variable α. The system $\boldsymbol{M}(\boldsymbol{DL})$ is thus normal.

The *canonical model* of the logic $\boldsymbol{DL}$ is the structure

$$\boldsymbol{Mo}(\boldsymbol{DL}) := (W(\boldsymbol{DL}), R, V_0, V_1), \tag{4.3.21}$$

where $W(\boldsymbol{DL})$ and R are defined as above, $V_0(p) := \{\nabla \in W(\boldsymbol{DL}) : p \in \nabla\}$ for every sentential variable p, and $V_1(\alpha) := A(\alpha)$ for every action variable α.

The following lemmas display the list of the more important properties of the canonical model.

Lemma 4.3.8

(1) *R is reflexive.*

Let α *be an action variable.*

(2) $\nabla\ R\ \nabla'$ *and* $\boldsymbol{O}\alpha \in \nabla$ *imply* $\boldsymbol{O}\alpha \in \nabla'$ *for all* ∇, ∇'.
(3) $\nabla\ R\ \nabla'$ *and* $\boldsymbol{O}\alpha \in \nabla$ *imply* $\nabla\ A(\alpha)\ \nabla'$ *for all* ∇, ∇'.
(4) *If* $\boldsymbol{P}\alpha \in \nabla$ *then there exists a maximal set* ∇' *such that* $\nabla\ R\ \nabla'$ *and* $\nabla\ A(\alpha)\ \nabla'$.

Proof (3). Suppose $\nabla\ R\ \nabla'$ and $\boldsymbol{O}\alpha \in \nabla$. By (2), $\boldsymbol{O}\alpha \in \nabla'$. In turn, in view of the axiom A4, $\boldsymbol{O}\alpha \in \nabla$ implies $\boldsymbol{P}\alpha \in \nabla$. Therefore $\nabla\ R\ \nabla'\ \ \&\ \ \boldsymbol{P}\alpha \in \nabla\ \ \&\ \ \boldsymbol{O}\alpha \in \nabla'$. So $\nabla\ A(\alpha)\ \nabla'$.

(4). Assume $\boldsymbol{P}\alpha \in \nabla$. By Corollary 4.3.7 there exists a maximal set ∇' such that $\boldsymbol{O}\alpha \in \nabla'$ and $\{\boldsymbol{O}\beta : \boldsymbol{O}\beta \in \nabla\} \subseteq \nabla'$. Thus $\nabla\ R\ \nabla'\ \ \&\ \ \boldsymbol{P}\alpha \in \nabla\ \ \&\ \ \boldsymbol{O}\alpha \in \nabla'$ which proves that $\nabla\ A(\alpha)\ \nabla'$. □

Lemma 4.3.9 *Let* ∇ *be a maximal set. For any variable* α,

$$\boldsymbol{O}\alpha \in \nabla \quad \textit{if and only if} \quad (\forall \nabla')(\nabla\ R\ \nabla' \Rightarrow \nabla\ A(\alpha)\ \nabla'). \tag{4.3.22}$$

Proof ($\Rightarrow$). Suppose $\boldsymbol{O}\alpha \in \nabla$. Let $\nabla\ R\ \nabla'$. Then $\nabla\ A(\alpha)\ \nabla'$, by Lemma 4.3.8.(3).

($\Leftarrow$). Assume the right-hand side of (4.3.22). The definition of $A(\alpha)$ yields $(\forall\nabla')$ $(\nabla\ A(\alpha)\ \nabla' \Rightarrow \boldsymbol{O}\alpha \in \nabla')$. So $(\forall\nabla')(\nabla\ R\ \nabla' \Rightarrow \boldsymbol{O}\alpha \in \nabla')$ by the transitivity of $\Rightarrow$. Thus, $\{\boldsymbol{O}\beta : \boldsymbol{O}\beta \in \nabla\} \subseteq \nabla'$ implies $\boldsymbol{O}\alpha \in \nabla'$, for all ∇', by the definition of R. Since trivially $\{\boldsymbol{O}\beta : \boldsymbol{O}\beta \in \nabla\} \subseteq \nabla$, we obtain that $\boldsymbol{O}\alpha \in \nabla$. □

Lemma 4.3.10 *Let* ∇ *be a maximal set. For any variable* α,

$$\boldsymbol{P}\alpha \in \nabla \quad \textit{if and only if} \quad (\exists \nabla')(\nabla\ R\ \nabla'\ \ \&\ \ \nabla\ A(\alpha)\ \nabla'). \tag{4.3.23}$$

Proof The implication ($\Rightarrow$) is a paraphrase of Lemma 4.3.8.(4). To prove the reverse implication, suppose the right-hand side of (4.3.23) is true for some ∇'. Since $\nabla\ A(\alpha)\ \nabla'$, (4.3.20) yields $\boldsymbol{P}\alpha \in \nabla$. □

Corollary 4.3.11 *Let* ∇ *be a maximal set. Then, for any* α,

$$\boldsymbol{F}\alpha \in \nabla \quad \textit{if and only if} \quad \neg(\exists \nabla')(\nabla\ R\ \nabla'\ \ \&\ \ \nabla\ A(\alpha)\ \nabla'). \tag{4.3.24}$$

□

The fundamental property of the canonical model $\boldsymbol{Mo}(\boldsymbol{DL})$ is that for any $\phi \in Sent$, and any maximal set ∇,

$$\boldsymbol{Mo}(\boldsymbol{DL}) \models_{\nabla} \phi \quad \text{if and only if} \quad \phi \in \nabla. \tag{4.3.25}$$

This is proved by induction on the length of ϕ, with (4.3.22)–(4.3.24) being involved to show that

$$\begin{aligned}
\boldsymbol{Mo}(\boldsymbol{DL}) \models_{\nabla} \boldsymbol{O}\alpha \quad &\text{if and only if} \quad \boldsymbol{O}\alpha \in \nabla,\\
\boldsymbol{Mo}(\boldsymbol{DL}) \models_{\nabla} \boldsymbol{P}\alpha \quad &\text{if and only if} \quad \boldsymbol{P}\alpha \in \nabla,\\
\boldsymbol{Mo}(\boldsymbol{DL}) \models_{\nabla} \boldsymbol{F}\alpha \quad &\text{if and only if} \quad \boldsymbol{F}\alpha \in \nabla.
\end{aligned}$$

From (4.3.25) and Lemma 4.3.5 we obtain that in general

$$\models \phi \quad \text{if and only if} \quad \boldsymbol{Mo}(\boldsymbol{DL}) \models \phi,$$

i.e., the sentences true in $\boldsymbol{Mo}(\boldsymbol{DL})$ are precisely $\boldsymbol{DL}$ theorems. From this observation Theorem 4.3.2 follows. □

The above Completeness Theorem can be more succinctly expressed in terms of admissible valuations.

An *admissible valuation* for $\boldsymbol{DL}$ is any mapping $h : Sent \to \{0, 1\}$ such that:

(1) h is a Boolean homomorphism for the classical connectives of *Sent*, i.e., for any $\phi, \psi \in Sent$

$$h(\phi \wedge \psi) = \min\big(h(\phi), h(\psi)\big)$$
$$h(\phi \vee \psi) = \max\big(h(\phi), h(\psi)\big)$$
$$h(\neg\phi) = \begin{cases} 1 & \text{if} \quad h(\phi) = 0 \\ 0 & \text{otherwise,} \end{cases}$$

(2) for any action variable α

$$h(\boldsymbol{O}\alpha) \leqslant h(\boldsymbol{P}\alpha)$$

and

$$h(\boldsymbol{F}\alpha) = \begin{cases} 1 & \text{if} \quad h(\boldsymbol{P}\alpha) = 0 \\ 0 & \text{otherwise.} \end{cases}$$

Each mapping h_0 defined on the union of the set of sentential variables and the set of formulas of the form $\boldsymbol{P}\alpha$, where α ranges over action variables, and with values in $\{0, 1\}$, extends to an admissible valuation h. This extension is not unique because in case of $h_0(\boldsymbol{P}\alpha) = 1$, one may assign any of the values 0, 1 to the formula $\boldsymbol{O}\alpha$. Then, unambiguously extend the mapping onto the remaining formulae.

Let H be the set of all admissible valuations.

Theorem 4.3.12 *For all ϕ, $\phi \in \boldsymbol{DL}$ if and only if ϕ is validated by all $h \in H$.*

Proof It is easy to see that every admissible valuation validates the axioms of $\boldsymbol{DL}$ and the Rule of Detachment. Consequently, $h(\phi) = 1$ for all $\phi \in \boldsymbol{DL}$ and all $h \in H$. This gives the "⇒"—part of the theorem.

To prove the reverse implication, we take the canonical model $\boldsymbol{Mo}(\boldsymbol{DL})$. Let, for each maximal set $\nabla \in W(\boldsymbol{DL})$, h_∇ be the characteristic function of ∇, i.e., for every $\phi \in Sent$,

$$h_\nabla(\phi) = \begin{cases} 1 & \text{if} \quad \phi \in \nabla \\ 0 & \text{otherwise.} \end{cases}$$

In view of Lemmas 4.3.4 and 4.3.9–4.3.11, h_∇ is an admissible valuation. Conversely, for every admissible valuation h there exists a unique maximal set ∇ such that $h = h_\nabla$, viz. the set $\nabla := \{\phi \in Sent : h(\phi) = 1\}$ is maximal. (The proof of this fact is left as an exercise.)

It follows from the above remarks and Theorem 4.3.2 that if ϕ is validated by all $h \in H$, then $\phi \in \boldsymbol{DL}$. So the "$\Leftarrow$"—part of the theorem also holds. □

It follows from the above theorem that the system ***DL*** is decidable. Moreover, Theorem 4.3.2 can be strengthened to a strong completeness theorem. We say that a set Γ *semantically entails* ϕ if, for every model M, if every sentence of Γ is true in this model then so is ϕ.

Theorem 4.3.13 (The Strong Completeness Theorem) *Let Γ be any set of sentences and ϕ any sentence. ϕ is deducible from Γ if and only if Γ semantically entails ϕ.*

We shall sketch the proof. The implication ($\Rightarrow$) can easily be established by induction on the length of the Γ-proof of ϕ. To prove the reverse implication, suppose the sentence ϕ is not deducible from Γ. We shall construct a model for *Sent*, in which the sentences of Γ are true and ϕ is not.

Let $W(\Gamma)$ be the family of all maximal sets that include Γ. Since Γ is consistent, $W(\Gamma)$ is nonempty by Lemma 4.3.5.(a). We then define the relation R, the actions $A(\alpha)$ on $W(\Gamma)$, and the assignments V_0, V_1 in the same way as in the proof of Theorem 4.3.2.

The quadruple $\boldsymbol{Mo}(\Gamma) := (W(\Gamma), R, V_0, V_1)$ is thus a model for *Sent* which obeys the following condition. Let $\psi \in Sent$ and $\nabla \in W(\Gamma)$. Then:

$$\boldsymbol{Mo}(\Gamma) \models_\nabla \psi \quad \textit{if and only if} \quad \psi \in \nabla. \tag{4.3.26}$$

(4.3.26) and Lemma 4.3.5.(b) imply that

$$\boldsymbol{Mo}(\Gamma) \models \psi \quad \textit{if and only if} \quad \Gamma \vdash \psi. \tag{4.3.27}$$

As $\boldsymbol{Mo}(\Gamma) \models_\nabla \Gamma$ for all $\nabla \in W(\Gamma)$, we also have that

$$\boldsymbol{Mo}(\Gamma) \models \Gamma. \tag{4.3.28}$$

Since the sentence ϕ is not deducible from Γ, (4.3.27) yields

$$\phi \textit{ is not true in } \boldsymbol{Mo}(\Gamma). \tag{4.3.29}$$

(4.3.28) and (4.3.29) prove that Γ does not entail semantically ϕ. □

The expressive power of the above deontic language is obviously very limited. One cannot express in it various interdependencies holding between different actions. For example, one semantically infers the sentence "I must not eat ice cream" from the sentence "I must not eat sweets." The sentence "If one eats ice-cream, then one eats sweets" underlines the subordinate nature of the action of eating ice cream with respect to the action of eating sweets. Similarly, if one fixes the oil pump in a car, one fixes this car. The action of fixing a car is superior to the action of fixing the oil pump in this car. In order to make comparisons of this sort, it suffices to extend the above deontic language by adjoining a binary functor $\angle$, forming a formula from action variables

$$\alpha \angle \beta \tag{4.3.30}$$

with the intended reading 'α is subordinated to β' or 'β is overriding the action α.' Semantically, for any frame (W, R) and any binary relations A, B on W:

$$A \angle B := \{u \in W : f_A(u) \subseteq f_B(u)\}. \tag{4.3.31}$$

The logical[1] value of 'β is overriding α' is therefore state dependent. (4.3.31) defines the meaning of a phrase of form (4.3.30) provided that the meaning of the action variables α and β has already been established. Note that $A \subseteq B$ if and only if $A \angle B = W$.

By changing the grammar of the deontic language by incorporating into its body a new formation rule allowing for expressions (4.3.30) for any action variables α and β, we see that every model $M = (W, R, V_0, V_1)$ validates sentences of the form

$$\boldsymbol{P}\alpha \wedge \alpha \angle \beta \to \boldsymbol{P}\beta \tag{4.3.32}$$

$$\boldsymbol{F}\beta \wedge \alpha \angle \beta \to \boldsymbol{F}\alpha. \tag{4.3.33}$$

(4.3.32) and (4.3.33)[2] immediately follow from the above semantic conditions that characterize $\boldsymbol{P}$ and $\boldsymbol{F}$ in arbitrary models M and the fact that

$$M \models_u \alpha \angle \beta \quad \text{if and only if} \quad (\forall w \in W)(u\, V_1(\alpha)\, w \Rightarrow u\, V_1(\beta)\, w). \tag{4.3.34}$$

It is an open problem whether the above relational semantics adequately characterizes the logical system that arises from $\boldsymbol{DL}$ by adjoining the axioms (4.3.32) and (4.3.33).[3]

[1] The discussion of the meaning of the connective $\angle$ for compound actions is more intricate. This problem will not be tackled here, though.

[2] If β represents the act of 'killing,' and α represents 'killing in self-defense,' then according to (4.3.33), in each case if killing is forbidden, then it is also forbidden to kill even if in self-defense. We cannot see any deontic paradox here. This conclusion is justified when α and β are treated as atomic actions (or the resultant relations of composite actions). But if the definition of $\angle$ is extended onto compound actions, (4.3.32) and (4.3.33) are no longer true; see Sect. 4.5.

[3] One can undermine the justifiability of the axioms (4.3.32) and (4.3.33) for compound actions. If a compound action is forbidden, we can see no reason why each of its 'parts' should be forbidden; see Sect. 4.5.

4.4 The Closure Principle

Every instance of the following principle

$$\boldsymbol{P}\alpha \to \neg \boldsymbol{F}\alpha, \tag{4.4.1}$$

which is a 'half' of the axiom (4.3.8), is clearly a tautological sentence. The principle states that every permitted action is not forbidden. Its converse

$$\neg \boldsymbol{F}\alpha \to \boldsymbol{P}\alpha, \tag{4.4.2}$$

which is called in jurisprudence the *closure principle*, is also a schema of tautological sentences with respect to the above relational semantics. The closure principle states that if an action is not forbidden, it is permitted.

(The closure principle also holds for the strong deontic operators $\boldsymbol{P}_S$ and $\boldsymbol{F}_S$: if A is an atomic action of a system $(W, R, \boldsymbol{\mathcal{A}})$, then for any $u \in W$, $u \in \boldsymbol{P}_S A$ if and only if $u \notin \boldsymbol{F}_S A$.)

Principle (4.4.1) is commonly accepted in the known approaches to deontic logic. It is logically equivalent to the tautology $\neg(\boldsymbol{F}\alpha \wedge \boldsymbol{P}\alpha)$ that says that no action is prohibited and permitted. The closure principle, however, is sometimes questioned as it is argued that systems of norms, including legal ones, are never complete in the sense that they do not codify all conceivable and potentially legally significant actions. For example, outside the criminal code there may remain groups of actions that norm-fixers cannot in advance decide whether they should be prohibited or not. Such a code is called *open*.

Tautologies (4.4.1) and (4.4.2) together make it possible to define the operator $\boldsymbol{F}$ in terms of $\boldsymbol{P}$ since the conjunction of (4.4.1) and (4.4.2) is equivalent to $\boldsymbol{F}\alpha \leftrightarrow \neg \boldsymbol{P}\alpha$. One may attempt to modify the above semantics in order to exclude the closure principle from the set of tautologies. One of the possible approaches changes the definition of an elementary action system so as to take into account, on a larger scale, knowledge limitations of the agents operating the system. This means accepting a more realistic definition of an action system.

The *open* action systems are identified with quadruples

$$(W, R^+, R^-, \boldsymbol{\mathcal{A}}), \tag{4.4.3}$$

where W is the nonempty set of *states* of the system, R^+ and R^- are binary, disjoint relations on W, and $\boldsymbol{\mathcal{A}}$ is a family of binary relations on W. The members of $\boldsymbol{\mathcal{A}}$ are called *atomic actions*. The triple

$$(W, R^+, R^-), \tag{4.4.4}$$

which is a reduct of (4.4.3), is called a *discrete system* or space. Both the relations R^+ and R^- impose some limitations on the possibilities of the transition from some states to others.

$R^+(u, w)$ says that the direct transition from u to w is certain while $R^-(u, w)$ says that the direct transition from u to w is excluded with certainty. The limitations expressed by R^+ and R^- arise both from the intrinsic order the system is based on and from the epistemic status of the agent which can be far from omniscience. (We assume, for the sake of simplicity, that the system is operated by only one agent.) These factors are more explicitly articulated by means of the following assumptions:

(a) The agent knows which transitions are excluded (these are the elements of R^-);
(b) The agent knows with certainty which transitions take place (these are the elements of R^+);
(c) The agent is unable to decide on some of pairs of states (u, w) if the system admits the direct transition from u to w or not.

Therefore $R^+ \cup R^-$ need not be the full relation $W \times W$.

We assume that the agent knows the elements of the family $\boldsymbol{\mathcal{A}}$, i.e., he/she can make the list of all his/her (atomic) actions and, for any atomic action $A \in \boldsymbol{\mathcal{A}}$, the agent knows the elements of A and the elements of R^+ and R^- as well.

Let $(u, w) \in A \in \boldsymbol{\mathcal{A}}$. If $(u, w) \in R^+$, the agent knows with certainty that the possible performance (u, w) is realizable. Consequently, the agent knows with certainty that A is performable in u. (But he may not know that u is the *current* state of the system; this fact may be hidden from him.) If $(u, w) \in R^-$, the agent knows with certainty that the possible performance (u, w) is not realizable. Since $R^+ \cup R^-$ need not be equal to $W \times W$, it may also happen that $(u, w) \notin R^+ \cup R^-$. In this case the agent *does not know* if the possible performance (u, w) is realizable or not—the knowledge available to him does not allow him to decide this issue.

We shall present a certain formal semantics for norms, which does not yield the closure principle. Before entering into the discussion of this topic, suppose that for some action A, the set $\delta_A(u)$ of possible effects of this action in the state u is empty. Then, in view of the condition (4.2.11) of Sect. 4.2, the action A is not permitted in u. The acceptance of the closure principle, then, immediately implies that A is forbidden in this state. However, if the closure principle is rejected, one may argue that A need not be regarded as forbidden in u. For instance, let A be the set of all possible moves of a particular type of piece in a game of chess (so A is one of the actions as described in Sect. 2.1). Suppose u is a possible configuration on the chessboard in which the pieces of a given type do not take part, i.e., they have been eliminated from the game and remain outside the board. Then, $\delta_A(u)$ is empty. In such a situation, we cannot say, obviously, that any move of any piece of the given type is admitted, since—physically—this piece is not on the chessboard. For the same reason we cannot say, however, that any movement of any of the pieces is forbidden in the position u.

These comments point to the fact that rejecting the closure principle requires in consequence an examination of the case when the set $\delta_A(u)$ is empty. We assume that in such states, action A is neither permitted nor forbidden.

Definition 4.4.1 Let the system (4.4.3) be fixed and let $A \in \mathcal{A}$.

(i) The action A is *permitted* in a state $u \in W$ if and only if $(\exists w \in W)\big(R^+(u,w) \;\&\; A(u,w)\big)$.

(ii) A is *forbidden* in u if and only if $(\exists w)\big(A(u,w)\big)$ & $(\forall w \in W)\big(A(u,w) \Rightarrow R^-(u,w)\big)$.

(iii) A is *obligatory* in u if and only if $(\exists w \in W)\big(R^+(u,w)\big)$ & $(\forall w \in W)\big(R^+(u,w) \Rightarrow A(u,w)\big) \;\&\; \neg(\exists w \in W)\big(R^-(u,w) \;\&\; A(u,w)\big)$.

Thus, A is obligatory in u if and only if every R^+-transition from u to a state w, i.e., such that $R^+(u,w)$, is accomplished by A (it is assumed that at least one such a transition exists) *and* there does not exist a possible performance (u,w) of A which—with certainty—is not realizable. If A is obligatory in u, then A is permitted in u and A is not forbidden in u. If the set $\delta_A(u)$ is empty, the action A is neither permitted nor forbidden in u.

As $R^+ \cap R^- = \emptyset$, every permitted action in u is not forbidden in this state. However, if A is not forbidden in u, then A need not be permitted in u. This is certainly the case if the set $\delta_A(u)$ of possible effects of A in u is empty. The second reason is that the condition $(\exists w \in W)\big(A(u,w) \;\&\; \neg R^-(u,w)\big)$ need not imply $(\exists w \in W)\big(A(u,w) \;\&\; R^+(u,w)\big)$. Thus, it may happen that the proposition $\neg \boldsymbol{P}_A \cap \neg \boldsymbol{F}_A$ is nonempty.

The above remarks give rise to a formal semantics for the language *Sent*. This semantics, whose construction is analogous to that presented in the previous section, determines a logical system in which the closure principle is annulled. More specifically, a *model* for *Sent* is now defined to be a quintuple

$$\boldsymbol{Mo} = (W, R^+, R^-, V_0, V_1), \tag{4.4.5}$$

where (W, R^+, R^-) is a space, V_0 is a mapping which to each sentential variable assigns a subset of W, and V_1 is a mapping which to each action variable assigns a binary relation on W.

Let $\boldsymbol{DL}^+$ be the logic which is defined by Detachment, the axioms (4.3.4)–(4.3.7) of $\boldsymbol{DL}$, and the axiom

$$\boldsymbol{P}\alpha \to \neg \boldsymbol{F}\alpha. \tag{4.4.6}$$

The above semantics adequately characterizes the logic $\boldsymbol{DL}^+$.

Theorem 4.4.2 (The Completeness Theorem) *Let α be a sentence of Sent. α is deducible in $\boldsymbol{DL}^+$ if and only if it is true in every model.*

Proof We shall omit the simple check that the logic $\boldsymbol{DL}^+$ is sound in every model (4.4.5). The crucial part of the proof is the construction of the canonical model for the logic $\boldsymbol{DL}^+$.

The *canonical model* of the logic $\boldsymbol{DL}^+$ is the structure

$$\boldsymbol{Mo}(\boldsymbol{DL}^+) := (W(\boldsymbol{DL}^+), R^+, R^-, V_0, V_1), \tag{4.4.7}$$

where $W(\boldsymbol{DL}^+)$ is the set of all maximal consistent sets of $\boldsymbol{DL}^+$, R^+ and R^- are the relations on $W(\boldsymbol{DL}^+)$ defined by the conditions:

$$R^+(\nabla, \nabla') \quad \text{if and only if} \quad \{\boldsymbol{O}\alpha : \boldsymbol{O}\alpha \in \nabla\} \subseteq \nabla', \tag{4.4.8}$$

$$R^-(\nabla, \nabla') \quad \text{if and only if} \quad \{\boldsymbol{F}\alpha : \boldsymbol{O}\alpha \in \nabla\} \cap \nabla' \neq \emptyset, \tag{4.4.9}$$

respectively.

V_0 is the mapping which to each sentential variable p assigns the set $\{\nabla \in W(\boldsymbol{DL}^+) : p \in \nabla\}$, and V_1 is the mapping assigning to each action variable α the following atomic action $A(\alpha)$ on $W(\boldsymbol{DL}^+)$:

$$A(\alpha)(\nabla, \nabla') \quad \text{if and only if either} \quad \text{(i)}\ R^+(\nabla, \nabla')\ \&\ \boldsymbol{O}\alpha \in \nabla'\ \&\ \boldsymbol{P}\alpha \in \nabla \tag{4.4.10}$$
$$\text{or} \quad \text{(ii)}\ R^-(\nabla, \nabla')\ \&\ \boldsymbol{O}\alpha \in \nabla'\ \&\ \boldsymbol{F}\alpha \in \nabla.$$

The proof of the following lemma can easily be established:

Lemma 4.4.3

(i) R^+ *is reflexive.*
(ii) $R^+ \cap R^- = \emptyset$.

Let α *be an action variable and* ∇ *a maximal set. Then:*

(iii) There exists a maximal set ∇' *such that* $R^+(\nabla, \nabla')$ *and* $\boldsymbol{O}\alpha \in \nabla'$.
(iv) If $\boldsymbol{F}\alpha \in \nabla$ *then there exists a maximal set* ∇' *such that* $R^-(\nabla, \nabla')$ *and* $\boldsymbol{O}\alpha \in \nabla'$. □

The key property of the canonical model is that for any sentence ϕ, and any maximal set ∇,

$$\boldsymbol{Mo}(\boldsymbol{DL}^+) \models_\nabla \phi \quad \text{if and only if} \quad \phi \in \nabla. \tag{4.4.11}$$

This is proved by induction on the length of ϕ. The following lemma makes it possible to show that

$$\boldsymbol{Mo}(\boldsymbol{DL}^+) \models_\nabla \boldsymbol{O}\alpha \quad \text{if and only if} \quad \boldsymbol{O}\alpha \in \nabla,$$
$$\boldsymbol{Mo}(\boldsymbol{DL}^+) \models_\nabla \boldsymbol{P}\alpha \quad \text{if and only if} \quad \boldsymbol{P}\alpha \in \nabla,$$
$$\boldsymbol{Mo}(\boldsymbol{DL}^+) \models_\nabla \boldsymbol{F}\alpha \quad \text{if and only if} \quad \boldsymbol{F}\alpha \in \nabla.$$

Lemma 4.4.4 *Let* ∇ *be a maximal set and* α *an action variable. Then:*

$\boldsymbol{O}\alpha \in \nabla$ *if and only if*

$$(\forall\nabla')\big(R^+(\nabla,\nabla') \Rightarrow A(\alpha)(\nabla,\nabla')\big) \;\&\; \neg(\exists\nabla')\big(R^-(\nabla,\nabla') \;\&\; A(\alpha)(\nabla,\nabla')\big); \tag{4.4.12}$$

$\boldsymbol{P}\alpha \in \nabla$ *if and only if* $(\exists\nabla')\big(R^+(\nabla,\nabla') \;\&\; A(\alpha)(\nabla,\nabla')\big)$; (4.4.13)

$\boldsymbol{F}\alpha \in \nabla$ *if and only if*

$$(\exists\nabla')\big(A(\alpha)(\nabla,\nabla')\big) \;\&\; (\forall\nabla')\big(A(\alpha)(\nabla,\nabla') \Rightarrow R^-(\nabla,\nabla')\big). \tag{4.4.14}$$

Proof The right-hand sides (RHS, for short) of (4.4.12)–(4.4.14) are paraphrases of conditions (i)–(iii) of Definition 4.4.1. (In (4.4.12) the condition $(\exists\nabla')\big(R^+(\nabla,\nabla')\big)$ is dropped since it follows from the reflexivity of R^+.)

(4.4.12). ($\Rightarrow$). Assume $\boldsymbol{O}\alpha \in \nabla$. Then, clearly $\boldsymbol{P}\alpha \in \nabla$. To prove the first conjunct of RHS of (4.4.12), assume $R^+(\nabla,\nabla')$. Then, $\boldsymbol{O}\alpha \in \nabla'$. So $R^+(\nabla,\nabla')$ & $\boldsymbol{O}\alpha \in \nabla'$ & $\boldsymbol{P} \in \nabla$. Thus, (4.4.10)(i) holds proving $A(\alpha)(\nabla,\nabla')$.

To prove the second conjunct on RHS of (4.4.12), suppose $R^-(\nabla,\nabla')$ & $A(\alpha)(\nabla,\nabla')$ holds for some ∇'. This implies that of the disjuncts (4.4.12)(i)–(ii) characterizing $A(\alpha)$, only (4.4.12)(ii) is true. So $\boldsymbol{F}\alpha \in \nabla$, which implies $\neg\boldsymbol{O}\alpha \in \nabla$. A contradiction.

($\Leftarrow$). Assume RHS of (4.4.12). Then, $(\forall\nabla')(R^+(\nabla,\nabla') \Rightarrow A(\alpha)(\nabla,\nabla'))$ is true. But it follows from the definition of $A(\alpha)$ that the sentence $(\forall\nabla')(A(\alpha)(\nabla,\nabla') \Rightarrow \boldsymbol{O}\alpha \in \nabla')$ is true. So $(\forall\nabla')(R^+(\nabla,\nabla') \Rightarrow \boldsymbol{O}\alpha \in \nabla')$ is true. Since $R^+(\nabla,\nabla)$, we thus have that $\boldsymbol{O}\alpha \in \nabla$.

(4.4.13). ($\Rightarrow$). Assume $\boldsymbol{P}\alpha \in \nabla$. By Lemma 4.4.3.(iii), there exists a maximal set ∇' such that $R^+(\nabla,\nabla')$ and $\boldsymbol{O}\alpha \in \nabla'$. So (4.4.10)(i) is true, which gives that $R^+(\nabla,\nabla')$ & $A(\alpha)(\nabla,\nabla')$.

($\Leftarrow$). Suppose $R^+(\nabla,\nabla')$ & $A(\alpha)(\nabla,\nabla')$ holds for some ∇'. This implies that of the clauses (4.4.10)(i)–(ii) characterizing $A(\alpha)$ only (4.4.10)(i) is true. So $\boldsymbol{P}\alpha \in \nabla$.

(4.4.14). ($\Rightarrow$). Assume $\boldsymbol{F}\alpha \in \nabla$, and suppose $A(\alpha)(\nabla,\nabla')$ holds. Clause (4.4.10)(i) cannot be true for it would imply that $\boldsymbol{P}\alpha \in \nabla$, which, in turn, would contradict the fact that $\boldsymbol{F}\alpha \in \nabla$. So (4.4.10)(ii) is true. Thus, $R^-(\nabla,\nabla')$ holds. It remains to show that $(\exists\nabla')(A(\alpha)(\nabla,\nabla'))$. But this is a direct consequence of Lemma 4.4.3.(iv).

($\Leftarrow$). Assume RHS.

Claim $(\forall\nabla')\big(A(\alpha)(\nabla,\nabla') \Rightarrow (9)(ii)\big)$.

Proof of the claim. Suppose that $A(\alpha)(\nabla,\nabla')$ & (4.4.10)(i) holds for some ∇'. Since (4.4.10)(i) is true, we have $R^+(\nabla,\nabla')$ and so it is not the case that $R^-(\nabla,\nabla')$. This contradicts RHS, proving the claim. □

Now, by RHS, we have that $A(\alpha)(\nabla,\nabla')$ holds for some ∇'. Hence, by Claim, we obtain that (4.4.10)(ii) is true, i.e., $R^-(\nabla,\nabla')$ & $\boldsymbol{O}\alpha \in \nabla'$ & $\boldsymbol{F}\alpha \in \nabla'$ for some ∇'. So $\boldsymbol{F}\alpha \in \nabla$.

This completes the proof of Lemma 4.4.4, proving at the same time formula (4.4.11). □

The Completeness Theorem, thus, follows from (4.4.11) and the fact that the intersection of all maximal sets of $\boldsymbol{DL}^+$ gives exactly the set of $\boldsymbol{DL}^+$ theorems. □

Remarks 4.4.5

(1). The atomic actions $A(\alpha)$ in the atomic model (4.4.7) are very modestly tailored—the intersection $A(\alpha) \cap (R^+)^c \cap (R^-)^c$ is the empty relation. In consequence, the conjunction of the negation of the permission of the action α and the negation of its prohibition reduces to the following equivalence:

$$\boldsymbol{Mo}(\boldsymbol{DL}^+) \models_\nabla \neg\boldsymbol{P}\alpha \wedge \neg\boldsymbol{F}\alpha \quad \text{if and only if} \quad \neg(\exists\nabla')\big(A(\alpha)(\nabla, \nabla')\big),$$

which may seem to be a somewhat trivial condition. We can prevent this from happening, though, by changing the definition of $A(\alpha)$, in the following way: the right-hand side of (4.4.10) is supplemented with the third disjunct, viz.,

$$\text{not } R^+(\nabla, \nabla') \ \&\ \text{not } R^-(\nabla, \nabla') \ \&\ \boldsymbol{O}\alpha \in \nabla' \ \&\ \neg\boldsymbol{P}\alpha \in \nabla \ \&\ \neg\boldsymbol{F}\alpha \in \nabla. \quad (4.4.15)$$

The thus defined action $A(\alpha)$ also satisfies the equivalences (4.4.12)–(4.4.14). The condition $\boldsymbol{Mo}(\boldsymbol{DL}^+) \models_\nabla \neg\boldsymbol{P}\alpha \wedge \neg\boldsymbol{F}\alpha$ is then fairly nontrivial.

(2). The strong completeness theorem, formulated below, can also be proved by means of a slight modification of the above proof. (The details are left to the reader.)

Theorem 4.4.6 (Strong Completeness Theorem) *Let Γ be a set of sentences and ϕ any sentence. Then, ϕ is deducible from Γ on the ground of* $\mathbf{DL}^+$ *if and only if Γ semantically entails ϕ (i.e., for any model (4.4.11), if every sentence of Γ is true in this model, then so is ϕ.)* □

4.5 Compound Actions and Deontology

The analytical theory of obligation presented thus far relates to atomic actions. Now let us expand the theory to encompass compound actions. This will allow us to include structural dependences occurring between the algebraic properties of compound actions and metalogical properties of deontic operators.

This section is concerned with the issue of defining for a given compound action $\mathbf{A}$ (over an elementary action system $\boldsymbol{M} = (W, R, \mathcal{A})$), the values of the deontic operators for $\mathbf{A}$. We shall begin with the operators $\boldsymbol{P}$ and $\boldsymbol{F}$.

Suppose that a compound action $\mathbf{A}$ is a singleton, which means that it consists of only one finite sequence of atomic actions, $\mathbf{A} = \{A_1 A_2 \ldots A_n\}$. Let u be a state. Intuitively, $\mathbf{A}$ is permitted in u if the relation R enables the agent(s) to perform the string of consecutive actions $A_1, A_2, \ldots, A_n$ commencing with the state u. Formally, $\mathbf{A}$ is permitted in u if there exists a nonempty string $u_1, \ldots, u_n$ of states such that $u\, R\, A_1\, u_1 \ldots u_{n-1}\, R\, A_n\, u_n$. We extend this definition to arbitrary compound actions as follows.

Definition 4.5.1 Let $\boldsymbol{M} = (W, R, \mathcal{A})$ be an elementary action system. Let $\mathbf{A}$ be a nonempty compound action for $\boldsymbol{M}$ and u a state of W. The action $\mathbf{A}$ is *permitted* in the state u, symbolically:

(a) $u \in \boldsymbol{P}\mathbf{A}$, if and only if there exist a nonempty string $u_1, \ldots, u_n$ of states and a sequence $A_1 A_2 \ldots A_n \in \mathbf{A}$ such that $u\, R A_1\, u_1 \ldots u_{n-1}\, R A_n\, u_n$.

(In other words, $u \in \boldsymbol{P}\mathbf{A} \;\Leftrightarrow_{df}\; (\exists w \in W)\; u\, Res\mathbf{A}\, w$.)

For the empty compound action $\boldsymbol{\emptyset}$ it is assumed that $\boldsymbol{P\emptyset} = \emptyset$. □

Thus, **A** is permitted in u if and only if for *at least one* sequence $A_1 A_2 \ldots A_n \in \mathbf{A}$, the action $\{A_1 A_2 \ldots A_n\}$ is permitted in u; equivalently, **A** is performable in u (see Definition 1.7.3.(i)). This is a weak form of permission. For example, drug use is permitted in this sense as some drugs, such as alcohol or cannabis, are permitted. (We abstract here from the precise definition of a narcotic.) A stronger definition would require that for *all* sequences $A_1 A_2 \ldots A_n \in \mathbf{A}$, the action $\{A_1 A_2 \ldots A_n\}$ is permitted in u. This option is not investigated here.

We want the closure principle to be rendered in the object language. The following scheme seems to be the best candidate: $\boldsymbol{F}\mathbf{A} = \neg \boldsymbol{P}\mathbf{A}$ for every compound action **A**. Accordingly, we arrive at the following definition:

Definition 4.5.2 Let $\boldsymbol{M} = (W, R, \mathcal{A})$ be an elementary action system. Let **A** be a nonempty compound action for $\boldsymbol{M}$ and u a state of W. The action **A** is forbidden in the state u, symbolically:

(b) $u \in \boldsymbol{F}\mathbf{A}$, if and only if there *does not* exist a nonempty string $u_1, \ldots, u_n$ of states and a sequence $A_1 A_2 \ldots A_n \in \mathbf{A}$ such that $u\, R A_1\, u_1 \ldots u_{n-1}\, R A_n\, u_n$.

(Equivalently, $u \in \boldsymbol{F}\mathrm{A} \;\Leftrightarrow_{df}\; \neg(\exists w \in W)\; u\, Res\mathbf{A}\, w$.) □

In other words, **A** is forbidden in u if and only if, for *every string* $A_1 A_2 \ldots A_n \in \mathbf{A}$, the action $\{A_1 A_2 \ldots A_n\}$ is forbidden in u. So, homicide is forbidden if and only if all its forms are forbidden.

It follows from Definition 4.5.2 that for the empty action $\boldsymbol{\emptyset}$, we have that $\boldsymbol{F\emptyset} = W$, i.e., $\boldsymbol{\emptyset}$ is forbidden everywhere. It also follows from the above two definitions that any compound action **A** is not permitted in every terminal state and hence it is forbidden in all terminal states.

We now turn to the issue of defining, for a given compound action **A** (over an elementary action system $\boldsymbol{M} = (W, R, \mathcal{A})$), the propositions $\boldsymbol{O}\mathbf{A}$, '**A** is obligatory.' We first assume that **A** consists of a single sequence of atomic actions, $\mathbf{A} = \{A_1 A_2 \ldots A_n\}$. Let u be a state. Intuitively, **A** is obligatory in u if and only if *at least one* atomic action occurring in the sequence of consecutive actions $A_1, A_2, \ldots, A_n$ becomes obligatory whenever one starts performing this string in the state u. Formally, there exists at least one sequence of states $u_1, \ldots, u_n$ such that $u\, R A_1\, u_1 \ldots u_{n-1}\, R A_n\, u_n$ *and* for every sequence $u_1, \ldots, u_n$ of states such that $u R A_1\, u_1 \ldots u_{n-1}\, R A_n\, u_n$ it is the case that for at least one i $(1 \leq i \leq n)$, the atomic action A_i is obligatory in u_{i-1}. (We adopt that $u_0 = u$.) The first conjunct of the above definition guarantees that the obligatoriness of **A** does not trivialize., i.e., the action $\{A_1 A_2 \ldots A_n\}$ is actually permitted in u. This is a weak form of obligatoriness. It is not assumed that all actions $A_1, A_2, \ldots, A_n$ are obligatory when one performs them starting at u.

The above remarks are encapsulated in the following definition:

Definition 4.5.3 Let $\boldsymbol{M} = (W, R, \mathcal{A})$ be an elementary action system. Let $\mathbf{A}$ be a nonempty compound action for $\boldsymbol{M}$ and u a state of W.

(c) If $\mathbf{A}$ is a singleton, $\mathbf{A} = \{A_1\, A_2 \ldots A_n\}$, then
$u \in \boldsymbol{O}\mathbf{A}$ if and only if $u \in \boldsymbol{P}\mathbf{A}$ *and* for every sequence $u_1, \ldots, u_n$ of states such that $u\, R\, A_1\, u_1 \ldots u_{n-1}\, R\, A_n\, u_n$ it is the case that for at least one i $(1 \leq i < n)$, the atomic action A_i is obligatory in u_{i-1}. (Here $u_0 = u$.)

(d) Generally, for an arbitrary $\mathbf{A}$,
$u \in \boldsymbol{O}\mathbf{A}$ if and only if $u \in \boldsymbol{P}\mathbf{A}$ and for every sequence $A_1 \ldots A_n \in \mathbf{A}$, if the action $\{A_1\, A_2 \ldots A_n\}$ is permitted in u then it is obligatory in u in the sense of (c).
For the empty compound action $\boldsymbol{\emptyset}$ it is assumed that $\boldsymbol{O}\boldsymbol{\emptyset} := \emptyset$. □

If $\mathbf{A}$ is an atomic action—more precisely, $\mathbf{A} = \{\langle A\rangle\}$ for some atomic action A—then the above definition gets reduced to the definition of the proposition $\boldsymbol{O}A$, as given in the formula (4.2.10) of Sect. 4.2. As $u \notin \boldsymbol{O}\mathbf{A}$ for any nonterminal state, $\mathbf{A}$ is not obligatory in such terminal states, which seems to be natural.

Here is a list of simple observations about the propositions $\boldsymbol{O}\mathbf{A}$ and $\boldsymbol{P}\mathbf{A}$.

Proposition 4.5.4 *Let* $\mathbf{A}$ *and* $\mathbf{B}$ *be nonempty compound actions.*

(1) $\boldsymbol{P}(\mathbf{A} \cup \mathbf{B}) = \boldsymbol{P}\mathbf{A} \cup \boldsymbol{P}\mathbf{B}$ *and* $\boldsymbol{P}(\mathbf{A} \cap \mathbf{B}) \subseteq \boldsymbol{P}\mathbf{A} \cap \boldsymbol{P}\mathbf{B}$.
(2) $\boldsymbol{O}\mathbf{A} \subseteq \boldsymbol{P}\mathbf{A}$.
(3) $\boldsymbol{O}(\mathbf{A} \cup \mathbf{B}) = \boldsymbol{O}\mathbf{A} \cup \boldsymbol{O}\mathbf{B}$ *and* $\boldsymbol{O}(\mathbf{A} \cap \mathbf{B}) \subseteq \boldsymbol{O}\mathbf{A} \cap \boldsymbol{O}\mathbf{B}$.
(4) If $\mathbf{A} \subseteq \mathbf{B}$, *then* $\boldsymbol{P}\mathbf{A} \cap \boldsymbol{O}\mathbf{B} = \boldsymbol{O}\mathbf{A}$.
(5) $\boldsymbol{P}(\mathbf{A} \circ \mathbf{B}) \subseteq \boldsymbol{P}\mathbf{A}$ *and* $\boldsymbol{O}\mathbf{A} \cap \boldsymbol{P}(\mathbf{A} \circ \mathbf{B}) \subseteq \boldsymbol{O}(\mathbf{A} \circ \boldsymbol{B})$.

The proofs are simple and are omitted. □

Note One may also consider other meanings of the operator $\boldsymbol{O}$ than that provided by Definition 4.5.3. For instance, one may adopt the following definition. Let $\mathbf{A}$ be a fixed compound action and u an arbitrary state of W. Suppose that there exists a path

$$u\, R\, u_1\, R\, u_2\, R \ldots R\, u_n, \tag{4.5.1}$$

for some states $u_1, \ldots, u_n$ $(n \geqslant 1)$, starting with the state u. There are several ways (if any) of accomplishing the direct transition (4.5.1)—there may exist atomic actions $A_1, \ldots, A_n$ such that $uA_1\, u_1 A_2 u_2 \ldots A_n u_n$. The string $A_1 A_2 \ldots A_n$ need not belong to $\mathbf{A}$. The fact that $\mathbf{A}$ is obligatory in u means that for every transition (4.5.1) there exist a number m, $1 \leq m \leq n$, and atomic actions $A_1, \ldots, A_m$ such that (i) $A_1 \ldots A_m \in \mathbf{A}$ and (ii) $u\, A_1\, u_1\, A_2 \ldots A_m\, u_m$. In other words, each possible exit from the state u must be initiated by a certain performance of the compound action $\mathbf{A}$, which will lead the system to a state u_m for some $m \leq n$. After reaching the state u_m, it is possible, most probably, to perform actions, in this state, which differ from $\mathbf{A}$, and which will, in turn, lead to the state u_n. It is vital to note here that a certain nonempty initial segment of (4.5.1) must be accomplished by means of the action $\mathbf{A}$.

This meaning of $\boldsymbol{O}$ is not discussed in this book. □

We shall define in this section the semantics of the *basic deontic logic of compound regular actions*. The logic is denoted by ***DLREG***. Its syntax is divided into the following four categories.

The category *SV* of sentential variables: $p_0, p_1, \ldots$
The category *AAV* of atomic action variables: $\alpha_0, \alpha_1, \ldots$
The category *RAT* of regular action terms
The category *SENT* of sentences.

SV and *AAV* are assumed to be primitive, disjoint categories. The category *RAT* is defined as follows:

(ri) $\underline{\emptyset}$ and $\underline{\varepsilon}$ are regular compound action terms ($\underline{\emptyset}$ and $\underline{\varepsilon}$ represents the empty composite action and the action $\boldsymbol{\varepsilon}$, respectively);
(rii) For every $n \geqslant 1$ and $\alpha_1, \ldots, \alpha_n \in AAV$, the expression $\{\alpha_1 \ldots \alpha_n\}$ is a regular compound action term;
(riii) If $\boldsymbol{\alpha}$ and $\boldsymbol{\beta}$ belong to *RAT*, then so do $\boldsymbol{\alpha} \cup \boldsymbol{\beta}$ and $\boldsymbol{\alpha} \circ \boldsymbol{\beta}$;
(riv) If $\boldsymbol{\alpha}$ belongs to *RAT*, then so does $\boldsymbol{\alpha}^*$.
(rv) Nothing else is a regular compound action term.

The category *SENT* of sentences is defined as follows:

(si) $SV \subseteq SENT$, i.e., every sentential variable is a sentence;
(sii) If $\boldsymbol{\alpha}$ is a regular compound action term, then $\boldsymbol{O\alpha}$, $\boldsymbol{P\alpha}$ and $\boldsymbol{F\alpha}$ are sentences;
(siii) If ϕ and ψ are sentences, then $\phi \wedge \psi$, $\phi \vee \psi$, and $\neg\phi$ are sentences.

A *model for SENT* is a quintuple

$$\boldsymbol{Mo} = (W, R, V_0, V_1, \boldsymbol{V}_2),$$

where

(i) (W, R) is a discrete system
(ii) V_1 is a mapping assigning to each atomic action variable α an atomic action $V_1(\alpha) \subseteq W \times W$
(iii) $\boldsymbol{V}_2$ is a mapping which assigns to every compound action term a compound action over $\{V_1(\alpha_n) : n \in \mathbb{N}\}$ in the following way:

$$\boldsymbol{V}_2(\underline{\emptyset}) = \boldsymbol{\emptyset} \text{ and } \boldsymbol{V}_2(\underline{\emptyset}) = \boldsymbol{\varepsilon},$$
$$\boldsymbol{V}_2(\{\alpha_1 \ldots \alpha_n\}) = \{\langle V_1(\alpha_1), \ldots, V_1(\alpha_n)\rangle\}$$

(i.e., $\boldsymbol{V}_2(\{\alpha_1 \ldots \alpha_n\})$ is a singleton whose only element is the n-tuple of relations $\langle V_1(\alpha_1), \ldots, V_1(\alpha_n)\rangle$. We recall that $\boldsymbol{\emptyset}$ is the empty set and $\boldsymbol{\varepsilon} = \{\varepsilon\}$.)

$$\boldsymbol{V}_2(\boldsymbol{\alpha} \cup \boldsymbol{\beta}) = \boldsymbol{V}_2(\boldsymbol{\alpha}) \cup \boldsymbol{V}_2(\boldsymbol{\beta})$$

(= the union of the compound actions $\boldsymbol{V}_2(\boldsymbol{\alpha})$ and $\boldsymbol{V}_2(\boldsymbol{\beta})$.)

$$\boldsymbol{V}_2(\boldsymbol{\alpha} \circ \boldsymbol{\beta}) = \boldsymbol{V}_2(\boldsymbol{\alpha}) \circ \boldsymbol{V}_2(\boldsymbol{\beta})$$

(= the composition of the compound actions $\boldsymbol{V}_2(\boldsymbol{\alpha})$ and $\boldsymbol{V}_2(\boldsymbol{\beta})$.)

$$\boldsymbol{V}_2(\boldsymbol{\alpha}^*) = (\boldsymbol{V}_2(\boldsymbol{\alpha}))^*$$

(= the iterative closure of the compound action $\boldsymbol{V}_2(\boldsymbol{\alpha})$.)
(The above clauses imply that $\boldsymbol{V}_2(\{\alpha_1 \ldots \alpha_n\}) = \boldsymbol{V}_2(\{\alpha_1\}) \circ \ldots \circ \boldsymbol{V}_2(\{\alpha_n\})$.)

V_0 is a mapping assigning to each sentential variable $p \in SV$ a subset $V_0(p) \subseteq W$. V_0 is then extended onto the entire set *SENT* by assuming that:

$$V_0(\boldsymbol{O\alpha}) := \boldsymbol{O}(\boldsymbol{V}_2(\boldsymbol{\alpha})) \text{ for any compound action term } \boldsymbol{\alpha} \tag{4.5.2}$$

(The right-hand side of 4.5.2 is defined as in Definition 4.5.3.)

$$V_0(\boldsymbol{P\alpha}) := \boldsymbol{P}(\boldsymbol{V}_2(\boldsymbol{\alpha})) \tag{4.5.3}$$

$$\begin{aligned} V_0(\boldsymbol{F}\alpha) &:= \boldsymbol{F}(\boldsymbol{V}_2(\alpha)) \\ V_0(\phi \wedge \psi) &:= V_0(\phi) \cap V_0(\psi) \\ V_0(\phi \vee \psi) &:= V_0(\phi) \cup V_0(\psi) \\ V_0(\neg\phi) &:= W \setminus V_0(\phi) \end{aligned} \tag{4.5.4}$$

(Note that

$$\begin{aligned} V_0(\phi \to \psi) &= (W \setminus V_0(\phi)) \cup V_0(\psi) \\ V_0(\phi \leftrightarrow \psi) &= V_0(\phi \to \psi) \cap V_0(\psi \to \phi).) \end{aligned}$$

A sentence ϕ is *true at u* in the model ***Mo***, in symbols:

$$\boldsymbol{Mo} \models_u \phi,$$

if $u \in V_0(\phi)$.

ϕ is *true in* ***Mo*** if and only if ϕ is true at every $u \in W$.

ϕ is *tautological* if it is true in every model ***Mo***.

Problem Axiomatize the set of all tautological sentences of ***DLREG***.

4.6 Righteous Systems of Atomic Norms

For a family N of atomic norms for an action system $\boldsymbol{M} = (W, R, \boldsymbol{\mathcal{A}})$, we let $\mathrm{N}_{\boldsymbol{P}}$ ($\mathrm{N}_{\boldsymbol{F}}$, $\mathrm{N}_{\boldsymbol{O}}$, respectively) denote the subfamily of N consisting of all permissive norms (prohibitive norms, obligatory norms, respectively).

A set N of atomic norms for $\boldsymbol{M}$ is said to be *righteous* for the system $\boldsymbol{M}$ if the following three conditions are satisfied:

(i)$_{\boldsymbol{P}}$ for every permissive norm $(\Phi, A, +) \in \mathrm{N}_{\boldsymbol{P}}$ and every $u \in \Phi$ there exists a state $w \in \delta_A(u)$ such that uRw.

(i)$_{\boldsymbol{F}}$ for every prohibitive norm $(\Phi, A, -) \in \mathrm{N}_{\boldsymbol{F}}$ and every state $u \in \Phi$ there does *not* exist a state w such that $u\,A\,w$ and $u\,R\,w$.

(i)$_{\boldsymbol{O}}$ for every obligatory norm $(\Phi, A, !) \in \mathrm{N}_{\boldsymbol{O}}$ and for every state $u \in \Phi$, $\emptyset \neq \delta_R(u) \subseteq \delta_A(u)$.

(i)$_{\boldsymbol{P}}$ says that for any permissive norm $(\Phi, A, +) \in \mathrm{N}_{\boldsymbol{P}}$ the action A is performable in every state $u \in \Phi$, and (i)$_{\boldsymbol{F}}$ says that for every prohibitive norm $(\Phi, A, -) \in \mathrm{N}_{\boldsymbol{F}}$ the action A is totally unperformable in any state $u \in \Phi$. In turn, (i)$_{\boldsymbol{O}}$ says that A is obligatory in any state $u \in \Phi$.

The relation R determines the 'freedom sphere' of the system, for R determines which actions are permitted and which are not. Righteous norms imposed on the action system $\boldsymbol{M}$ should not violate the limits of this freedom sphere. What permissive norms allow should not be less restricted than what the relation R admits, and in an analogous way, what norms prohibit should not be more restricted than what R excludes. In short, permissive norms cannot be more tolerant or liberal than R, and prohibitive norms—more restrictive than R. On the other hand, the 'obligatoriness' of a norm arises from the compulsion imposed by R.

The following proposition directly follows from the above remarks.

Proposition 4.6.1 *Let* N *be a family of atomic norms for* $\boldsymbol{M} = (W, R, \boldsymbol{\mathcal{A}})$. N *is righteous for* $\boldsymbol{M}$ *if and only if*

(ii)$_{\boldsymbol{P}}$ *for every* $(\Phi, A, +) \in \mathrm{N}_{\boldsymbol{P}}$, *the proposition* $\Phi \to \boldsymbol{P}A$ *is equal to* W;
(ii)$_{\boldsymbol{F}}$ *for every* $(\Phi, A, -) \in \mathrm{N}_{\boldsymbol{F}}$, *the proposition* $\Phi \to \boldsymbol{F}A$ *is equal to* W;
(ii)$_{\boldsymbol{O}}$ *for every* $(\Phi, A, !) \in \mathrm{N}_{\boldsymbol{O}}$, *the proposition* $\Phi \to \boldsymbol{O}A$ *is equal to* W. □

For every action system $\boldsymbol{M} = (W, R, \boldsymbol{\mathcal{A}})$ there exist sets $\mathrm{N}_{\boldsymbol{P}}$, $\mathrm{N}_{\boldsymbol{F}}$ and $\mathrm{N}_{\boldsymbol{O}}$ of permissive, prohibitive and obligatory atomic norms such that the family $\mathrm{N} := \mathrm{N}_{\boldsymbol{P}} \cup \mathrm{N}_{\boldsymbol{F}} \cup \mathrm{N}_{\boldsymbol{O}}$ is righteous for $\boldsymbol{M}$. For example, as $\mathrm{N}_{\boldsymbol{P}}$, $\mathrm{N}_{\boldsymbol{F}}$ and $\mathrm{N}_{\boldsymbol{O}}$ one can take the following sets of norms:

$$\mathrm{N}_{\boldsymbol{P}} := \{(\Phi, A, +) : (\forall u \in \Phi)(\exists w \in W)\big(A(u, w)\ \&\ R(u, w)\big)\},$$
$$\mathrm{N}_{\boldsymbol{F}} := \{(\Phi, A, -) : (\forall u \in \Phi)\big(\neg(\exists w \in W)(A(u, w)\ \&\ R(u, w))\big)\},$$
$$\mathrm{N}_{\boldsymbol{O}} := \{(\Phi, A, !) : (\forall u \in \Phi)(\exists w \in W)\big(R(u, w)\ \&\ (\forall u \in \Phi)(\forall w \in W)(R(u, w) \Rightarrow A(u, w))\big)\},$$

Remark If the figure $(\Phi, A, +)$ is regarded as a strongly permissive norm, read as "Φ, therefore A is strongly permitted," then the proposition corresponding to it is equivalent to $\Phi \to \boldsymbol{P}_S A$. In this case we can also speak of systems of norms righteous in the strong sense. This means, in particular, that for every positive norm $(\Phi, A, +) \in \mathrm{N}$ the action A is totally performable in every state $u \in \Phi$. Equivalently, N is righteous in the strong sense if it is righteous and $\Phi \subseteq \boldsymbol{P}_S A$ for every positive norm $(\Phi, A, +) \in \mathrm{N}$, $\Phi \subseteq \boldsymbol{F}_S A$ for every negative norm $(\Phi, A, -) \in \mathrm{N}$, and $\Phi \subseteq \boldsymbol{O}_S A$ for every obligatory norm $(\Phi, A, !) \in \mathrm{N}$. □

Norms righteous in one model need not be righteous in another. In order to illustrate this, suppose that $\boldsymbol{M} = (W, R, \mathcal{A})$ and $\boldsymbol{M}' = (W, R', \mathcal{A})$ are action systems with the same sets of states and atomic actions. The models differ merely with relations of a direct transition. Each set of interpreted (i.e., semantic) norms N for $\boldsymbol{M}$, is also a set of norms for $\boldsymbol{M}'$ and *vice versa*. The set N may be righteous for $\boldsymbol{M}$, but not necessarily for $\boldsymbol{M}'$. Conversely, N may be righteous for $\boldsymbol{M}'$, but this may be not the case for the system of $\boldsymbol{M}$.

In this context the question arises whether a given set N of norms for an action system $\boldsymbol{M}$ can determine, in a conceivable way, the relation R of direct transition in the system. In order to facilitate the discussion we shall confine ourselves to families of permissive norms only.

A set $\mathrm{N}_{\boldsymbol{P}}$ of permissive atomic norms for $\boldsymbol{M} = (W, R, \mathcal{A})$ is said to be *adequate (for* $\boldsymbol{M}$*)* if and only if it is righteous for $\boldsymbol{M}$ and for every pair $(u, w) \in R$ there exists a norm $(\Phi, A, +) \in \mathrm{N}_{\boldsymbol{P}}$ such that $u \in \Phi$ and $u\,A\,w$.

The adequacy of a family $\mathrm{N}_{\boldsymbol{P}}$ of atomic norms for $\boldsymbol{M}$ thus ascertains that every direct transition $u\,R\,w$ is in the range of a suitable norm of $\mathrm{N}_{\boldsymbol{P}}$, i.e., the transition $u R w$ is accomplished by an action that is permitted by a norm of $\mathrm{N}_{\boldsymbol{P}}$. It immediately follows from the definition of adequacy that the inclusion $R \subseteq \bigcup \mathcal{A}$ is a necessary condition for a set of norms to be adequate for $\boldsymbol{M}$.

Proposition 4.6.2 *For every action system* $\boldsymbol{M} = (W, R, \mathcal{A})$ *such that* $R \subseteq \bigcup \mathcal{A}$ *there exists a family* $\mathrm{N}_{\boldsymbol{P}}$ *of permissive atomic norms which is adequate for* $\boldsymbol{M}$.

Proof For each $A \in \mathcal{A}$ define

$$\Phi_A := \{u \in W : (\exists w \in W)(R(u, w)\ \&\ A(u, w))\}.$$

Let

$$\mathrm{N}_{\boldsymbol{P}} := \{(\Phi_A, A, +) : A \in \mathcal{A}\}.$$

$\mathrm{N}_{\boldsymbol{P}}$ is righteous for $\boldsymbol{M}$. Let $(u, w) \in R$. As $R \subseteq \bigcup \mathcal{A}$, there exists an action $A \in \mathcal{A}$ such that $(u, w) \in A$. Thus, $u \in \Phi_A$ and so $\mathrm{N}_{\boldsymbol{P}}$ is adequate for $\boldsymbol{M}$. □

Proposition 4.6.3 *Let* $\boldsymbol{M}_i = (W, R_i, \mathcal{A})$, $i = 1, 2$, *be two elementary action systems with the same sets* W *of states and* $\mathcal{A}$ *of atomic actions. Let* $\mathrm{N}_{\boldsymbol{P}}$ *be a family of permissive atomic norms (for both systems). If* $\mathrm{N}_{\boldsymbol{P}}$ *is adequate for* $\boldsymbol{M}_1$ *and* $\boldsymbol{M}_2$, *then the relations* R_1 *and* R_2 *satisfy the following condition:*

$$(\forall u \in W)[(\exists w \in W)\ R_1(u, w) \Leftrightarrow (\exists w \in W)\ R_2(u, w)].$$

Proof Let $u \in W$ and suppose $R_1(u, w)$ for some w. By the adequacy of $\mathrm{N}_{\boldsymbol{P}}$ for $\boldsymbol{M}_1$ there exists a norm $(\Phi, A, +) \in \mathrm{N}_{\boldsymbol{P}}$ such that $u \in \Phi$ and $u\,A\,w$. As $\mathrm{N}_{\boldsymbol{P}}$ is righteous for $\boldsymbol{M}_2$, there exists a state $w' \in f_A(u)$ such that $R_2(u, w')$ holds. By a symmetric argument one shows that $(\exists w \in W)\ R_2(u, w)$ implies $(\exists w \in W)\ R_1(u, w)$, for all $u \in W$. □

The conclusion which can be drawn from the above proposition is that a set $\mathrm{N}_{\boldsymbol{P}}$ of norms, adequate for $\boldsymbol{M}$, need not unambiguously determine the relation R of direct transition in $\boldsymbol{M}$. A given permissive norm $(\Phi, A, +)$ allows, in an arbitrary state $u \in \Phi$, for the performance of the action A; the norm, however, does not *precisely* define the effect of this action. The norm only requires that it has to be an element of $f_A(u)$.

We recall the formal language defined in Sect. 4.3. Any triple of the form

$$(\phi, \alpha, +), \quad (\phi, \alpha, -), \quad (\phi, \alpha, !) \tag{4.6.1}$$

where[4] $\phi \in Sent$ and α is an action variable, is a permissive, prohibitive and obligatory atomic norms in the *syntactic* form, respectively. They are interpreted in elementary action systems $\boldsymbol{M} = (W, R, \boldsymbol{\mathcal{A}})$ as triples

$$(\Phi, A, +), \quad (\Phi, A, -) \ \text{ or } \ (\Phi, A, !) \tag{4.6.2}$$

respectively, where A is an atomic action of $\boldsymbol{\mathcal{A}}$ and Φ is an elementary proposition, i.e., $\Phi \subseteq W$. (Strictly speaking, formal norms (4.6.1) are interpreted in $\boldsymbol{M}$ by means of a pair of mappings (V_0, V_1), where V_0 assigns to each sentential variable a subset of W and V_1 assigns to each action variable an atomic action in $\boldsymbol{\mathcal{A}}$—see Sect. 4.3.)

We may consider sets N of atomic norms (4.6.1) and ask about being righteous on the syntactic level; that is, asking whether various interpretations (4.6.2) of the norms of N lead to righteous sets in the semantic sense, as expounded above. Another obvious question is about consistency of a set of formal norms. How are we to understand consistency in this context? Intuitively, two formal norms $(\phi, \alpha, +)$ and $(\phi, \alpha, -)$, (where the same sentence ϕ and the same action variable α occur in both norms), are inconsistent—the same action in identical circumstances is permitted and forbidden. The solution we propose is to define consistency of norms in terms of normative sentences associated with norms. Accordingly, the following definition of consistency of norms is adopted. Given a set N of formal norms, we define:

$$\begin{aligned} \mathrm{N}_s := &\{\phi \to \boldsymbol{P}\alpha : (\phi, \alpha, +) \in \mathrm{N}\} \cup \\ &\{\phi \to \boldsymbol{F}\alpha : (\phi, \alpha, -) \in \mathrm{N}\} \cup \{\phi \to \boldsymbol{O}\alpha : (\phi, \alpha, !) \in \mathrm{N}\}. \end{aligned}$$

---

[4] The obligatory norm $(\phi, \alpha, !)$ may be also interpreted as the command "ϕ, therefore perform α!".

Postulate of Consistency of Elementary Norms:

A set of formal norms N *is consistent if and only if the corresponding set* N_s *of norm sentences is consistent in the logical system* ***DL*** *defined in Sect.* 4.3. □

If an action system $\boldsymbol{M} = (W, R, \boldsymbol{\mathcal{A}})$ provides a righteous interpretation for a set N of formal norms (by means of a pair of mappings (V_0, V_1)), then N is consistent (because this 'righteous' interpretation validates the sentences of N_s at every state $u \in W$).

Consistency is an aspect which characterizes 'good' sets of formal norms. There are other aspects that come to light:

(1) the incompatibility of two obligatory norms with the same hypothesis Φ occurs when the actions required by them cannot be jointly performed in Φ.
(2) praxiological incompatibility occurs when the implementation of the action of one norm annihilates the results of the action of the other norm.

The notion of a righteous interpretation of a system of formal norms excludes such pathologies from sets of atomic norms.

Every righteous set of formal norms is consistent but not vice versa—consistency is a weaker notion. The consistency of a set N of formal norms is equivalent to the fact that there is an interpretation (V_0, V_1) of *Sent* in an action system $\boldsymbol{M}$ validating the set N_s of normative sentences corresponding to N in *some* state of the model. In turn, Proposition 4.6.1 implies that a set N of formal norms is righteous if and only if there is an interpretation of *Sent* in an action system $\boldsymbol{M}$ validating the set N_s in *every* state of the model (because $V_0(\psi) = W$ for every normative sentence ψ of N_s.)

The above remarks may give rise to a theory of justice based on the semantic notion of a righteous set of norms. But the meaning of the term 'justice,' inherent in this approach, relativizes it to a model of action: a set of formal norms is righteous *with respect* to a model $\boldsymbol{M}$.

Definition 4.6.4 Let N be a righteous set of (semantic, i.e., interpreted) atomic norms for an action system $\boldsymbol{M} = (W, R, \boldsymbol{\mathcal{A}})$ and let

$$u_0, u_1, u_2, \ldots, u_{n-2}, u_{n-1}, u_n \tag{4.6.3}$$

be a finite[5] nonempty sequence of states of W. We say that (4.6.3) *complies* with the rules of N if for every i $(0 \leq i \leq n-1)$:

$(!)_i$ for every obligatory norm $(\Phi, A, !) \in$ N, if $u_i \in \Phi$, then $u_i \, A \, u_{i+1}$,
$(+)_i$ there is a permissive norm $(\Phi, A, +) \in$ N such that $u_i \in \Phi$ and $u_i \, A \, u_{i+1}$,
$(-)_i$ there is no prohibitive norm $(\Phi, A, -) \in$ N such that $u_i \in \Phi$ and $u_i \, A \, u_{i+1}$.

□

[5]Infinite runs of states (in both directions) also make sense. Some remarks on infinite runs in situational action systems are presented in Sect. 2.2.

The sequence (4.6.3) is therefore a result of performing a string of atomic actions

$$A_1 \ldots A_{n-1} A_n,$$

each action being a component of some norm of N, so that $u_0\, A_1\, u_1\, A_2\, u_2 \ldots u_{n-2} A_{n-1}\, u_{n-1} A_n\, u_n$ is an operation of $\boldsymbol{M}$ and the obligatory norms in N have priority. This string of actions is not unambigously defined but its elements occur in the norms of N. Since N is righteous, the operation $u_0\, A_1\, u_1\, A_2\, u_2 \ldots u_{n-2}\, A_{n-1}\, u_{n-1}\, A_n\, u_n$ is realizable.

Conditions $(!)_i$ and $(+)_i$ are not contradictory. For instance, if $u_0, u_1, u_2, \ldots, u_{n-2}, u_{n-1}, u_n$ represent 'states' of a car while driving it (cf. Example 4.2.4), where each state is determined by the street location of the car and its velocity and the norms enforcing two actions—speed limit A and right-hand traffic B—are obviously obligatory. (This system is situational—time matters here, but we shall disregard the situational envelope.) Therefore $(!)_i$ holds for these two obligatory norms for *every* state u_i $(0 \leq i \leq n-1)$. In other words, the driver performs these two actions (jointly!) at every state u_i. (There are other obligatory actions as (e.g.) the car must stop at red traffic lights. But this action may be obligatory in some points u_i.) These two norms do not exclude the possibility of performing other permitted action at some u_i, such as turning right. The norms of N may also forbid turning left at some u_i.

Definition 4.6.4 formulates concisely the factual principles of conduct which complies with norms. The above comments allow for the building a theory of justice and liberty supported on the notion of a righteous system of norms (it will not matter here whether they are specifically formally or informally). The meanings of the terms 'justice' and 'freedom' are not of an absolute character but are relativized to a definite model. One can speak about a righteous (or just) system of norms inside the given model or from the point of view of the model. Different models allow different interpretations of the same set of formal norms. As mentioned above, norms that are righteous in one system will not be just in another one. By analogy, we can say that in the case of norms of law, the term 'righteous' bears a different meaning to a follower of Mohammad, who is subjected to the norms of sharia, and another one to a citizen of Europe, where laws stemming from the ancient Rome are in force. Islam does not separate the secular life from the religious one and regulates religious customs and the organization of religious authority, as well as the everyday life of a Moslem. Islam operates by means of a different deontology which derives from a blending of religious law and morality. Sharia is based on the assumption that law is meant to provide all norms necessary for the spiritual and physical development of the individual. Moslem's actions are divided into five categories—necessary, glorious, permitted, reprehensible, and prohibited. (It should be noted that there are various schools of law whose interpretations of sharia differ: the Hanafi, the Maliki, the Shafi'i, or the highly conservative Hanbali school.)

A penal law includes a set of norms of a deontological character—orders and prohibitions, the violation of which is punishable with a sanction. It makes use of basically two types of norms: sanctioned and sanctioning. A sanctioned norm orders, for example, punishment for committing a theft. A sanctioning norm imposes on the court the duty of inflicting a suitable penalty in compliance with the code. Therefore, sanctioning norms define various sanctions as a consequence of violating a sanctioned norm.

Inasmuch as in different places different people behave in a similar way and commit similar crimes (e.g., they steal), it is differences between individual penal codes that are clear-cut if one takes into account the following two aspects: the range of sanctioned norms determined through the enumeration of punishable acts; and the size of sanctions. The latter determines differences in sanctioning norms. Certain norms of sharia, which contain such sanctions as corporal punishment, the amputations of hands, and stoning, are regarded as alien to contemporary western legal systems. On the other hand, European law has not worked out uniform norms relating, for example, to the ritual killing of animals. The following picture, taken from the Internet, is very instructive, because it ideographically depicts sanctioned and sanctioning norms in a nutshell:

Fig. 4.4 Picture on the internet: podroze.onet.pl

Systems of norms that regulate various aspects of collective and individual life are dynamic quantities changing over time. The direction and the character of the changes are determined by different factors and their analysis depends more on sociology and social psychology rather than on logic. Various groups, such as lobbies, opinion-forming environments, newspapers, the television, the Internet, trade unions, business circles, and the authorities of different levels and power ranges exert a primary influence on the evolution of systems of norms, especially legal ones. The order of things is generally such that groups bring about a change of the legal *status quo*. An example here is the ultimately successful pressure to abolish the capital punishment in Poland at the end of the 1980s. A second example is provided by current discussions in the press concerning the legal recognition of homosexual relationships. As regards the latter, there are a number of arguments for and against being brought forward. It is claimed that a change in the law will make it more just overall, as it will not ignore issues relating to the life and well-being of minority groups. Some liberal lawyers and economists (Friedrich Hayek, amongst others) reject the notion of *social justice* as impossible to define and argue that attempts lead to overinterpretation, these being driven different pressure groups whose goal is to realize their particular interests.

The evolution of a system of norms leads to the abandonment of certain old forms and to the working out and implementation of a new legal framework. In consequence, systems of social and individual action become altered (obviously they are not atomic systems of action in view of their degree of complexity). In such systems, new norms will be righteous, inasmuch as they are actually abided by. (Abiding by norms is, of course, something regulated by norms of a higher order, which are related to, e.g., the functioning of the judicature, the police, or the army.)

In this discussion of righteous systems of norms, one aspect should be underlined. The accepted definition relativizes justice to a model of action: a righteous system of norms is one that is compliant with the model of action. The norms which are binding in the model precisely reflect the range of liberty resulting from the construction (structure) of the model; in particular the norms do not violate the limitations imposed on by the transition relation between states. Actual systems of action, which people of flesh and blood are entangled in, are multi-rung, often hierarchized, structures, with a complicated knot of mutual links between parts of the structures. They are clearly not atomic systems. One may advance the thesis, however, that the basic components of the actual system which has been formally reconstructed can be identified with complexes of atomic systems in which, additionally, the situational envelope has been taken into account. In general, in real systems of action, there is no concordance between the factual functioning of the system and the accepted norms as regulators of activity: norms are simply violated. The theory above relates to an ideal situation which is characterized by a full harmonizing of norms and actions.

The second question deals with the *sense* of justice. Norms—at least the fundamental ones included in constitutions and legal codes of states—are derived from supreme values adhered to by given communities. According to the Judeo-Christian tradition, these basic norms should not violate the commandments of the Decalogue. The Age of Enlightenment brought new suggestions for supreme principles; for example, Kant's categorical imperative, the conception of social contract and the ideas of freedom and egalitarianism. Every conscious human being has some norms of conduct instilled in their childhoods, ones that shape their sense of justice and the sensitivity to right and wrong. These principles can change over time as the forms of social life in which individuals find themselves change.

In 2013, the press reported on the drastic form which the evictions of tenants in Spain took. These evictions were the consequence of the strict mortgage law which in force since 1909. In recent years half a million people have been turned out onto the street, often because of just one or two overdue monthly mortgage repayments.

The inability to pay typically resulted from a loss of work, whereupon the immediate repayment of the whole debt was ordered or usurious interest rates were introduced. People who had for years been repaying their mortgage were suddenly deprived of the right to remain where they lived. According to the press, this was an instance of the execution of anachronistic legal norms that departed from a contemporary sense of justice. Following the decision of the European Court of Justice concerning one case of this type, the Spanish government decided to change the law. The action is pending.

Looking at the issue more deeply, one can say that from the legal perspective of *the beginning of the 20th century*, the system of mortgage law which has been in force until today, is *just*: the law was introduced in compliance with civilized procedures; the law is respected; the activities of banks and tenants are in agreement with the letter of law; and there are no violations of it. Everyone who applies for a credit is familiar with the regulations of the law and knows what documents they are signing. The essence of the matter lies in the fact that these norms do not take into account the change in the dominant form of residential ownership which occurred during the century. Properties used to be largely rented with the option of buying them just for the well-off. Nowadays properties are more readily purchased in the real estate market. Owning one's own property is common in Spain and in many other countries.

The above remarks situate the problem of justice on two levels; the first level is constituted by the action system a given set of sanctioned norms refers to, the other level is occupied by the close-coupled action system whose functioning is determined by the totality of sanctioning norms (more on sanctioned and sanctioning forms in the next paragraph). Roughly, the first one concerns various legal aspects of the social life; the other regulates the work of justice. Both the systems may be unjust. In the discussed example, the two systems regulate various mortgage issues. Mortgage law, together with complimentary regulations of various banks, is a peculiar melange of principles (or norms) regulating the mode of granting loans and of sanctioning norms, i.e., norms that impose various forms of control (and punishments) for defiance of mortgage repayments. The heart of the matter is the lack of equilibrium between the two groups of norms, viz. the sanctioned and sanctioning norms. The regulations which settle the principles of granting loans were, with time, stretched in a natural manner to cover a broad spectrum of potential (and often actual) clients who earn their living by doing contract work. On the other hand, the sanctioning regulations have not been altered; they are anachronistic and do not take into account the specific character of the economic situation of many contemporary clients of banks. It can be said that we have come to deal here with a situation where certain norms which are adjusted to the system of procedures, i.e., the obligatory norm ordering to make immediate payment of an installment of the granted credit, together with the coupled norm which imposes defined sanctions, are too restrictive—the loan-taker is not able to perform the ordered action, having lost his or her job. (The system of actions is created here by all of the potential clients who submit loan applications.)

The inadequacy of the binding legal system in relation to actions (the model) actually taken by citizens consists in not making provision, by the relevant regulations, for states where purchasers of flats lose the so-called credit repayment capacity (are unable to repay the loans), e.g., in consequence of their losing work and means of obtaining income. (It is indeed hardly possible to imagine that legal norms in the capitalist reality should guarantee full and stable employment that is unlimited over time). Yet, from the perspective of the theory of action systems it is vital that mortgage law should take into account cases of bitter random instances affecting loan-takers, and, consequently, the obligatory norms of this law should admit more flexible forms of mortgage repayment schedules. (Because otherwise, the obligatory action of repayments of installments may be unperformable in the aforementioned states.) The essence of the matter lies—on the one hand—in transition to a new model of action (one of a more complex set of states and more sophisticated legal actions), on the other one—in the acceptance of more subtle procedures of granting and repaying credits, formulated within a new and fair system of norms for this particular system.

4.7 More About Norms

The conditions in which a given atomic norm, i.e., any triple of one of the following forms

$$(\Phi, A, +), \quad (\Phi, A, -), \quad (\Phi, A, !), \tag{4.7.1}$$

is applicable—in other words the range of a norm—are fully determined by the elementary proposition Φ. If the system is in a state u belonging to Φ, the norm $(\Phi, A, +)$ allows the agent of A to perform this action in the state u. Analogously, the norm $(\Phi, A, -)$ forbids him to perform A in u, and $(\Phi, A, !)$ commands him to perform A in u.

In modern civilized societies, it is a principle that the validity of a norm is established by a sovereign decision taken by the legislature. Democratic legislation determines the conditions and the detailed procedures of inclusion and exclusion of norms into and from the system of law in force. There is the commonly accepted principle that a norm does not hold forwards or backwards in force. Not holding forwards means that the norm is not applied either before the date of its taking effect or—in the case of an annulment of the norm—after the date of its exclusion from the legal system. Not holding backwards means that the legalized norm does not cover actions taken before the date of its legalization. (A norm may however be introduced at a time but be decreed to apply from a time earlier than the time of introduction. In this case, though, civilized legislature requires that the intention to introduce a norm with an earlier date of effect should be made known to whom it will be applicable *prior to* this date.) The principles above, likewise the Closure Principle, ought to be treated here as meta-norms that regulate the functioning of norms within the system of law. The presence of these principles causes atomic norms to be integrated into a determined situational envelope each time we follow the effectiveness of a norm. One of the elements of the situational envelope of an elementary norm is the time parameter which determines whether the norm is valid or has been canceled by the legislator's decision. The second element is the presence of the legislator, the author and the guardian of the norm, equipped with sanctions enforcing respect of the norm. The third and last element is the presence of the defined addressee of the norm. Certain behaviors are expected from the addressee of the norm in the conditions defined in the assumptions of the norm. We assume that the addressee of an elementary norm is identical with the agent of the action specified in the consequent of the norm.

The addressee can be defined, for instance, through reference to a certain general or common name. A good example here is the norm which determines the principles of the use of arms by police officers. The word 'police officer' is, obviously, a general name, yet the addressee of the norm is not a 'police officer' in general, i.e., a certain notion, but the designate of this name, that is a functionary of the police forces, each time mentioned by their name and surname, say, Kowalski or Smith, who find themselves in the situation (state) that is defined in the antecedent of the norm, thus, e.g., in a given place at a given time, chasing a criminal. In the case of obligatory norms $(\Phi, A, !)$ the addressee is to perform action A which is mentioned in the consequent as long as the antecedent is satisfied. On the other hand, when the norm has the form $(\Phi, A, -)$, it is expected that the addressee will not undertake action A in the conditions determined by Φ. (The fact that the agent of action A may never do this action in the circumstances described by Φ is not paradoxical. In a game of chess, White may never move a given figure, e.g. the king. This does not change the fact he is still the *agent* of this action. This remark signals the need for an in-depth analysis of the essence of the relation 'agent—action.'

The notion of an atomic norm distinguishes—in the elementary formula of the norm—the antecedent or *assumption* (also referred to as the hypothesis of the norm), which defines the conditions in which the norm applies, and also the consequent, also referred to as the *disposition* of the norm, specifying the determined actions related to this norm, as well as determining the status of this action: obligatory or forbidden, or permissive. The disposition of the norm also marks out each time the addressee of this norm through a situational framing. The disposition determines the form of the addressee's behavior in the conditions assumed by the norm; if the norm is obligatory, the addressee *has to* carry out the action contained in the disposition in the circumstances in which the assumption of the norm is satisfied. We have not introduced here an additional division into general and individual norms due to the fact whether the addressee is defined as a class of people or as an individual. These questions are settled by the form of the elementary system of acting, to which the given norm is fitted, and the situational surrounding of the system. The latter determines the agents of the actions, and consequently the addressees of the norm. We have also omitted here the problem of classifying norms by the rank the norm has in hierarchical structure of laws and the branch of law to which the norm belongs.

As mentioned, jurisprudence introduces conjugate norms, viz. sanctioned and sanctioning norms. The former determines the manner in which the addressee should behave in the conditions set out by the hypothesis (assumption) of the norm. The sanctioning norm, conjugate with the sanctioned one, defines the type of behavior on the part of the legislator in the case when the addressee of the sanctioned norm does not abide by it. The principle 'nullum crimen sine lege' is in force here. The area of our interest here solely covers norms occupying the lowest level of the hierarchy of the law. The assumption can be made that each obligatory or forbidden atomic norm is a sanctioned one. The question arises whether and when the sanctioning norm interrelated with the latter is atomic as well. Let us assume that $N := (\Phi, A, !)$ is an obligatory norm for an elementary action system $\boldsymbol{M} = (W, R, \boldsymbol{\mathcal{A}})$. The norm N regulates the behavior of the agent (agents) of the action A in the system $\boldsymbol{M}$, in the conditions determined by the proposition Φ. The norm N^*, interrelated with N, regulates the behavior of the legislator (or an organ empowered by the legislator) *toward* the addressee of the norm N. The specific character of the interrelated norm manifests itself in the fact that such a norm commands the punishment of the addressee of the sanctioning norm N, when, in the conditions determined by the hypothesis of the norm N, the addressee of the norm has or has not performed an improper action specified in the disposition of the norm. For instance, N^* can command to that a driver be imprisoned if the latter has not given first aid to an injured passenger. (We assume, of course, that the driver himself/herself has not suffered in the accident.) The norm itself (N) commands that the aid be given to the passenger. The assumption of the norm N can be fully specified. Let us agree to accept, then, the cautious thesis that sanctioning norms may be—in many cases—treated as atomic norms. The sanctioning norm N^* *is not*, however, the atomic norm for the system $\boldsymbol{M}$. The system of action to which the norm N^* is adjusted, differs fundamentally from the system $\boldsymbol{M}$. The space of states is here defined by determining the scale of violations of the norm N and the gradation of penalties associated with this scale. Let us suppose a simple framework in which penalties have the following format:

(n) "The addressee of the norm N is punished with n years' imprisonment.",

where n is a small positive natural number. The legislator performs one action: inflicting the punishment. This action can be identified in the above model with the set consisting of pairs of states $(*_N, n_N)$, where $*_N$ is the state described by the sentence stating that the addressee of the norm N has violated this norm and n_N is the state in which the addressee of the norm N is punished with n years' imprisonment according with (n). (In civilized criminal codes there is the principle that the punishment be commensurable with the crime, with the duration of the sentence reflecting the type and magnitude of the crime.) If we accept that in the output elementary system $\boldsymbol{M}$, the family $\boldsymbol{\mathcal{A}}$ of atomic actions includes only one action, the action A, specified in the norm N, then the above-defined action system, on the set of states of the form $*_N$ and n_N, with n ranging over a finite set of consecutive positive natural numbers, can be called a system *interconnected* with the system $\boldsymbol{M}$. The interconnected norm N^* is thus adjusted to the interconnected system; this causes the norm N^* to be reduced to an atomic one in many cases.

We shall now consider norms with a more complex structure. These norms, apart from the notion of a state of a system, take into account (to a very modest extent), certain elements of the situational envelope of the system. More precisely, only one purely situational aspect of action is considered here. This is the atomic action directly preceding the performance of the current action. The notion of a possible situation, therefore, is reduced to the pair

$$(u, A), \tag{4.7.2}$$

where u is a state of the system and A is an arbitrary atomic action. (4.7.2) is read:

(1)* 'u is a state of the system and this state is the result of performing the action A.'

The norms which will be discussed here in brief will also be adjusted to the type of situation above. These norms will comply with the rules of transformation of situations of type (4.7.2). The norms being considered here are quadruples of the form $(A, \Phi, B, +)$, or $(A, \Phi, B, -)$, or $(A, \Phi, B, !)$, where A and B are atomic actions of some fixed action system $\boldsymbol{M} = (W, R, \boldsymbol{\mathcal{A}})$ and Φ is an elementary proposition, i.e., Φ is a subset of W.

The norm

$$(A, \Phi, B, +) \tag{4.7.3}$$

is read:

$(2)_*^+$ 'It is the case that Φ after performing the action A, therefore the action B is permitted',

or in short: 'Φ after performing the action A, therefore the action B is permitted.' The norms of the above shape will be called *positive* (alias *permissive*).

Quadruples of the form

$$(A, \Phi, B, -) \tag{4.7.4}$$

are called *negative* (or *prohibitive*) norms. (4.7.4) reads:

$(2)_*^-$ 'Φ after performing the action A, therefore the action B is prohibited.'

In turn, the quadruple

$$(A, \Phi, B, !) \tag{4.7.5}$$

which is called an *obligatory* (or *imperative*) norm, reads:

$(2)_*^!$ 'Φ after performing the action A, therefore the action B is obligatory.'

One may extend the above definition of a norm by allowing additional situational parameters in it; in particular one may take into account a proposition Ψ referring not to the present state of the system itself but to a course of previous actions on $\boldsymbol{M}$. This proposition, which would enter into the norm, may represent, for example, what sequences of actions have already been undertaken by the agents or how many times particular actions were performed. This additional factor has to be taken into consideration, for example, in the obligatory norms that govern castling in a game of chess.

The range of the norms (4.7.1) depends only on two factors:

(i) what action was performed last;

(ii) the extent of knowledge about the actual state of the system represented by the proposition Φ.

The moment the factors (i) and (ii) are established, a positive norm allows the performance of an action from $\mathcal{A}$, while a negative norm forbids the performance of a certain action from $\mathcal{A}$. Because of factor (i), triples (4.7.3)–(4.7.5) may be regarded as norms with a limited 'memory.' The memory is not associated with a history of the system itself (and so the sequence of states through which the system passes) but requires only recognizing the recently undertaken action.

As with elementary norms, (4.7.3) does not prejudge the possibility of performing action B when the antecedent of the norm is satisfied. If action A has been performed and, just after performing A, it is true that Φ, it may still happen that in each state $u \in \Phi$ the action B is unperformable.

It is possible to refine the shape of norms by taking other, more complex factors constituting the history of the system such as those referring to particular sequences of previously performed actions and their agents or taking into account the order of some or all the actions performed before. A game of chess proceeds according to norms with memory—White's move has to be followed by Black's.

As is well known, an important problem of deontic logic is that of how to properly represent conditional obligations such as *If you smoke, then you ought to use an ashtray*. We do not tackle this problem here. But *dyadic norms* of the form: *B is obligatory, given A* or: *B is forbidden, given A*, where A and B are actions, are obviously well defined. In deontic logic they are marked by $\boldsymbol{O}(B|A)$ and $\boldsymbol{F}(B|A)$,, respectively. Assuming that A and B are atomic actions in some action system $\boldsymbol{M} = (W, R, \mathcal{A})$, we may identify $\boldsymbol{O}(B|A)$ and $\boldsymbol{F}(B|A)$ with $(A, W, B, !)$ and $(A, W, B, -)$, respectively. (Note, however, that smoking a cigar is not an atomic action!) $\boldsymbol{O}(B|A)$ is paraphrased as "After performing the action A, the action B is obligatory." Similarly, $\boldsymbol{F}(B|A)$ is paraphrased as "After performing the action A, the action B is forbidden." Of course, a more adequate rendering would be: "While performing the action A and just after that, the action B is obligatory" or "While performing the action A, the action B is forbidden," but representing of such norms within the here formalism outlined here would require passing to situational action systems endowed with tense components.

Chapter 5
Stit Frames as Action Systems

Abstract Stit semantics gives an account of action from a certain perspective: actions are seen not as operations performed in action systems and yielding new states of affairs, but rather as selections of preexistent trajectories of the system in time. Main problems of stit semantics are recapitulated. The interrelations between stit semantics and the approach based on ordered action systems are discussed more fully.

5.1 Stit Semantics

Action theory has considerably broadened the scope of traditional semantics by way of introducing agents on the stage. Agents design various action plans, perform actions, carry out reasoning about the effects of actions and, last but not least, they bring about the desirable states of the world within which they act.

One of the crucial problems which a theory of action faces is that of the formal representation of actions. It is still an area where there is a good deal of disagreement about fundamental issues and lack of generally-acknowledged mathematical results in central area of concerns. Traditional formal semantics is based on (fragments of) set theory as an organizational method: the majority of semantic concepts (frames, possible worlds, truth-valuations, satisfaction, meaning, etc.) are reconstructible within some parts of contemporary set theory. How then to describe actions? How to model them? The theory of action is still lacking a well-entrenched paradigm. As a result of many attempts undertaken to base the theory on a few plausible principles, many competing formal systems have been developed, each of which is imperfectly applicable to real-life situations in various, say, moral, legal, or praxeological contexts.

It was the work on deontic logic that initiated a deeper reflection on action and provided various formal resources for expressing the idea that certain actions are obligatory, permitted, or forbidden (von Wright)—"that closing the window, for example, is obligatory." However, the dominant line of research allows deontic operators (functors) to apply not to actions directly, but to arbitrary sentences. This means that certain states of affairs are qualified as obligatory, forbidden, or that they ought or ought not to be. It has been argued that the second perspective is more general.

J. Czelakowski, *Freedom and Enforcement in Action*, Trends in Logic 42,
DOI 10.1007/978-94-017-9855-6_5

Thus, *that an agent ought to close the window* is interpreted as (in a presumably logically equivalent way) *that it ought to be that the agent closes the window*. In other words, "the study of what agents ought to do is subsumed under the broader study of what ought to be, and among the things that ought or ought not to be are the actions agents perform or refrain from" (Horthy 2001). The emphasis in deontic logic on the notion of what ought to be has been criticized, e.g., by Geach and von Wright, since it leads to distortions when the existent formalisms are applied to the task of analyzing an agent's "oughts."

In this chapter, we shall outline an approach to action developed mainly by American researchers. This approach is conventionally referred to as the *Pittsburgh School*. (The name derives from the fact that the leading researchers, who made important contributions to action theory, are professionally affiliated to Pittsburgh University. But, of course, this group of researchers is much broader and includes logicians and philosophers from other universities.) In fact, our goal is twofold. First, we shall present the main ideas developed by logicians, computer scientists, and philosophers belonging to the above school in a concise, but undistorted way, thus giving the reader the opportunity of becoming acquainted with their approach to action. On the other hand, we wish to reconstruct the main concepts of this approach within the framework presented in previous chapters. This requires establishing the exact relationship between the two approaches. The main semantic tool in the approach we wish to present is that of a *stit* frame. We advocate here the thesis that stit models are reconstructible as a special class of situational action systems, as defined in Chap. 2. Generally, stit semantics, worked out by Chellas, gives an account of action from a certain perspective: actions are seen here not simply as operations performed in action systems and yielding new states of affairs, but rather as selections of preexistent histories or trajectories of the system in time. Actions available to an agent in a given state (moment) u are simply the cells of a partition of the set of possible histories passing through u. Stit semantics is therefore time oriented and time, being a situational component of action, plays a privileged role in this approach. We shall recapitulate the main principles of stit semantics after Horthy (2001) and relate them to the concepts of action presented here.

One of the crucial problems the theory of action faces is the issue of representing and reasoning about what agents ought to do, a notion that is distinguished from that of what ought to be the case. The analysis of this notion is based on the treatment of action in *stit* semantics. Stit semantics is the best known account of agency in the theory of action. This semantic approach concentrates on the meaning of formal constructions of the form

$$\text{“}a \text{ (an agent) sees to it that } \phi\text{”},$$

usually abbreviated simply as

$$[a\ stit : \phi].$$

These constructions provide a semantic account of various stit operators within a temporal framework in which the future is open or indeterminate. From the set-theoretic perspective, the above constructions are developed within the framework of the theory of trees. As is well-known, this framework permits the introduction of the 'standard' deontic operation $\bigcirc$ with the intended meaning "It ought to be that". It is then reasonable to propose syntactic complexes of the form $\bigcirc[a\ stit : \phi]$—with the intended meaning "It ought to be that a sees to it that ϕ"—as an analysis of the idea that seeing to it that ϕ is something a ought to do. Such an attitude, explicitly expounded, e.g., in Horthy's book (2001), agrees with the thesis which identifies the notion of what an agent ought to do with the notion of what it ought to be that the agent does. The analysis of the notion of what an agent ought to do, proposed by Horthy, is based on "a loose parallel between action in indeterministic time and choice under certainty, as it is studied in decision theory." This requires introducing a certain (quasi) ordering—a kind of dominance ordering—in the study of choice under uncertainty. This preference ordering then defines both the optimal actions which an agent performs and the proposition whose truth the agent should guarantee.

Definition 5.1.1 A *branching system* is a pair of the form $\boldsymbol{W} = (W, \leq)$, where W is a nonempty set and $\leq$ is a tree-like order on W. Thus, the binary relation $\leqslant$ is an order (i.e., $\leq$ is reflexive, transitive, and antisymmetric[1]) with the following *tree-like property*:

$$\text{for any } u, v, w \in W, \text{ if } u \leqslant w \text{ and } v \leqslant w, \text{ then either } u \leqslant v \text{ or } v \leqslant u. \qquad \square$$

The above condition states that for any $w \in W$, the set $\downarrow w = \{u \in W : u \leqslant w\}$ is linearly ordered. The *tree property* is stronger. We recall that a poset $(W, \leqslant)$ with zero $\mathbf{0}$ is a tree if for any $w \in W$, the set $\downarrow w$ is well ordered. As it is customary, the elements of W are called *states*. We shall often interchangeably use the terms: *events*, *moments*, etc. Following the terminology adopted in the theory of trees, the elements of W are also called *nodes*. Intuitively, the fact that $u \leqslant v$ means here that the state u is *equal or earlier* than v. The notation $u < v$ means that $u \leqslant v$ and $u \neq v$. In branching systems, forward branching represents the openness or indeterminacy of the future, and the absence of backward branching represents the determinacy of the past. Branching systems form the indeterministic framework of the theory of branching time, originally presented in Prior's book (1967), and then developed by Thomason (1970, 1984).

We recall that a set of states $h \subseteq W$ is a *chain* (or: h is linearly ordered) if it satisfies the trichotomy postulate, that is, for any states u and v belonging to h, either $u < v$ or $v < u$ or $u = v$. The set h is a maximal chain of states whenever it is linearly ordered and there is no chain g that properly includes h.

The term "tree-like property" may seem to be a bit puzzling. Some branching structures need not look like trees. For example, two linearly ordered sets put vertically in parallel form a branching system. But, of course, we will be mainly concerned with structures, which are depicted as trees.

---

[1] Stit semantics assumes even a weaker condition that $\leqslant$ is a quasi-order. Here we shall confine the discussion to order relations.

Any maximal chain in $(W, \leqslant)$ is called a (possible) *history*. (The existence of histories follow from the Axiom of Choice (AC). The fact that in every poset, each chain can be extended to a maximal chain is equivalent to AC and is referred to as *Kuratowski's Lemma*.) The tree property postulates that for every $w \in W$, the set $\downarrow w := \{u \in W : u \leqslant w\}$ is a chain in $(W, \leqslant)$. It is *not* assumed here that $\downarrow w$ is well-ordered by $\leqslant$. This last, stronger property is assumed in the theory of trees.

Intuitively, each history h represents a complete evolution of the system in time, "one possible way in which things might work out" (Horthy 2001, p. 6). If u is a state and h is a history, the fact that $u \in h$ intuitively means that u occurs at some place of the trajectory h, or that h passes through u. For each $u \in W$, we define

$$H_u := \{h : h \text{ is a history and } u \in h\}.$$

H_u is the set of histories passing through u.

The set of possible worlds accessible at a state $u \in W$ is identified with the set H_u of histories passing through u. The propositions at u are identified with sets of accessible histories, i.e., subsets of H_u.

The crucial role in the stit approach is played by the notion of an index. Each *index* is a pair (u, h) consisting of a state u, together with a history h throughout that state, i.e., $h \in H_u$. Following the notation adopted, e.g., in Horthy's book (2001), each such pair is written as u/h.

Branching systems are a semantic tool that enables one to develop a tense logic of branching time. We shall sketch the main ideas.

Sentences are evaluated with respect to indices. Let $\mathbf{W} = (W, \leqslant)$ be a branching system. A *valuation* (or an *assignment*) is a function V assigning a set of indices $V(x)$ to each sentential variable x. The mapping V is then inductively extended onto the set of all sentences in the following way:

$$V(\phi \wedge \psi) := V(\phi) \cap V(\psi),$$
$$V(\neg\phi) := \text{ the complement of } V(\phi).$$

The language is usually furnished with some other connectives. It is usually assumed that the language contains temporal connectives P and F, as well as the modal connectives $\Diamond$ and $\Box$.

The connective P represents simple past tense and is read "It was the case that." The connective F represents simple future tense and is read "It will be the case that." The intended meanings of the above connectives are provided by the following conditions:

$$V(P\phi) := \{u/h : (\exists u' \in h)\ u' < u \ \ \& \ \ u'/h \in V(\phi)\},$$
$$V(F\phi) := \{u/h : (\exists u' \in h)\ u < u' \ \ \& \ \ u'/h \in V(\phi)\},$$

The above framework makes it possible to fix the meanings of temporal connectives as well as the meanings of the above modal connectives. In this context, the formula $\Box\phi$ means that ϕ is settled, or historically necessary. Intuitively, $\Box\phi$ is true at some state (moment) if ϕ is true at that state no matter how the future turns out. Analogously, $\Diamond\phi$ means that ϕ is still open as a possibility. $\Diamond\phi$ is true at some state (moment) if "there is still some way in which the future might evolve that would lead to the truth of ϕ" (Horthy 2001, p. 9). Formally,

$$V(\Box\phi) := \{u/h : (\forall h' \in H_u)\ u/h' \in V(\phi)\},$$
$$V(\Diamond\phi) := \{u/h : (\exists h' \in H_u)\ u/h' \in V(\phi)\}.$$

The fact that $u/h \in V(\phi)$ is read 'ϕ is satisfied at an index u/h in the branching system $\boldsymbol{W}$' and denoted by $\boldsymbol{W}, u/h \models \phi$, i.e.,

$$W, u/h \models \phi \quad \textit{if and only if} \quad u/h \in V(\phi).$$

The satisfaction relation $\models$ is thus a binary relation between the set of indices and the set of formulas. It is easy to see that the satisfaction relation preserves the validity of classical logic. The pair $(\boldsymbol{W}, \models)$ is then called a *model.*

$|\phi|_u$ is the *proposition* expressed by the sentence ϕ at the state u in the model $(\boldsymbol{W}, \models)$. Thus,

$$|\phi|_u = \{h \in H_u : \boldsymbol{W}, u/h \models \phi\}.$$

Equivalently, $|\phi|_u = \{h \in H_u : u/h \in V(\phi)\}$.

In what follows, however, we will be not much concerned with purely logical aspects of these semantic constructions. Instead, we will present further constructions strictly linked with the theory of action.

Given a branching system $\boldsymbol{W} = (W, \leqslant)$ and $u \in W$, we say that two histories h_1 and h_2 are *undivided at* u if there is a state v such that $u < v$ (i.e., v is properly later than u) and $v \in h_1 \cap h_2$. (Due to the maximality of histories as chains in $(W, \leqslant)$, we also have that $u \in h_1 \cap h_2$.) The fact that histories h_1 and h_2 are undivided at u is pictorially represented by the following diagram:

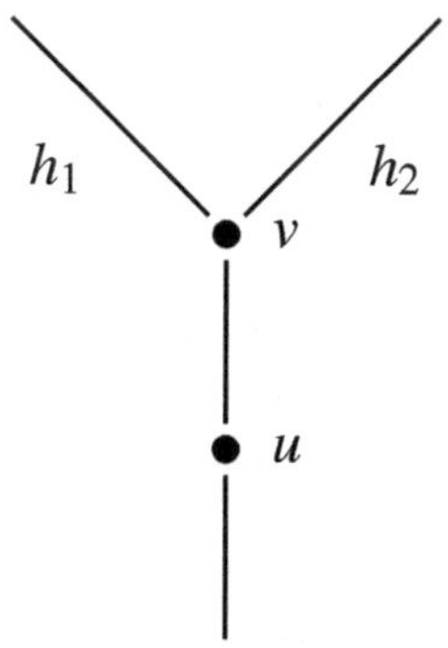

Fig. 5.1

We arrive at the crucial definition for this section:

Definition 5.1.2 A *stit frame* is a structure of the form

$$\boldsymbol{M} = (W, \leqslant, Agent, Choice),$$

where $(W, \leqslant)$ is a branching system, $Agent$ is a nonempty set of agents, and $Choice$ is a function defined on the Cartesian product $W \times Agent$. The function $Choice$ assigns to each agent a and a state u a partition, denoted by $Choice^u_a$, of the set H_u. (We may equivalently say that $Choice^u_a$ is an equivalence relation on the set H_u, for all $u \in W$ and $a \in Agent$.)

The elements of the partition $Choice^u_a$ are called the *actions available to the agent a at u*.

Furthermore, $Choice$ is subject to two requirements of

(1) no choice between undivided histories, and
(2) the independence of agents.

If V is a valuation in the branching system $(W, \leqslant)$, then the extended system

$$M = \langle W, \leqslant, Agent, Choice, V\rangle$$

is called a *stit model*. □

We shall now define the meaning of the above two requirements.

The requirement of no choice between undivided histories says that for each agent a and each state u, *any* two histories, which are undivided at u, belong to the same choice cell of the partition $Choice^u_a$ (equivalently, they belong to the same equivalence class of the equivalence relation corresponding to $Choice^u_a$.

The requirement of the independence of agents says, intuitively, that at any given state, the particular equivalence class that is selected by one agent cannot affect the choices available to another. The formal definition of this condition is more involved. Let u be a state. An *action selection function* at u is a mapping s assigning to each agent a an action available to a at u, i.e., $s(a) \in Choice^u_a$ for every $a \in Agent$. Each such selection function "represents a possible pattern of action, a selection for each agent of some action available to that agent at u" (Horthy 2001, p. 31). We let $Select_u$ denote the set of all selection functions at u. The condition of independence of agents says that, for any state $u \in W$ and for each $s \in Select_u$, the intersection $\bigcap\{s(a) : a \in Agent\}$ is nonempty.

The agent a is said to *perform the action $K \in Choice^u_a$ at the index u/h* just in the case h is a history belonging to K. It makes no sense to say that an agent performs an action at a state u, but only at a state/history pair u/h. The histories belonging to K are called *possible outcomes* of K.

Let $h \in H_u$. There is then a unique equivalence class in the range of $Choice^u_a$ which contains h. This class is denoted by $Choice^u_a(h)$. It represents the particular action performed by the agent at the index u/h.

The stit operator (connective), playing the central role, is referred to as the *Chellas stit* and marked by *cstit*, because it is an analogue of the operator introduced in the sixties by Chellas.

Valuations are extended onto stit sentences in the following way. A statement of the form $[a\ cstit : \phi]$, which expresses the fact that the agent a sees to it that ϕ, is defined as true at an index u/h just in the case the action performed by a at u/h guarantees the truth of ϕ, which means that $Choice_a^u(h) \subseteq |\phi|_u$. Thus,

$$V([a\ cstit : \phi]) = \{u/h : Choice_a^u(h) \subseteq |\phi|_u\}.$$

(There is also another operator closely related to the Chellas stit, marked by *dstit*, and called the *deliberative stit*. The cstit and dstit operators are interdefinable in the presence of the necessity functor—see Horthy (2001, p.16).

The semantic constructions above offer a number of logical schemes involving the necessity operator, the temporal, and stit operators. They are not discussed here. We remark only that various proof systems for logical calculi based on stit semantics or its fragments have been proposed. We mention in these context the names of Xu, Wölfl, Wansing, Herzig, Balbiani et al. (2007) provides a more extensive bibliography of the work done in this area. The above framework for studying agency allows for a natural treatment of the related concepts of individual ability, as distinguished from that of impersonal possibility. Even though it is possible for it to rain tomorrow, no agent has the ability to see it that it will rain tomorrow. The notion of what an agent is able to do is identified with the notion of what it is possible that the agent does, expressed by the modal formula

$$\Diamond[a\ cstit : \phi].$$

Other topics discussed within the framework of the above formalism are those of refraining, group agency and group ability. Kenny's objections to the thesis that the notion of ability cannot be formalized by means of modal concepts and Brown's analysis of agency are also discussed.

5.2 Stit Frames as Action Systems

Stit frames can be viewed as action systems as defined in the previous chapters. Our goal is to reconstruct stit frames within the setting of ordered action systems.

Let

(1) $\boldsymbol{M} = (W, \leqslant, Agent, Choice)$

be a stit frame. We shall assign to (1) a certain ordered (and constructive) action system

(2) $\boldsymbol{M}^* = (W, \leqslant, R, \boldsymbol{\mathcal{A}})$.

(This assignment will not be one-to-one.) $\boldsymbol{M}$ and $\boldsymbol{M}^*$ have the same set of states and the same order relation $\leqslant$. Consequently, $(W, \leqslant)$ is a reduct of both the systems $\boldsymbol{M}$ and $\boldsymbol{M}^*$. The transition relation R between states in the system $\boldsymbol{M}^*$ coincides with the order relation $\leqslant$, i.e., $R = \leqslant$. The definition of elementary actions is more involved.

We recall that for any $u \in W$ and any $a \in Agent$, $Choice_a^u$ is a partition of the set H_u, where H_u is the set of histories passing through u.

Let Sel be a set of functions f defined on the Cartesian product $W \times Agent$ such that, for all $u, v \in W$ and all $a \in Agent$:

(3) $f(u, a) \in Choice_a^u$, i.e., $f(u, a)$ is a partition cell in $Choice_a^u$;
(4) if $u \leqslant v$ then $f(v, a) \subseteq f(u, a)$;

Furthermore, it is assumed that

(5) for every pair $(u, a) \in W \times Agent$ and for every choice cell $K \in Choice_a^u$, there exists a function $f \in Sel$ such that $f(u, a) = K$.

Thus, (5) says that the values of the functions of Sel exhaust the class of all partition cells.

Condition (3) is motivated by the following observation. Suppose $u, v \in W$ are states such that $u \leqslant v$. Then, the tree property for $(W, \leqslant)$ implies that $H_v \subseteq H_u$. Furthermore, the requirement of no choice between undivided histories means that for any $a \in Agent$ and for any choice cells $K \in Choice_a^u$, $L \in Choice_a^v$, $K \cap L \neq \emptyset$ implies that $L \subseteq K$. Consequently, every choice cell $K \in Choice_a^u$ includes the union $\bigcup\{L \in Choice_a^v : K \cap L \neq \emptyset\}$ as a subset. Thus, (4) postulates that if $f(u, a) = K$, then the value $f(v, a)$ belongs to the family $\{L \in Choice_a^v : K \cap L \neq \emptyset\}$. This is a kind of consistency between the actions $f(u, a)$ available for a at u, and the action $f(v, a)$, which is available for a at v.

Given a set Sel of functions, defined as above, and a function $f \in Sel$, we then proceed to define, for each agent $a \in Agent$, a binary relation $A_a^f \subseteq W \times W$. For $u, v \in W$, we put:

(6) $u \; A_a^f \; v$ if and only if $u \leqslant v$ and there exists a history $h \in f(u, a)$ such that $v \in h$.

The relations from the family

$$\boldsymbol{\mathcal{A}}_a := \{A_a^f : f \in Sel\}$$

are called *actions available to the agent* a.

Thus, $u \; A_a^f \; v$ means that by performing the action A_a^f at u, the agent a carries out the system to a later state v, belonging to some history $h \in f(u, a)$. In other words, by performing the action A_a^f at u, the agent a can guarantee that the history to be realized will lie among those belonging to $f(u, a)$. This is in accordance with the "philosophy" of action that underlines stit semantics: an action performed by an agent at u results in choosing a history passing through u.

Let

$$\mathcal{A} := \bigcup\{\mathcal{A}_a : a \in Agent\}.$$

$\mathcal{A}$ is the totality of all actions available to the agents $a \in Agent$. $\mathcal{A}$ is the set of atomic actions of the system (2) we wished to define. This makes the definition of the system (2) complete.

It immediately follows from (6) that every action A_a^f is included in the order $\leqslant$. But since $\leqslant$ is also the transition relation R, it follows that the resulting action system is constructive and normal.

5.3 Ought Operators

The crucial task is to incorporate the deontic operator $\bigcirc$, read as

"It ought to be that",

into the above framework based on branching time. Technically, *Ought* is defined as a function assigning a nonempty subset of H_u to each state (moment) u. The set $Ought_u$ is intuitively thought as containing the *ideal* histories through u, those in which things turn out as they ought to. A sentence of the form $\bigcirc\phi$ is then defined as true at u/h just in case ϕ is true in each of these ideal possibilities, i.e., ϕ is true at u/h' for each history $h' \in Ought_u$. The resulting structure

$$\langle Tree, Agent, Choice, Ought\rangle$$

is called a *standard deontic* stit frame. Deontic systems based on standard stit frames model normative theories that can do no more than classify situations as either ideal or nonideal. The above formalism is then generalized to accommodate a broader range of normative theories, allowing for more than two values. Technically, this is achieved by replacing the primitive *Ought* in the standard deontic stit frames by a function *Value* assigning to each state u a mapping $Value_u$ of the set of histories from H_u into some fixed set of values. Thus, each history h through a state, rather than being classified as ideal or nonideal, is instead assigned a certain value at that moment. These values represent the worth or desirability of histories. The set of values is assumed to be partially ordered. $Value_u(h) \leqslant Value_u(h')$ means that the value assigned to the history h' at u is at least as great as that assigned to h at u. The structure

$$\langle Tree, Agent, Choice, Value\rangle$$

is called a *general deontic* stit frame. This class of frames subsumes standard frames as a special case and enables one to represent more involved normative theories. Of particular concern are *utilitarian* theories in which the value functions are subject to two additional constraints. The first condition says that the range of the function $Value_u$ is a set of real numbers, for all u. The second constraint—*uniformity along histories*—requires that the numbers assigned to a single history do not vary from state to state; formally, $Value_u(h) = Value_{u'}(h)$ for any states u, u' belonging to h. Utilitarian stit frames form, therefore, a subclass of the class of general deontic stit frames. Horthy's book (2001) is concerned with an exposition of the basic conceptual issues related to deontic stit frames, and the focus is on the utilitarian deontic framework.

As expected, general deontic models and utilitarian deontic models determine two deontic logical systems. These logics are both impersonal, offering accounts only of what ought to be, abstracting from any issues of agency. A certain explication of the notion of agency can be obtained by combining the agency operator with the above impersonal account of what ought to be. Technically, this results in a proposal according to which the notion that an agent a ought to see it that ϕ is analyzed through a formula of the form $\bigcirc[a\ cstit : \phi]$. This proposal is referred to as the *Meinong/Chisholm analysis*. Thus, the general idea underlying the Meinong/Chisholm analysis is that what an agent ought to do is identified with what it ought to be that the agent does. A number of objections against the Meinong/Chisholm analysis have been raised, e.g., by Geach, Harman and Humberstone. Their analysis of the problem shows that the above proposal is vulnerable to a simple objection called the *gambling problem* in the literature. In conclusion, as Horthy points out, although the above utilitarian framework carries some plausibility, the Meinong/Chisholm analysis yields incorrect results and therefore must be rejected as a representation of the purely utilitarian notion of what an agent ought to do.

Horthy (2001) writes: "...the general goal of any utilitarian theory is to specify standards for classifying actions as right or wrong; and in its usual formulation, act utilitarianism defines an agent's action in some situation as right just in case the consequences of that action are at least as great in value as those of any of the alternatives open to the agent, and wrong otherwise" (p. 70). In the above analysis of agency, the alternative actions open to an agent a in a given situation are identified with the actions belonging to $Choice_a^u$ for an appropriate u. However, in the framework of indeterminate time, the matter of specifying the consequences of an action presents some difficulties. In decision theory, a situation in which the actions available to an agent lead to various possible outcomes with known probability is described as a case of *risk*, while a situation in which the probability with which the available actions might lead to various outcomes is either unknown or meaningless, is described as a case of *uncertainty*. Horthy is mainly concerned with the case of uncertainty. Consequently, the above framework excludes *expected value act utilitarianism* because the probabilistic information necessary to define expected values of actions is, generally, unavailable. Instead, a theory called *dominance act utilitarianism* is formulated. It is a form of act utilitarianism applicable in the presence of both indeterminism and uncertainty and based on the dominance quasi-order among actions.

Let K and K' be members of $Choice_a^u$. Then K' *weakly dominates* K (in symbols: $K \oslash K'$) whenever, roughly, the results of performing K' are at least as good as those of performing K in every state, and K' *strongly dominates* K (in symbols: $K \otimes K'$) whenever K' weakly dominates K but it is not weakly dominated by K. According to Horthy, an action available to an agent a at a state u is classified as *right* at an index u/h exactly when this action belongs to certain subset $Optimal_a^u \subseteq Choice_a^u(h)$—the set of the *optimal* actions available to that agent; and the action is classified as *wrong* at m/h otherwise.

Although the classification of actions as right or wrong offered by the dominance theory is relativized to full indices—moment/history pairs—the classification does not depend on histories: any action that is classified as right at u/h is also right at u/h' for each history $h' \in H_u$. The classification of actions offered by the dominance theory is also referred to as *state* (or *moment*) *determinate*.

In order to avoid the difficulties inherent in the Meinong/Chisholm analysis, a new two-argument functor $\odot[\ldots$ *cstit* $: \ldots]$ is introduced, allowing for the construction of statements of the form $\odot[a$ *cstit* $: \phi]$ with the intended meaning: 'a ought to see it that ϕ'. The evaluation rule for this functor is rather intricate but the underlying idea is rather straightforward. A sentence ϕ is *safely guaranteed* by an action K available to an agent a whenever the truth of ϕ is guaranteed by K and also by any other action available to a that weakly dominates K. Thus, $\odot[a$ *cstit* $: \phi]$ is true at u/h if and only if, for every action K available to a at u that *does not* guarantee the truth of ϕ, there is another action K' available to a at u that both strongly dominates K and safely guarantees A.

The semantics of the deontic operator $\odot$ simplifies in the case of finite choice situations, where the set of optimal actions available to the agent is finite (and non-empty). The above semantics, based upon a dominance of semantics among actions, is strictly linked with the notion of causal independence. A proposition is *causally independent* of the actions available to a particular agent whenever the truth or falsity of this proposition is guaranteed by a source of causality other than the actions of that agent. The notion of independence is correlated with counterfactuals. We shall not discuss here the formal representation of these correlations.

It is worth noticing that the notion of what an agent ought to do is extended to yield an account of the agent's conditional oughts as well. This requires introducing sets $Choice_a^u/X$ as containing those actions available to a at u that are consistent with $X \subseteq H_u$. Conditional analogues of earlier concepts of weak and strong dominance are also defined. Combining these concepts, one arrives at the notion of conditional optimality. For example, Horthy's book (2001) turns to the task of analyzing statements of the form $\odot[a$ *cstit* $: \phi/\psi]$ expressing what an agent a ought to do under specified background conditions ψ. The notion of conditional optimality is useful in the explication of a version of act utilitarianism different from the theory of dominance act utilitarianism sketched above. This version is referred to as *orthodox act utilitarianism*. Roughly, while the dominance theory associates with any *state* a single classification of the actions available to an agent as right or wrong, the orthodox theory allows an action to be classified as right at some *histories* throughout a state but wrong at others. In order to represent the claim that a ought to see it that ϕ in the orthodox sense, a new operator $\oplus$ representing the orthodox perspective of an agent's oughts is defined. Informally, the statement $\oplus[a\ cstit : \phi]$ is true at u/h if and only if the truth of ϕ is guaranteed by each of the actions available to a that are optimal given the circumstances; that is, the particular state in which a finds himself at this index.

An account governing groups of agents, as well as individuals, and an account governing individual agents viewed as acting in co-operation with the members of a group are also analyzed. The formal machinery enabling one to systematically articulate various groups of normative concepts is developed in Horthy's book. This formal apparatus largely parallels the individual case. The key role is played here by the partition $Choice^u_\Gamma$ representing the actions available to the group Γ at the state u. The statement $[\Gamma\ cstit : \phi]$ means that, through the choice of one or another of its available actions, the group Γ guarantees the truth of ϕ.

Having defined dominance act utilitarianism for both individuals and groups, the relations holding between these two versions are explored. The crucial issues are whether the satisfaction of dominance act utilitarianism by each member of the group entails that the group itself satisfies this theory, and conversely, whether satisfaction by a group entails satisfaction by its individual members. The answer to these two questions in the case of dominance theory is: no—individual and group satisfaction are independent notions. But the orthodox theory can be extended to cover groups as well. In the case of the orthodox version of act utilitarianism, while satisfaction by each individual in a group is again possible even when the group itself violates the theory, satisfaction of the orthodox theory by a group does, in fact, *entail* satisfaction by the individuals the group contains.

The theory of the dominance and orthodox ought operators for individual agents is extended to group agents, so as to allow for sentences of the forms $\odot[\Gamma\ cstit : \phi]$ and $\oplus[\Gamma\ cstit : \phi]$ expressing, respectively, the dominance and orthodox meanings of the statement: "The group Γ ought to see to it that ϕ".

Group deontic operators enable one to express, in a formal way, a number of issues concerning the relations among the oughts of different groups of agents, and also the oughts governing collectives of agents and their individual members. The matter of *inheritance* (both downward and upward) of whether individuals inherit oughts from the groups to which they belong, and conversely, of whether groups inherit oughts from their individual members is an example of such an issue.

The utility of the above framework is illustrated in the context of the discussion on *rule utilitarianism*. According to rule utilitarianism an action available for an individual is classified as right whenever that action is one that the individual would perform in the best overall pattern of action.

The above framework, by placing the discussion in a context that distinguishes explicitly between the orthodox and dominance forms of act utilitarianism, allows us to relate rule utilitarianism to the two forms of act utilitarianism defined above. The tenor of the discussion of these issues is that the conflict between rule utilitarianism and the dominance theory of act utilitarianism is not severe. In situations allowing for multiple optimal patterns of action, these two forms of utilitarianism are consistent.

The above theories, only outlined here, specify the normative concepts governing an agent on the basis of the actions available at a definite *state*, ignoring actions that might be available later on, being (e.g.) a result of certain implemented action plans. To avoid such a distorted picture of the agent's real normative situation, the above theory is also focussed on elaborating a general framework that allows an agent's choices to be "evaluated against the background of later possibilities." This analysis is based on the key technical concept of a *strategy*. This is a partial function σ assigning to each state u in the domain of the strategy a certain action available to the agent at u; formally, σ is a partial function on T such that $\sigma(u) \in Choice_a^u$ for each u in the domain of σ. Considerations are limited to certain classes $Strategy_a^M$ of structurally ideal strategies. $Strategy_a^M$ is the set of strategies for a in a field M that are defined at u, complete in M, and nonredundant. $Strategy_a^M$ is a kind of generalization of $Choice_a^u$. Just as $Choice_a^u$ represents the options available to the agent a at u, $Strategy_a^M$ represents the options available to a throughout the entire field $M \subseteq \{u' : u \leqslant u'\}$. The problem of defining a strategic analogue of the standard cstit operator, say $scstit$, for "strategic Chellas stit," leads, however, to serious semantic difficulties. The reason is that although an index u/h provides enough information to identify the action performed by a, it does not provide enough information to identify the strategy being executed by a, thus not allow us to determine whether or not that strategy guarantees the truth of ϕ.

In order to resolve these difficulties, some ways are mentioned in which these issues might be approached. One of them is similar to the (Ockhamist) approach to the future tense statements adopted in tense logic—since a state (moment) u alone does not provide sufficient information for the evaluation of statements involving *scstit*, this approach recommends that these statements be evaluated rather at *triples* of the form $u/h/J$, where $h \in J$ and $J \subseteq H_u$ than at indices consisting of pairs m/h only. Leaving aside the issue of finding solutions to the strategic *cstit* problem, the above difficulties can be bypassed in the treatment of the *strategic ability* by way of introducing of a fused operator $\Diamond[\ldots\ scstit : \ldots]$. The intuitive meaning of $\Diamond[a\ scstit : \phi]$ is that a has the ability to guarantee the truth of ϕ by carrying out an available strategy. Such statements are evaluated at indices being triples $u/h/M$, where u/h is an ordinary index and M is a field at u. (In fact, the truth or falsity of a strategic ability statement is independent of the h-component of such an extended index.)

The strategic ought operator allows for the construction of statements of the form $\odot[a\ scstit : \phi]$. Such a statement, when evaluated at an extended index $u/h/M$, conveys the meaning that the agent a ought to guarantee the truth of ϕ as a result of his/her actions throughout the field M. The semantics of the strategic ought operator validates many logical principles, analogous of theses of the momentary cstit operators. This issue is not pursued here.

We have presented several models of the deontology of the ought operators. These 'oughtological' models are based on tree-like structures, in which a distinguished situational parameter—the time factor—plays a special role. In fact, they are situational models. Each index $s = u/h$ can be viewed as a situation and u is the state of this situation. A similar remark applies to 'strategic' triples $u/h/M$, where u/h is an ordinary index and M is a field at u.

It seems that at least two ways in which possible modifications of the above theory might be approached. First, the range of the theory might be broadened by considering models founded rather on directed graphs than trees only; that is, the models presented in Parts I and II of this book. These graph-like models are based on two constituents, separating actions from their situational components: the notion of an action system and the notion of a situational envelope of an action system. By appealing to such graph-like models rather than to tree-like structures, it is possible to represent some simple action systems, like a game of chess or a production line in a factory, in an undistorted way. (In the case of a game of chess, instead of moments of time, a more significant role is played by the distributions of pieces on the chessboard and the order in which the two players, White and Black, perform their moves.) This, of course, requires incorporating the concept of a *situation* as a certain set-theoretic entity into the general framework of the theory of action.

Chapter 6
Epistemic Aspects of Action Systems

Abstract The theory of action conventionally distinguishes real actions and doxastic (or epistemic) actions. Real actions (or as we put it—praxeological actions) bring about changes in material objects of the environment external to the agent. Epistemic actions concern mental states of agents—they bring about changes of agents' knowledge or beliefs about the environment as well as about other beliefs. Some logical issues concerning knowledge, action, truth, and the epistemic status of agents are discussed. In this context the frame and ramification problems are also analyzed. The key issue raised in this chapter is that of non-monotonicity of reasoning. A reasoning is non-monotonic if some conclusions are invalidated by adding more knowledge. In this chapter a semantic approach to non-monotonic forms of reasoning, combining them with action theory, is presented. It is based on the notions of a frame and of a (tree-like) rule of conduct for an action system.

6.1 Knowledge Models

6.1.1 Introduction

This book[1] does not cover some in-depth aspects of reasoning about action and change such as the ramification problem and the frame problem. A significant amount of work has been done by many researchers to shed light on these issues from a more formal perspective. The problem of linking non-monotonic reasoning with action deserves special attention. All these problems are inherent to any formalization of dynamic systems. Yet another issue is defeasibility in normative systems and the problem of regulation consistency. (This is different from the simple consistency of the set of norms.) As to the defeasibility problem, we refer here to the work by

[1]The author is deeply indebted to Dr. Matthew Carmody for many insightful comments to this paragraph. Many of them have been taken into account.

J. Czelakowski, *Freedom and Enforcement in Action*, Trends in Logic 42,
DOI 10.1007/978-94-017-9855-6_6

Leon van der Torre (1997, 1999) and his colleagues from the 1990s.[2] The problem of regulation consistency is discussed by Cholvy (1999). Other papers that are relevant are: Castilho et al. (2002), Eiter et al. (2005), Thielscher (1997, 2011), Varzinczak (2006, 2010), Zhang and Foo (2001, 2002), Zhang et al. (2002).

We shall present some preliminary remarks with the intention of better understanding some of these difficult problems from the point of view of the formalism developed in this book. This especially concerns the problem of non-monotonic reasoning, which is examined here in more detail from the semantic and pragmatic perspectives.

The theory of action conventionally distinguishes *real* actions and *doxastic* (or *epistemic*) actions. Real actions (or as we put it—*praxeological* actions) bring about changes in material objects; these changes manifest themselves in changes of "the environment external to the agent." Epistemic actions concern mental states of agents—they bring about changes of agents' knowledge or beliefs "about the environment as well as about other beliefs."[3] As for real actions, there are two main approaches. According to the first approach, actions are as *events* of a certain kind. Actions are done with an intention or for some reason. As Segerberg (1985) writes: "To understand action both are needed, agent and world, intention and change." The other approach views material actions as changes of states of affairs brought about by agents. Actions are described here by their results/outcomes. This view on action and agency is close to modal and dynamic logic. Doxastic actions are often divided into two classes: belief revision and belief update. It is interesting to note that in the study of doxastic actions, the tools borrowed from (multi)modal logic and computer science are extensively used to express epistemic operators. The introduction to Trypuz (2014) gives a brief account of the main philosophical ideas that underlie the theory of action. This book is intended to unify various views on praxeological and epistemic actions. 'Bare' actions are modeled by elementary action systems. By introducing a situational envelope of an elementary system (and hence passing to situational action systems), it is possible to model various issues pertaining to agency, intentionality, belief update, and so on.

---

[2] Van der Torre (1997) writes: "There does not seem to be an agreement in deontic logic literature on the definition of 'defeasible deontic logic'". It is generally accepted that a defeasible deontic logic has to formalize reasoning about the following two issues.

1. *Resolving conflicts*. Defeasibility becomes relevant when there is a (potential) conflict between two obligations. In a defeasible deontic logic a conflict can be *resolved*, because one of the obligations overrides, in some sense, the other one.

2. *Diagnosing violations*. Consider the obligation "normally, you should do p." Now the problem is what to conclude about somebody who does not do p. Is this an exception to the normality claim, or is it a violation of the obligation to do p?"

[3] See Introduction in Trypuz (2014).

6.1.2 Propositional Epistemic Logic

The language of propositional epistemic logic results from augmenting the language of classical propositional logic (CPC) with a countable set of unary epistemic connectives K_a $(a \in A)$. The set A represents agents. $K_a\phi$ states 'Agent a knows that ϕ', for arbitrary sentence ϕ. One may also add the doxastic connectives B_a $(a \in A)$. $B_a\phi$ then states 'Agent a believes that ϕ'. A standard semantic interpretation of epistemic and doxastic operators will shortly be provided in terms of multimodal Kripke frames called *knowledge models*. Before presenting them, we briefly discuss some issues concerning knowledge and truth.

6.1.3 The Idea of Knowledge

Wikipedia provides a brief account of the term 'knowledge': "Knowledge is a familiarity, awareness or understanding of someone or something, such as facts, information, description, or skills, which are acquired through experience or education by perceiving, discovering, or learning. Knowledge can refer to a theoretical or practical understanding of a subject." Philosophers have tended to focus on 'knowledge-that' or propositional knowledge. A traditional concern of the epistemologist has been to provide an analysis of the concept of knowledge. Attempts to do this stretch back to Plato, Meno, Theaetetus but have enjoyed a renaissance in the last 50 years because of 'Gettier cases'.

6.1.4 Gettier

Gettier presented the traditional analysis of knowledge consisting of three necessary and sufficient conditions. Let a be an agent and ϕ a sentence. The tripartite analysis says that a knows that ϕ if and only if

(a) ϕ is true
(b) a believes that ϕ
(c) a is justified in believing ϕ.

Let us admit and ignore that the conditions (especially (b) and (c)) are potentially to some extent vague. Gettier argued that there were cases in which a has a justified true belief but—it strikes us—not knowledge. For example, Piotr has a clock on his office wall which has worked perfectly for the past 10 years. One Monday morning, he looks up at his clock and forms the true belief that it is 10.15 a.m. His belief is justified in view of the past history of the clock. However, the clock happened to stop working on Sunday at 10.15 a.m. Had he looked up at any other time he would have formed a false belief. He does not know it is 10.15 a.m. (The example is a variation of that from Bertrand Russell.)

Many philosophers have attempted to explain the strong intuition that he does not know by inserting clauses to ensure (for example) that an agent's beliefs are not defeasible or have been arrived at through reasoning through a false lemma. But the opinion is that good solutions are not available—the definition of knowledge is a matter of ongoing debate among philosophers.

Concerning the knowledge operator in epistemic logic, one may pose two questions here

(M) What is the meaning of the formula $K_a\phi$? How should it be explicated in simpler terms?

and

(K) What is the logic and semantics of this operator? How are truth-values to be assigned to $K_a\phi$?

These questions are clearly related. Nevertheless, one need not complete the traditional epistemic project of answering (M) before addressing (K). For we have a sufficiently clear understanding of the concept of knowledge to undertake answering (K). (Had we not, we could of course attempt (M); for a reasonable grasp of the concept must be presumed in order to engage in its analysis.) For our purposes, it is (K) that matters. Of course, in exploring the logic of knowledge, we will be exploring the concept of knowledge but, as it were, indirectly.

6.1.5 Exploring (K)

Let us therefore explore the concept of knowledge a little more in order to address (K). First of all, knowledge is factive and logically grounded—*a* knows only true facts. Epistemic logic rejects various forms of false knowledge which are not grasped by models. So, epistemic logic assumes the scheme:

T: $$K_a\phi \rightarrow \phi,$$

because the agent knows only true facts (but of course not all of them); hence in particular, if agent *a* knows that ϕ is true,[4] then ϕ is true. *T* is the axiom of truth—the requirement that whatever is known is true.[5]

---

[4]Formally, in the developed form one would write: "*a* knows that the sentence 'ϕ' is true." Here 'ϕ' is the name of ϕ in the metalanguage and "*a* knows that 'ϕ' is true" is shorthand for "*a* knows that the sentence bearing the name 'ϕ' is true." To simplify the notation, we shall assume that the particle 'that' plays the role of the quotation marks 'and'. We therefore write "*a* knows that ϕ is true".

[5]One distinguishes between a sentence ϕ and the metalanguage parlance *about* ϕ using appropriate quotation marks. Thus the expressions "The sentence '*a* knows ϕ' is true" and "*a* knows that ϕ is true" bear different meanings.

The schemes

$$K_a\phi \rightarrow K_a(\phi \vee \psi) \quad \text{and} \quad K_a\psi \rightarrow K_a(\phi \vee \psi)$$

are also valid but their converses $K_a(\phi \vee \psi) \rightarrow K_a\phi$ and $K_a(\phi \vee \psi) \rightarrow K_a\psi$ are not.

The axiom T and CPC imply that

(1) $$K_a(\neg\phi) \rightarrow \neg K_a(\phi)$$

is logically valid for all ϕ. Indeed, $K_a(\neg\phi) \rightarrow \neg\phi$ and $\neg\phi \rightarrow \neg K_a(\phi)$ are valid by T and the contraposition of T, respectively. Hence, $K_a(\neg\phi) \rightarrow \neg K_a(\phi)$ holds, by the transitivity of $\rightarrow$.

6.1.6 Extensionality

Epistemic operators are not extensional. The following example is instructive. When the unary epistemic operator 'Newton knew that…' is applied to the true sentence '$8 = 5 + 3$' we obtain the true sentence 'Newton knew that $8 = 5 + 3$'. The truth-value of this sentence is not necessarily preserved if we swap the embedded sentence with one that has the same extension nor if we swap any term in it for a coextensive one. For example, neither 'Newton knew that Lincoln was president of the USA' nor 'Newton knew that the atomic number of oxygen = 8' is true. To capture this and other phenomena, Chapter 6 in Czelakowski (2001), provides a semantic approach to propositional attitudes based on referential frames.

6.1.7 What Is Involved in Knowing?

It is evident that knowledge of a proposition is *not* equivalent to knowledge of the truth of a sentence that expresses that proposition, as represented by (2):

(2) "Agent a knows that ϕ" is translated as the metasentence "a knows that ϕ is true."

The sentence on the right-hand side of (2) does not belong to the object language; it belongs to the metalanguage. If one accepts (2), then the sentences on both sides of the equivalence (2) carry the same logical value.

According to (2), the epistemic sentence "Adam knows that Walter Scott is the author of Rob Roy" is paraphrased as "Adam knows that the sentence 'Walter Scott is the author of Rob Roy' is true." It is also assumed that classical logic is valid in metalanguage and the usual conventions for truth as, e.g., "The sentence '$\neg\phi$' is true" is equivalent to "The sentence 'ϕ' is false" are adopted. (That is, we suppose Adam has also a certain basic understanding of truth, falsity, and negation.) If one additionally assumes that the connective 'a knows that…' is extensional for truth conventions, one gets from (2) that "a knows that $\neg\phi$ is true" is equivalent to "a knows that ϕ is false." Accordingly, each negative fact of the kind "Agent a knows that $\neg\phi$," that is, the formula $K_a(\neg\phi)$, is then under (2) equivalent in the metalanguage to "Agent a knows that ϕ is false." For example, the sentence "Adam knows that Goethe is not the author of Rob Roy" receives, under paraphrase (2), the equivalent form "Adam knows that the sentence 'Goethe is the author of Rob Roy' is false."

Note, however, that (1) becomes paradoxical under paraphrase (2), because (1) then equivalently states that if a knows that $\neg\phi$ is true (and hence *a fortiori*, a knows that ϕ is false), then a does not know that ϕ is true, which is sheer nonsense (because if a knows that if ϕ is false, then a knows everything about the truth and falsity of ϕ, by bivalence). Thus the paraphrase (2) is untenable if a basic knowledge of classical logic is available to a.

But one may retain partial paraphrases:

(a) If a knows ϕ, then a knows that ϕ is true,

and

(b) If a knows $\neg\phi$, then a knows that ϕ is false,

without arriving at a contradiction.

6.1.8 Situational Semantics

On grounds of situational semantics, one may try to paraphrase the formula $K_a\phi$ in the metalanguage of epistemic discourse as follows:

(3) "Agent a knows that ϕ" is translated as "a knows the situation described by ϕ."

One may also say that a knows the fact or the event represented by ϕ. As in the case of paraphrase (2), the sentence on the right-hand side of (3) does not belong to the object language. Another question that is left open is the meaning of the term 'situation'. We may, of course, refer here to various theories of situations which are available in the literature. In some approaches, the meaning of a sentence is identified with the situation described by this sentence. Without dwelling deeply on situational semantics and its nuances, one may then argue that (3), the situational rendering of $K_a\phi$, yields that a knows the situation described by ϕ if and only a knows the situation described by the negation $\neg\phi$. Thus, from the viewpoint of Suszko's situational semantics (see Wójcicki 1984), if one knows the (positive) situation that Paris is the capital of France, one also knows the (negative) situation that Paris is not the capital of France and vice versa.

According to the Fregean standpoint, there are only two possible situations being the referent of each well-defined sentence—truth or falsity. This is the content of so-called Fregean Axiom—see Suszko (1975). As Suszko writes: "The semantical assumption that all true (and similarly all false) sentences describe the same, i.e., have a common referent (Bedeutung) is called the Fregean axiom." (In the older literature, the word 'Bedeutung' was translated as 'meaning' but nowadays it is translated as 'referent' while the word 'Sinn' is translated as 'meaning'.) Thus, from the Fregean perspective, if one accepts (3), yet another metalinguistic semantic translation of the epistemic connective K_a is conceivable:

"Agent a knows that ϕ" is translated as "a knows the referent of ϕ" (that is, "a knows the truth-value of ϕ").

But this interpretation also yields nonsense. We may therefore conclude that if one assumes (3), then epistemic logic rejects the Fregean Axiom. (In fact, epistemic logic abolishes the Fregean Axiom even without assuming (3).) As Suszko (1975) writes: "It is obvious today that the abyss of thinking in a natural language does not fit into the Fregean scheme."

Similar remarks apply to other propositional attitudes as "a believes that ϕ", "a accepts ϕ", "a maintains that ϕ," etc. (Beliefs are probably true, possibly true, or likely to be true.)

6.1.9 The Logical Framework

It is true that $K_a\phi \rightarrow \phi$ but not the converse, as there are many truths we do not know. There is no simple link between the truth of ϕ and the truth of $K_a\phi$. The traditional epistemic project—the answer to (M) seeks to specify the link. The logical project can to some extent sidestep it by helping itself to the catch-all *epistemic means* agents have to put themselves in different *epistemic states*. Contemporary epistemic logic offers a sophisticated semantic apparatus by means of which it answers (K), avoiding simplifications and paradoxes. The semantics is provided by multirelational Kripke frames called knowledge models. Thus epistemic logic is viewed as multimodal logic applied for knowledge representation.

In case of one-agent systems, a *model* is a triple $\boldsymbol{M} = (W, R, V)$, where R is a binary relation on a set W (the set of 'possible worlds') and V is a mapping assigning a subset of W to each propositional variable. $\boldsymbol{M}$ is called a *Kripke-model* and the resulting semantics *Kripke-semantics*. A propositional variable p is true in a world w in $\boldsymbol{M}$ (written $\boldsymbol{M}, w \models p$) if and only if w belongs to the set of possible worlds assigned to p, i.e., $\boldsymbol{M}, w \models p$ if and only if $w \in V(p)$, for all p. The formula $K\phi$ is true in a world w (i.e., $\boldsymbol{M}, w \models K\phi$) if and only if for all $w' \in W$, if $R(w, w')$, then $\boldsymbol{M}, w' \models \phi$. Thus $K\phi$ is true at the agent's state of knowledge represented by w, then $R(w, w')$ encodes all epistemic means available to the agent enabling him/her to establish the truth of ϕ at w'. The semantics for the Boolean connectives follow the usual recursive recipe. A modal formula is said to be *valid* in the model $\boldsymbol{M}$ if and only if this formula is true in all possible worlds of W.

6.1.10 Other Renditions

As for other syntactic renditions of $K_a\phi$, computer scientists have proposed that the epistemic operator $K_a\phi$ should be read as "Agent a knows implicitly ϕ," "ϕ follows from a's knowledge," "ϕ is agent a's possible knowledge," etc. Propositional attitudes like these should replace the usual "agent a knows ϕ."

Dynamic epistemic logic (DEL) represents changes in agents' epistemic status by transforming models. DEL takes into account interactions between agents or groups that basically consist in sending information (as announcements) and analyzes the consequences of the flow of information on agents' epistemic states. As Hendricks (2003) points out: "Inquiring agents are agents who read data, change their minds, interact or have common knowledge, act according to strategies and play games, have memory and act upon it, follow various methodological rules, expand, contract or revise their knowledge bases, etc. all in the pursuit of knowledge. Inquiring agents are *active agents*."

The theory of action raises interesting praxeological problems strictly linked with agents' abilities to perform actions. Knowledge of actions is genuine procedural knowledge and not always reducible to verbalisable propositional contexts. Consider for example, the action of driving a car. Although the agent must know certain facts—that one must put petrol in a car to make it work, that the brake pedal is to the left of the accelerator pedal—there are skills involved that cannot be conveyed in words, such as when to change gears or how to reverse park. So, the theory must discriminate between theoretical knowledge (*episteme*) and the skills that should come into use (*techne*).

We may extend the language of CPC by introducing action variables α, β in a similar way as in Chap. 4. One may then consider praxeological operator $Perf_a(\alpha)$ which read as: "Agent a has ability to perform the action α." It is an open problem how to define the semantics of this operator.

6.1.11 Knowledge Models as Action Systems

A *knowledge model* $(A, W, (A_a)_{a\in A})$ consists of a set of *agents* A, a set of *states* W, and, for each agent $a \in A$, an *accessibility relation* $A_a \subseteq W \times W$. Subsets of W are called *propositions*, *events* etc. depending on context. For a given $u \in W$, the event $\delta_a(u) := \{w \in W : A_a(u, w)\}$ is the set of states that is considered possible for a at u, while all other states are excluded by a at u. We say that *a knows a proposition Φ at u* if $\delta_a(u) \subseteq \Phi$. The event that *$a$ knows Φ*, denoted $K_a(\Phi)$, is the set of all states in which a knows Φ:

$$K_a(\Phi) = \{u \in W : \delta_a(u) \subseteq \Phi\}.$$

In other words, a knows that Φ at u if and only if, for every $w \in W$, $A_a(u, w)$ implies $w \in \Phi$. The function $K_a : \wp(W) \to \wp(W)$ thus defined is called the a's knowledge operator.

The following observation is immediate:

Lemma 6.1.1 *For any sets $\Phi, \Psi \subseteq W$, $\Phi \subseteq K_a(\Phi)$ if and only if $A_a[\Phi] \subseteq \Psi$.*

($A_a[\Phi]$ is the A_a-image of Φ, $A_a[\Phi] := \{w \in W : (\exists u \in \Phi)\ A_a(u, w)\}$.) Thus a knows Ψ at all states $u \in \Phi$ if and only if the epistemic 'action' A_a carries every state of Φ to a state in Ψ.

Proof ($\Rightarrow$). Assume $\Phi \subseteq K_a(\Psi)$ and $w \in A_a[\Phi]$. Hence $A_a(u, w)$ for some $u \in \Phi$. But $u \in \Phi$ and $\Phi \subseteq K_a(\Psi)$ give that $u \in K_a(\Psi)$. This immediately implies that $w \in \Psi$.

($\Leftarrow$). Assume $A_a[\Phi] \subseteq \Psi$ and $u \in \Phi$. We claim $u \in K_a(\Psi)$, i e., $\delta_a(u) \subseteq \Psi$. Let then w be an arbitrary element of $\delta_a(u)$. Hence $A_a(u, w)$ holds. But $u \in \Phi$, $A_a(u, w)$ and $A_a[\Phi] \subseteq \Psi$ imply that $w \in \Psi$. This shows that $\delta_a(u) \subseteq \Psi$, proving that $u \in K_a(\Psi)$. □

For each two agents a and b we put $K_{ab}(\Phi) := K_a(\Phi) \cap K_b(\Phi)$. The operator C_{ab} defined by

$$C_{ab}(\Phi) = \bigcap_{n \geqslant 1} K^n_{ab}(\Phi)$$

is called the *common knowledge* (between a and b) *operator*. (Here $K^1_{ab}(\Phi) := K_{ab}(\Phi)$ and $K^{n+1}_{ab}(\Phi) := K_{ab}\big(K^n_{ab}(\Phi)\big)$ for all $n \geqslant 1$.)

Formally, every epistemic model $\big(A, W, (A_a)_{a \in A}\big)$ falls under the definition of a normal action system in which R, the relation of direct transition between states, can be taken to be the union of the set of relations $\{A_a : a \in A\}$. Therefore, the epistemic model $(A, W, (A_a)_{a \in A})$ may be identified with the normal action system $(W, R, \{A_a : a \in A\})$. Note, however, that in epistemic models, the accessibility relations A_a should not be treated as atomic actions. Each A_a is rather the resultant relation (in the formal sense of this notion defined in Sect. 1.7) of a bunch of various simpler infallible epistemic and non-epistemic in character actions undertaken by the agent a at state u leading from his state of knowledge u to a state w (in our terminology: A_a is the resultant relation *Res* $\mathbf{A}_a$ of a certain compound 'epistemic' action $\mathbf{A}_a$ available to a). In short, the compound action $\mathbf{A}_a$ encompasses various ways of acquiring information by the agent a as learning, exchanging information, proving theorems, conducting experiments, etc., and A_a is the resultant relation of the action. In epistemic models one usually abstracts from the structure of ways of acquiring knowledge and takes into account the initial and final states only because only these two states matter. Dynamic logic is the logic of changing knowledge (DEL). The situation in dynamic epistemic logic is more involved, because the models defined there also take into consideration various forms of communication between the agents in time-dependent contexts.

Although the whole apparatus of epistemic logic is not introduced here, knowledge models serve for the interpretation of sentences in an epistemic language, where each sentence corresponds to an event, the propositional connectives correspond to set theoretic operations, and the language connective 'a knows' corresponds to the (semantically defined) operator K_a. The operator K_a is monotonic and each knowledge model validates the axiom of truth T. (For more information see e.g., van Benthem 2013a; van Ditmarsch et al. 2013; Kooi 2011; Dov Samet 2008.)

6.2 The Frame Problem

The term 'frame problem' derives from a technique used by cartoon makers. This technique, called 'framing', consists in superimposing of a sequence of different images to produce a foreground in motion on the "frame," which does not change, and depicts the background of the scene. In reference to action theory the frame problem can be articulated as the problem of identification and description of the conditions (states of affairs, etc.) which are irrelevant or not affected by the actually performed action. Every action changes some states of affairs. However, not all factors that make-up a definite state are changed during the course of an action. The factors that constitute situational aspects of an action, such as time or space localization, are undoubtedly fluid.

The ramification problem and the frame problem are two sides to the same coin. The ramification problem is concerned with the indirect consequences of an action. It addresses the questions: *How do we represent what happens implicitly due to an action? How do we control the secondary and tertiary effects of an action?*

The learned skills and extraordinary deftness in using these skills enable people to attain vitally important goals at minimal cost and power. The ability of humans, presumably refined over the millenia of our cognitive development, makes it possible to decide immediately which factors are relevant and which are not while undertaking an action. It is assumed that the common sense law of inertia "according to which properties of a situation are assumed by default not to change as the result of an action" underlies the efficient action (after Stanford Encyclopedia of Philosophy (SEP), the entry "Frame Problem").

The frame problem was initially formulated in the context of artificial intelligence in 1969 by John McCarthy and Patrick J. Hayes in their article *Some Philosophical Problems from the Standpoint of Artificial Intelligence*. The solutions that have been put forward by computer scientists normally refer to the common sense law of inertia. This yields the logical setting of the frame problem, where the heart of the matter is in defining a set of logical formulas describing the results of actions without the need for taking into account factors being "non-effects of those actions." The solution consists in adopting certain axioms called frame axioms that explicitly define the non-effects of actions. (For example, "painting an object will not affect its position, and moving an object will not affect its color"—see SEP). A meta-assumption is adopted ("the general rule-of thumb") that "an action can be assumed not to change a given property of a situation unless there is evidence to the contrary." This default assumption is therefore identified with the above law of inertia (in the common sense). However, there appear difficulties of logical character there—the consequence operation of the standard logical systems is monotone, i.e., the set of conclusions increases with the addition of further formulae as premises. This is not the case when one takes into account various results of actions. For e.g, the assumption "that moving an object will not affect its color" is invalidated when one moves the object into a vat of dye, or presses it against a newly-painted wall, or into a flame, and so on. The solution of the frame problem in the logical wording results in adopting various non-monotonic reasoning formalisms as circumscriptions (McCarthy 1986). We shall return to this issue.

Later the term 'frame problem' received a much broader meaning in philosophy, where, according to some approaches, it touches the essence of the problem of rationality. The contemporary literature differentiates the epistemological frame problem from its computational counterpart. After the Stanford Encyclopedia of Philosophy (the entry "Frame Problem") the first problem is formulated as follows: "How is it possible to holistic, open-ended, context sensitive relevance to be captured by a set of propositional (...) representations used in classical AI?" Whereas its computational counterpart is encapsulated in the form of the question: "How could an inference process tractably to be confined to just what is relevant, given that relevance is holistic, open-ended, and context-sensitive?" In this short introduction to the frame problem it is not even possible to telegraph many of the questions raised by the researchers as, the relevance problem and the threat of infinite regress there.

One may maintain that the frame problem may be situated not only at the epistemic level but also on the praxeological level. There, it is defined as the problem of isolating the factors that guarantee efficient action, as well as one of describing those factors that are irrelevant in the given situation (or state) when the action is taken, that is, those that do not affect the process of performability of the action. Praxeology describes humans' engagement in purposeful actions, as opposed to reflexive behavior caused by external factors, like sneezing or coughing. These external factors may be treated as actions (or rather disturbances) of certain metasystems led by the forces of nature, social groups, states, armies, etc.

In the formalism adopted in this book, where the basic semantic unit is that of an elementary action system, the frame problem can also be articulated in at least two ways that are parallel to the threads raised in the above discussion. In the epistemological setting, the frame problem concerns the limitations of the epistemic abilities of the agents of actions. Whereas each elementary or situational action system possesses explicitly defined components as extensional, set-theoretic entities such as states or atomic actions, only limited knowledge in this respect may be available to the agents assigned to the system. This topic is more thoroughly discussed in the section on ideal agents. In the praxeological setting the frame concerns executive rather than epistemic limitations of the agents: they may have a full knowledge of the system (therefore all conceivable scenarios of future events are available to them); but their executive abilities are limited so that they may be unable to run the courses of action in accordance with the strategies adopted by them. Then one may discern mixed situations, being the resultant of epistemic and praxeological components. We know from the preceding paragraph that knowledge is not reducible to the knowledge of truth. Episteme (theoretical knowledge) is a more involved concept. One may argue, however, that the epistemic setting of the frame and ramification problems is more fundamental than the praxeological one. The reason is that the praxeological limitations of the agents involved in various actions often stem from ignorance. The fact that someone is not able to drive a car is conditioned by the potential driver's lack of knowledge—he simply did not learn the art of steering a car, and thus does not know how to do it (despite the fact that he may possess a purely theoretical knowledge of the construction of vehicles). One may maintain that all sorts of praxeological limitations are secondary in relation to epistemic limitations. It seems, however, that there have been merged two types of knowledge here: theoretical (episteme) with practical (techne). Knowledge of efficient action is organized on two levels—a linguistic level (a theoretical one), and a practical level, where knowledge is, among others, the skill of using a tool, or a machine. Mutual dependences between the levels and types of knowledge are complex and we are not going to analyze them in-depth here.

As an interlude, it is worth quoting here a short passage relating to the production of Damascus steel. The fact that nowadays we produce Damascus steel is a derivative of our knowledge concerning its production; after centuries this knowledge has successfully been recreated (with approximation), both on the theoretical and practical levels. The very production process itself has ceased to pose difficulty. Thus, in this case, episteme, as solely theoretical knowledge harmonizes with techne, practical knowledge. The production of Damascus steel is an interesting example of a compound action system. Damscus steel was used in South Asian and Middle Eastern swordmaking.[6] European knights became acquainted with the power of Damscus-steel swords during the Crusades. The swords made of this steel surpassed the white steel manufactured at that time in Europe: the blades were reputed to be tough, resistant to shattering, and capable of being honed to a sharp, resilient edge. (One should add, however, that modern steel outperforms these swords.) The blades are characterized by distinctive patterns of banding and mottling reminiscent of flowing water. The production methods were meticulously concealed and never revealed. For unknown reasons the production of the patterned swords gradually declined, ceasing by around 1750, and the process was lost to metalsmiths. Several hypotheses have ventured to explain this decline as, e.g., "the breakdown of trade routes to supply the needed metals, the lack of trace impurities in the metals, the possible loss of knowledge on the crafting techniques through secrecy and lack of transmission, or a combination of all the above." Undoubtedly a vital factor could also be the fact that firearms became more and more popular, which resulted in a dramatic drop in orders for cold steel. The original method of producing Damascus steel is not known. For many years attempts were made to rediscover the methods in which Damascus steel was produced and, in consequence, to start its production on a larger scale. (Recreating Damascus steel is therefore a subfield of experimental archeology.)

Thus, the problem was, how—having necessary raw materials at one's disposal, such as ores and additions (e.g., carbon, glass)—to obtain a product identical in

[6]These remarks are based on the entry 'Damascus steel' from the Wikipedia. The citations are taken from this entry.

all structural and functional respects to that of ancient Damascus steel. Early efforts proved futile. Because of differences in raw materials and manufacturing techniques, modern attempts to duplicate the metal have not been entirely successful. The empirical data collected for years and historical records allowed, to some extent, researchers to establish the origin of the raw materials used in the production of the steel. They were special ores found in India, which contained key trace elements, like tungsten or vanadium. The crucial problem, which required solving, was to determine the structure of Damascus steel on a microscopic scale. The breakthrough came with the discovery of the presence of cementite nanowires and carbon nanotubes in the composition of Damascus steel by a team of researchers from Germany. It is now thought that "the precipitation of carbon nanotubes most likely resulted from a specific process that may be difficult to replicate should the production technique or raw materials used be significantly altered." It is claimed that these nanostructures are a result of the forging process.

From the perspective of the theory of action, the difficulty in reproducing Damascus steel consists in identification of the action system(s) pertinent to the original technology. The earlier attempts erroneously identified various components of this system. The actions undertaken resulted in steel which displayed different parameters from those of the original product. From the historical perspective, the output systems of actions, which were to recreate production of Damascus steel, were subjected to a series of modifications through a change in the sets of states and atomic actions, so that an action system could be worked out, in which steel with properties that are approximate or identical with those of Damascus steel could be obtained as a result.

Currently, Damascus steel having properties very close to the classical Damscus steel is manufactured in a process which most likely resembles the original one fairly closely. This is indicated by its simplicity and similar properties of products used.

Formal action systems $\boldsymbol{M} = (W, R, \boldsymbol{\mathcal{A}})$ may be viewed as approximations of the real state of affairs. The relation R mirrors the reality—the relation determines (at least when one takes into account only physical limitations) what is possible or what is not. The agent usually has a vague, approximate idea of the relation. A more competent agent has a better image of R. A much better approximation is achieved by top experts. For example, during a game of solitaire, it is an objective fact that at some step the game cannot be continued further. But when a game can indeed be prolongated, the poor player may stop it maintaining that further steps are unperformable. His conviction cannot be mixed with objective characteristics of this particular game. But we are reluctant to adopt 'A God's eye point of view' as an epistemic or pragmatic criterion of performability of actions. We are inclined rather to adopt the top experts' perspective on this matter.

The choice of the set W and the relation R, and also the shape of the actions of $\boldsymbol{\mathcal{A}}$ depends much on the 'depth' of the description of the real situation. A transition from one system $\boldsymbol{M} = (W, R, \boldsymbol{\mathcal{A}})$ to a more adequate model $\boldsymbol{M'} = (W', R', \boldsymbol{\mathcal{A}}')$ results in changing the set of states, the relation of direct transition, and the atomic actions. It seems that in the description of this transition one can distinguish two components. First, the states of the initial system $\boldsymbol{M}$, i.e., the elements of W, may be treated as sets of finite sequences of states of the system $\boldsymbol{M}'$. This means that a one-to-one function k from W to the power set of the set of finite sequences of elements of W', $k : W \to \wp\big((W')^*\big)$ is singled out. The sequence $k(u)'$ is called the *reconstruction* of the state $u \in W$ within the system $\boldsymbol{M}'$. Thus, the whole set W is reconstructed in the power set of the set $(W)^{*\prime}$ as the family of sets of finite sequences $\{k(u) : u \in W\}$. It is natural to require that if u and w are different states of W, then the sequences belonging to $k(u)$ and $k(w)$ do not contain a common element. Second, the atomic actions of $\boldsymbol{\mathcal{A}}$ and the transition relation R of $\boldsymbol{M}$ may be also reconstructed within the system $\boldsymbol{M}'$ by means of another map l. The transition from $\boldsymbol{M}$ to $\boldsymbol{M}'$ makes each atomic action of $\boldsymbol{\mathcal{A}}$ replaced by a compound action, i.e., by a set of finite strings of atomic actions of $\boldsymbol{\mathcal{A}}'$. Thus, each atomic action of $\boldsymbol{M}$ is replaced by a set of sequences of atomic actions on the deeper level provided by $\boldsymbol{M}'$. The simplified description of actions the system $\boldsymbol{M}$ offers, formulated in the 'input–output' terms of its atomic actions, gives way to a more refined description on the basis of the system $\boldsymbol{M}'$. The difficult problem of how to formally define such an interpretation of $\boldsymbol{M}$ in $\boldsymbol{M}'$ should come under further scrutiny; this is not tackled in this book.

The frame problem has already been signaled in this book by discriminating elementary action systems from situational action systems. The former represent 'hard' praxeological aspects of action originating, e.g., from the laws of mechanics or the engineering principles, while the latter represent 'soft' and 'fluent' ones. These volatile aspects of action are encapsulated in the concept of the situational envelope of an elementary action system. The situational envelope of an action system, though relevant to workflow or game rules, does not set 'hard' action components aside.

What factors/entities constitute the situational envelope of an elementary action system is often hard to foresee. In the case of the game of chess we include in them, apart from a distribution of pieces on the chessboard, also the order of moves; that is, whether the successive move is to be performed by the white or the black player. One can say that we have come here to deal with a certain *type* of situational action system determined by the general rules of the game of chess.

Let us imagine now that a concrete game of chess is being played within the tournament, where two opponents are competing for the title of world champion, held in a beautifully illuminated sports arena. The game is progressing smoothly in compliance with the rules and *White* is to make his move. Let us mark these situations with s. Suddenly, all the lights in the hall go out and it gets completely dark. A failure has occurred in the nearest power station (perhaps an accidental failure, or maybe a terrorist attack...). Despite the fact that the state on the chessboard and the rules of the game make it officially possible to continue the game, it is automatically broken off and *White* cannot make the move. Both the participants of the game and the audience have found themselves in a new, different, situation which will be marked with s_ω. Of course, s_ω is not included in the situational envelope of the game of chess presented in Sect. 2.1.

The above (fictitious) case suggests including chance situations (a sudden blackout, an earthquake, a terrorist attack, etc.) in the notion of a surrounding. They are extreme situations. In consequence, knowledge relating to the source of the situation and the relation Tr is here *ex post* knowledge; it is not fully articulated before the inclusion of the system into the move. Within the compass of situation action theory there are found—as we mentioned earlier—systems that are *a priori* 'foreseeable', in which a transition from one to the other situation falls completely under the control of agents included into the system. It is clear that if somebody plans a sports tournament in Japan, he should make provision for the possibility of occurrence of an earthquake. So as to minimize its influence on the course of the tournament, a variety of additional protective measures are used such as the reinforcement of the structure of the stadium and the number of medical services on duty during the event. The very tournament itself, being a situational system, is described in terms of the categories presented above, though without the need to assume that an unexpected chance incident will 'upset' it. No person with a sound sense of reality plans to conduct their life activities under the assumption that life on Earth will be destroyed by a meteorite. Various external sudden random incidents, as unpredictable or of low predictability, are not composed into the notional apparatus presented here. Probabilistic systems *sensu stricto* are an obvious exception. They are discussed in Chap. 1.

The above example illustrates yet another aspect of an action that may be pertinent to the ramification problem, viz., the distinction between ignorance and uncertainty. The ramification problem has therefore two sides. The first side concerns the ignorance and inability of agents to conduct purposeful actions in the intended way *within* a definite action system operated by them. The second side concerns uncertainty and it is linked to the presence of *external* forces influencing the action system. These external forces are framed in the form of meta-action systems (nature, political or military forces, big business, etc.). These metasystems manifest their action through earthquakes, floods, cosmic disasters, wars, long or short term economic or demographic processes, and so on. The given action system is subordinated to various metasystems; the first (along with its agents) is merely a pawn in the game run by the other. One can try to learn the rules of the game (modern natural science has made a spectacular success in this domain); the area of knowledge of metasytems is impressive. Metasystems are, however, often unpredictable and uncertainty remains. It is from here that the significance of probabilistic action systems stems. Obviously, we will not settle here the metaphysical question: What is the order of all things? Is it a Stoic order (God's reason and universal order), or an Epicurean one (chaos of atoms, fortuity)?

Let $\boldsymbol{M} = (W, R, \boldsymbol{\mathcal{A}})$ be an elementary action system. A map which is called the *deliberation operator* and denoted by $\boldsymbol{D_M}$ is associated with the system $\boldsymbol{M}$. The deliberation operator is a function from $\wp(W) \times \boldsymbol{\mathcal{A}}$ to $\wp(W)$. Thus, to each subset $\Phi \subseteq W$ and each atomic operation $A \in \boldsymbol{\mathcal{A}}$ the operator $\boldsymbol{D_M}$ assigns a subset of W. To simplify matters, let us assume that the system is normal, that is, $R \subseteq A$ for all $A \in \boldsymbol{\mathcal{A}}$. Then we define:

$$\boldsymbol{D_M}(\Phi, A) := \{w \in W : (\exists u \in \Phi)\ A(u, w)\},$$

for all $\Phi \subseteq W$, $A \in \boldsymbol{\mathcal{A}}$. $\boldsymbol{D_M}(\Phi, A)$ is the set of all possible results of the action A when performed in states belonging to Φ. $\boldsymbol{D_M}$ describes the dynamics of $\boldsymbol{M}$—the consecutively performed string of atomic actions $A_0, A_1, \ldots, A_n$ which has been initialized a Φ-state u_0, determines a path

$$u_0, u_1, \ldots, u_n, u_{n+1}, \tag{6.2.1}$$

in the space of states (we shall restrict considerations to finite paths), where $u_0\, A_0\, u_1\, A_1 \ldots u_n\, A_n\, u_{n+1}$. (It may happen that there is no possibility of the prolongation of (6.2.1), when for no atomic action A there is a state u such that $u_{n+1}\, A\, u$.)

Looking at the first argument Φ of $\boldsymbol{D_M}$, one may say that $\boldsymbol{D_M}$ assigns to the set Φ a family of subsets of W, viz., $\{\boldsymbol{D_M}(F, A) : A \in \boldsymbol{\mathcal{A}}\}$, representing the possible outcomes of all atomic actions when performed in Φ-states. The set Φ represents epistemic limitations of the agents 'operating' the system $\boldsymbol{M}$: they may not know exactly the state of the system, but they know that this state is localized in Φ. It is assumed that the agents know the description of atomic actions, that is, they know which pairs (u, w) belong to a given atomic action A and which do not. Hence they know that after performing an action A the system will pass to (an unspecified) state belonging to $\boldsymbol{D_M}(\Phi, A)$. The problem is that the set $\boldsymbol{D_M}(\Phi, A)$ may be empty and possibilities of performing of A in Φ-states are then illusive. Yet another situation is such that after performing the action A, the agents intend to perform another action, say B. If the set $\boldsymbol{D_M}(\Phi, A)$ is disjoint with $Dom(B)$, the domain of B, then there is no such possibility of such conduct—the action B is totally unperformable in any state of $\boldsymbol{D_M}(\Phi, A)$. Agents know about this! Epistemically, a more interesting situation is when $Dom(B)$ in a nonempty way intersects with $\boldsymbol{D_M}(\Phi, A)$ and $Dom(B)$ is not a superset of $\boldsymbol{D_M}(\Phi, A)$. After performing A, the system $\boldsymbol{M}$ passes to some state in $\boldsymbol{D_M}(\Phi, A)$. But the knowledge as to whether the achieved state belongs to the domain of B or not may be unavailable to the agents—they are at a loss. What can they do?

In light of the above remarks, the essence of the above problem is contained in the *epistemic* limitations of the agents operating a given action system. Even assuming that the action system is managed by ideal agents, profoundly knowledgeable of all the components of the system such as states, the transition relation, precise description of atomic actions, and so on, certain situational aspects (including the epistemic ones) can remain beyond their reach. The example of a game of chess proves instructive here again. A novice can know perfectly the rules of the game; nonetheless, many documents that influence winning of the game can rest beyond his epistemic powers. Although he knows the rules of the game of chess, his knowledge of strategies may be limited. He may not to be able to foresee, or simply remember some moves, or the like. There occurs here the information phenomenon of *informational explosion*, when the amount of information necessary to take the proper decision which guarantees success increases exponentially. On the other hand, it is impossible to exclude actions, often by other casual agents, external toward the given system, which disturb the functioning of the system such as forces of nature.

The deliberation operator restricts reasonings about possible events and scenarios to the world encapsulated by the action system $\boldsymbol{M}$. This operator does not take into account various external action systems and forces that may influence the dynamics of $\boldsymbol{M}$—only the states and action undertaken within the framework of $\boldsymbol{M}$ matter. In other words, the deliberation operator represents a form of the closed-world assumption in action theory.

On the side of mathematics, the definition of the elementary system $\boldsymbol{M} = (W, R, \boldsymbol{\mathcal{A}})$ can be strengthened by the so-called hidden parameters, which can be made into components of the state of things. In other words, instead of the set W we can consider the set of pairs $\{(u, \bullet_u) : u \in W\}$, where $\bullet_u$ marks a block of primitively undefinable or unknown parameters which, in the course of action, may turn out to be activated in the state u. Let $\boldsymbol{M} = (W, R, \boldsymbol{\mathcal{A}})$ represent the system of actions that describe the building of a skyscraper in Tokyo. ($\boldsymbol{M}$ is basically an ordered situational action system, but we disregard situational components and the order altogether.) Suddenly, an earthquake occurs yet the building being raised does not break up; there is, however, slight damage done to it, which can be repaired. As a result of the earthquake, that is nature's acting on the state of the system, there follows a transition from $\boldsymbol{M}$ to another system. What system? All the states that are vital to the further construction of the building, including this in which the earthquake followed, ought to be provided with an additional label in the place $\bullet_u$ notifying that there was an earthquake and further construction will follow in the conditions after the earthquake. (In the case of earlier states, as ones insignificant to the construction, marker $\bullet_u$ remains unfilled.)

In light of the above comments, it appears that there is not one recipe for solving the frame problem in the context of elementary or situational action systems. The above-presented comments point merely to a few alternative formal frameworks of the problem.

6.3 Action and Models for Non-monotonic Reasonings

6.3.1 Reasoning in Epistemic Contexts

We begin by recalling the well-known logical puzzle called *three cards*. The three cards are red, blue, and green, respectively. Darek, Jarek, and Marek are fair players (they do not lie and cheat), who have one card each. They only see their own card, and so they are ignorant about the other agent's card. (But they know that the cards are red, blue, and green.) They want to know the colors of the cards of their partners.

(a) Darek asks Jarek the question: *Do you have the blue card?*

Who amongst them right after Darek's question *and before* Jarek's answer to it, already knows the distribution of the cards? The answer is: the person (either Jarek or Marek) who does *not* have the blue card. Why? Darek, while asking the question, announces at the same time that he does not have the blue card (this is the presupposition of his question). Therefore, the blue card is in the hands of Jarek or Marek. Have a look then at Jarek and Marek. One of them (he who does *not have* the blue card) rightly infers that the blue card is in the hands of his colleague. Since he knows the color of his own (non-blue) card, and knowing that the card of his colleague is blue, he at the same time knows the color of Darek's card. *Ergo*: he knows the distribution of the cards. Yet each of the remaining two people do not know the distribution.

(b) Jarek sincerely answers: *I do not have the blue card.*

Who amongst them concretely *after* hearing Jarek's answer, knows the distribution of cards? The answer is: Darek and Jarek. Why? After hearing the answer, Darek rightly concludes that the blue card is in Marek's hands. Therefore, as Darek knows both his own card and Marek's card, Darek knows also the card of Jarek. Hence Darek knows the distribution of cards.

Right after Darek's question, Jarek knows that Darek does not have the blue card. Of course, Jarek knows his card (his card is not blue). Hence, Jarek deduces that the blue card is in Marek' hands. Therefore, Jarek also knows the color of Darek's card. Hence Jarek also knows the distribution of cards.

Marek has the blue card. (After public announcements (a) and (b), the remaining players also know that fact.) But Marek is unable to determine the particular color of the cards of Darek and Jarek. Therefore, Marek still does not know the distributions of cards.

One may generate alternate scenarios with different questions and answers addressed to other players. But here we have a specific action system operated by three agents: Darek, Jarek, and Marek. Each of the agents asks questions to other agents. The addressee of the question performs another action, namely that of answering the question.

The above logical exercise in public announcements can be seen from a wider perspective. The task consists in recognition of distribution of three cards in the hands of three players. The cards are marked by the letters b, g, r, respectively. Each possible distribution is therefore represented by a sequence of the above letters of length 3 (without repetitions). Let us agree that the first element of the sequence is the card belonging to Darek, the second card—to Jarek, and the third one—to Marek. There are altogether $3! = 6$ distributions.

The cards have been dealt out. The players want to identify this distribution. It is assumed that the situational envelope of the game contains three agents and an external, neutral observer. Each agent asks another player a question. The addressee responds giving an answer of type YES/NO only. More exactly, each question is treated as a figure Φ?, where Φ is a subset of $\{b, g, r\}$. (Usually Φ is a one-element set.) Φ? is read: *Is your card of a color belonging to* Φ? For example, $\{b, g\}$? is a query: *Is your card blue or green*? Possible answers are identified with figures of the form Φ!, where Φ is a subset of $\{b, g, r\}$. For instance, $\{g, r\}$! is the answer: *I have the green or the red card.* (As classical logic is assumed, in this model the above answer is equivalent to the sentence: *I do not have the blue card.*)

Both verbal actions, i.e., asking a question and answering it, result in a change of epistemic states of the players (that is, tightening suitable disjunctions from state descriptions by eliminating certain disjuncts). This follows from the fact that apart from the above labeled epistemic actions (asking questions and answering them), the players also perform other actions—they carry out reasoning according to the rules of classical logic on Boolean combinations of the sentences Darek(c), Jarek(d), Marek(e), for all permutations (c, d, e) of the set $\{b, g, r\}$. (E.g., the rule *modus tollendo ponens* is widely applied.) Thus a detailed description of the dynamics of the game would also require taking into account a succession of the rules of inference applied by each of the players because these rules enable each of them to acquire better knowledge of the distribution of the cards after performing the mentioned epistemic actions (asking and/or answering). The above description of the game leads to a rather complex situational action system. The process of gaining knowledge about cards distribution is of cumulative character, that is, by performing the above epistemic actions and drawing logical conclusions, the players get more detailed data about the localization of the actual distribution in the space of possible states. The paradigm of monotonic knowledge is not overruled here because the undertaken verbal and logical actions do not change the actual distribution of the cards.

Each state of the above action system is represented by three components: the individual knowledge of possible distributions of cards by Darek, Jarek, and Marek, respectively. We shall assume God's eye perspective in describing the arisen situations. God knows the actual distribution of cards; he/she also knows the (fluent) knowledge of the players concerning the distribution of the cards. From the above analysis it follows that the *actual* distribution is described either by the sentence Darek$(g) \wedge$ Jarek$(r) \wedge$ Marek(b) or by Darek$(r) \wedge$ Jarek$(g) \wedge$ Marek(b). Let us consider the first case:

$$\text{Darek}(g) \wedge \text{Jarek}(r) \wedge \text{Marek}(b) \tag{6.3.1}$$

The initial knowledge available to each of the agents is described by the sentences:

$$\begin{array}{ll}
\text{Darek(A1)}: & \text{Darek}(g) \wedge \big[\big(\text{Jarek}(r) \wedge \text{Marek}(b)\big) \vee \big(\text{Jarek}(b) \wedge \text{Marek}(r)\big)\big] \\
\text{Jarek(A1)}: & \text{Jarek}(r) \wedge \big[\big(\text{Darek}(g) \wedge \text{Marek}(b)\big) \vee \big(\text{Darek}(b) \wedge \text{Marek}(g)\big)\big] \\
\text{Marek(A1)}: & \text{Marek}(b) \wedge \big[\big(\text{Darek}(g) \wedge \text{Jarek}(r)\big) \vee \big(\text{Darek}(r) \wedge \text{Jarek}(g)\big)\big].
\end{array}$$

Immediately after asking the above question by Darek and after performing some logical derivations by the agents, the epistemic status of the agents turns into:

$$\begin{array}{ll}
\text{Darek(A2)}: & \text{Darek}(g) \wedge \big[\big(\text{Jarek}(r) \wedge \text{Marek}(b)\big) \vee \big(\text{Jarek}(b) \wedge \text{Marek}(r)\big)\big] \\
\text{Jarek(A2)}: & \text{Jarek}(r) \wedge \text{Darek}(g) \wedge \text{Marek}(b) \\
\text{Marek(A2)}: & \text{Marek}(b) \wedge \big[\big(\text{Darek}(g) \wedge \text{Jarek}(r)\big) \vee \big(\text{Darek}(r) \wedge \text{Jarek}(g)\big)\big].
\end{array}$$

(Note that under option (6.3.1), Jarek already knows the distribution of cards.) After answering the question by Jarek the above state turns into:

$$\begin{array}{ll}
\text{Darek(A3)}: & \text{Darek}(g) \wedge \text{Jarek}(r) \wedge \text{Marek}(b) \\
\text{Jarek(A3)}: & \text{Jarek}(r) \wedge \text{Darek}(g) \wedge \text{Marek}(b) \\
\text{Marek(A3)}: & \text{Marek}(b) \wedge \big[\big(\text{Darek}(g) \wedge \text{Jarek}(r)\big) \vee \big(\text{Darek}(r) \wedge \text{Jarek}(g)\big)\big].
\end{array}$$

(Here Jarek also knows the distribution of cards.)

Abstracting from the logical actions performed by the agents, we see that the above actions give rise to the transitions between epistemic states:

$$\begin{aligned}\langle \text{Darek(A1), Jarek(A1), Marek(A1)} \rangle &\rightarrow_R \langle \text{Darek(A2), Jarek(A2), Marek(A2)} \rangle \\ &\rightarrow_R \langle \text{Darek(A3), Jarek(A3), Marek(A3)} \rangle.\end{aligned}$$

Analogously, if the actual distribution of cards is

$$\text{Darek}(r) \wedge \text{Jarek}(g) \wedge \text{Marek}(b), \tag{6.3.2}$$

the following epistemic state is initial:

$$\begin{aligned}&\text{Darek(B1)}: && \text{Darek}(r) \wedge \big[\big(\text{Jarek}(g) \wedge \text{Marek}(b)\big) \vee \big(\text{Jarek}(b) \wedge \text{Marek}(g)\big)\big] \\ &\text{Jarek(B1)}: && \text{Jarek}(g) \wedge \big[\big(\text{Darek}(r) \wedge \text{Marek}(b)\big) \vee \big(\text{Darek}(b) \wedge \text{Marek}(r)\big)\big] \\ &\text{Marek(B1)}: && \text{Marek}(b) \wedge \big[\big(\text{Darek}(g) \wedge \text{Jarek}(r)\big) \vee \big(\text{Darek}(r) \wedge \text{Jarek}(g)\big)\big].\end{aligned}$$

Immediately after asking the question by Darek the above state turns into the state:

$$\begin{aligned}&\text{Darek(B2)}: && \text{Darek}(r) \wedge \big[\big(\text{Jarek}(g) \wedge \text{Marek}(b)\big) \vee \big(\text{Jarek}(b) \wedge \text{Marek}(g)\big)\big] \\ &\text{Jarek(B2)}: && \text{Jarek}(g) \wedge \text{Darek}(r) \wedge \text{Marek}(b) \\ &\text{Marek(B2)}: && \text{Marek}(b) \wedge \big[\big(\text{Darek}(g) \wedge \text{Jarek}(r)\big) \vee \big(\text{Darek}(r) \wedge \text{Jarek}(g)\big)\big].\end{aligned}$$

After Jarek's answer, the above state changes to:

$$\begin{aligned}&\text{Darek(B3)}: && \text{Darek}(r) \wedge \text{Jarek}(g) \wedge \text{Marek}(b) \\ &\text{Jarek(B3)}: && \text{Jarek}(g) \wedge \text{Darek}(r) \wedge \text{Marek}(b) \\ &\text{Marek(B3)}: && \text{Marek}(b) \wedge \big[\big(\text{Darek}(g) \wedge \text{Jarek}(r)\big) \vee \big(\text{Darek}(r) \wedge \text{Jarek}(g)\big)\big].\end{aligned}$$

The above transitions between states are depicted as:

$$\begin{aligned}\langle \text{Darek(B1), Jarek(B1), Marek(B1)} \rangle &\rightarrow_R \langle \text{Darek(B2), Jarek(B2), Marek(B2)} \rangle \\ &\rightarrow_R \langle \text{Darek(B3), Jarek(B3), Marek(B3)} \rangle.\end{aligned}$$

The above example merely plays an illustrative role. It shows that in action systems operated by many agents one may speak of global states of the system treated as sequences of the "private" states of particular agents. (In the above case these are sequences of length 3, because there are only three agents: Darek, Jarek, and Marek.) Each such string of private epistemic states makes up a global state. Knowledge of the global system is restricted for each agent. Thus, operating on global states requires adopting here an external perspective on the system by assuming the existence of an omniscient superagent who always knows the epistemic states of the particular agents operating the system (God's eye perspective). According to this standpoint, an epistemic model with n interacting individuals (agents) $a_1, \ldots, a_n$ is set-theoretically represented as an action system whose global states are sequences $w = (w_1, \ldots, w_n)$, where the ith component w_i encodes the 'private' knowledge of the agent a_i about the world for $i = 1, \ldots, n$. (The world here may be a distribution of cards.) The sequence w defines the knowledge of the superagent about the knowledge of the agents about the world. (This omniscient agent knows the epistemic states of each agent a_i). Let W be the set of global states; W is thus a set of sequences $(w_1, \ldots, w_n)$ of length n.

A simple model can be described as follows. Let $\boldsymbol{A}$ be a finite Boolean algebra. Each filter of $\boldsymbol{A}$ is principal and every element x of $\boldsymbol{A}$ is a combination of atoms of $\boldsymbol{A}$. W is the set of all n-tuples $(w_1, \ldots, w_n)$ of proper filters of $\boldsymbol{A}$. It is assumed that the filter generated by the union $w_1 \cup \cdots \cup w_n$ is proper; this corresponds to the fact that the union of the partial knowledge available to all agents $a_1, \ldots, a_n$ is consistent. The filter generated by the union $w_1 \cup \cdots \cup w_n$ is the knowledge of the superagent in $(w_1, \ldots, w_n)$. If it is an ultrafilter, we call it a *world*. If the filter generated by $w_1 \cup \cdots \cup w_n$ is an ultrafilter, the knowledge available to the superagent is complete. In other words, though the private knowledge w_i of each agent a_i need not be complete, the *joint* knowledge of the agents is complete (and the superagent knows that). Verbal actions performed by particular agents (queries and responses) are treated as atomic actions on the set of states. The agents want to describe the filter w generated by $w_1 \cup \cdots \cup w_n$. In other words, each of them wants to achieve the private state w. (Therefore the n-tuple $(w, \ldots, w)$ is the final global state.) Each agent a_i may publicly asks the agent a_j a question of the form Φ?, where Φ is an element of the algebra. Φ? reads as: *Does Φ belong to your knowledge*? In other words, in the state w_i the agent a_i asks a_j a whether Φ is an element of the filter w_j. (As $\boldsymbol{A}$ is finite, each Φ is the join of a finite number of atoms of $\boldsymbol{A}$.) We shall mark this question as a quadruple $(a_i, \Phi, a_j, ?)$, where $i \neq j$. The agent a_j responds to this question giving one of two answers: Yes, No. We mark these actions as (a_j, Yes), (a_j, No). Suppose w_j is a private state of a_j. The agent a_j answers Yes in w_j if Φ belongs to the filter w_j, i.e., the set of atoms that make Φ contain the atoms whose join is the generator of the principal filter w_j. a_j answers No if Φ is not in w_j, i.e., there is an atom in the generating element of w_j which is disjoint with Φ.

In the model we consider there is no need to revise the earlier knowledge gained by the agents—they simply enlarge their filters whenever possible to reach eventually the filter generated by the union $w_1 \cup \cdots \cup w_n$. (The possibility of knowledge contraction is excluded. This enables us to avoid well-known difficulties in the mathematical modeling of the dynamics of knowledge.) There is an order concerning asking questions. We may assume for simplicity that the last asked agent, after giving an answer, has the right to ask the next question to an agent selected according to a certain meta-rule adopted by all players. (The model we describe is therefore a simple situational action system.)

We make the following assumptions, which are known to the agents:

(a) The union $w_1 \cup \cdots \cup w_n$ generates an ultrafilter w of $\boldsymbol{A}$.
(b) Each question $(a_i, \Phi, a_j, ?)$, when publicly asked by a_i, assumes that Φ *is not* an element of the filter w_i.

The goal of the game:

(c) Each of the players wants to achieve w as his final private state.

(c) is adopted by the agents.

If a_i asks $(a_i, \Phi, a_j, ?)$, then according to (b), an atom of $\boldsymbol{A}$ that is included in the generator of the filter w_i is not included in Φ and all agents know that after the question has been asked. The action $(a_i, \Phi, a_j, ?)$ transforms the state $(w_1, \ldots, w_n)$ into a new global state $(v_1, \ldots, v_n)$. (The picture is more complicated here, because after hearing $(a_i, \Phi, a_j, ?)$ each agent sets rules of logic in action. Just these rules enable them to determine particular components of $(v_1, \ldots, v_n)$.) Here v_i coincides with w_i (the agent a_i does not update his private knowledge w_i when asking $(a_i, \Phi, a_j, ?)$.) But other agents a_k may do that by enlarging the filter w_k. As mentioned, hearing $(a_i, \Phi, a_j, ?)$, they rightly deduce that *some* of the atoms that make up Φ are not included into the generator of w_i. But it is often impossible for each of them to infer which particular atoms are relevant here. But if a_k can do that and he can identify the atoms of Φ which are not in the generator of w_i, he enlarges his filter w_k by deleting those atoms in the generator of w_k he can identify as not belonging to the set of atoms of Φ. If in the state w_j the agent a_j answers Yes to $(a_i, \Phi, a_j, ?)$, i.e., Φ is an element of the filter w_j, each of the agents deduce that all atoms in the generator of w_j are atoms of Φ. Then the agent a_k $(k \neq j)$ updates his knowledge by enlarging the filter w_k as follows: a_k deletes from the generator of w_k the atoms which are in Φ and not in the generator of w_j. If in the state w_j the agent a_j answers *No* to $(a_i, \Phi, a_j, ?)$, i.e., Φ is not an element of the filter w_j, each of the agents deduce that there are atoms in Φ that are not in the generator of w_j. Each agent a_k updates his knowledge by deleting from the generator of w_k these atoms (unless he can identify them).

How to achieve goal (c)? The agents may apply various strategies. For instance they may at the outset ask questions of the form $(a_i, \Phi, a_j, ?)$, where Φ is a coatom of $\boldsymbol{A}$. Since $w_1 \cup \cdots \cup w_n$ generates an ultrafilter w of $\boldsymbol{A}$, there exists *exactly one* coatom Φ_0 which is not a member of each of the filters $w_1, \ldots, w_n$. (w is generated by the atom being the complement of Φ_0). They may attempt to identify Φ_0. Here much depends on the organization of the work. (This belongs to the situational envelope of the system.) For example, if for every coatom Φ and any i, j $(i \neq j)$, the question $(a_i, \Phi, a_j, ?)$ can be eventually answered, the identification problem is solved.

The above remarks merely provide a sample of the problems epistemic logic raises. But, more importantly, these remarks point out that epistemic logic is tractable by the formal apparatus presented in this book.

The problem of building a formal apparatus to represent knowledge undergoing changes in time is discussed in the literature. The idea is to merge together epistemic logical systems applied in the description of knowledge and temporal logical systems. Basic systems of epistemic logic are split into two groups: the first group is concerned with a singular cognitive subject (an agent) and the second group takes into account the knowledge of many cooperating agents. But the most challenging problem concerns the description of the dynamic of knowledge when it is necessary to 'temporalize' the systems under question. In this context some systems of temporal-epistemic logic created by the use of the fusion method or the Finger-Gabbay method (1992) of temporalization of logical systems as well as alternating time temporal epistemic logic ATEL (van der Hoek and Wooldridge 2003; Herzig and Troquard 2006) should be mentioned.

The classical approach to common knowledge, originating from epistemology and epistemic logic, underlines the significant role of consensus among the members of a group of agents (the latter forming together a collective agent). Reaching common knowledge facilitates the process of reasoning in the group—the members of the group draw common conclusions from the universally known and accepted premises. Although this approach is useful in modeling various aspects of group conduct as in modeling the process of forming norms and conventions or building models of reasoning of other agents, the cost of reaching a consensus may turn out to be too high. In recent advancements concerning group knowledge, the role of group consensus is not emphasized and emphasis is placed on the following epistemic aspects of group cooperation:

1. Instead of the principle 'what everybody knows', collective knowledge is aimed at providing a *synthetic* body of information abstracted from the information provided by particular group members. This synthetic information may concern selected aspects (or fragments) of knowledge and is utilized in various forms of collective action.
2. The group conclusions are not treated as private conclusions accepted by particular members of the group. The members of the group comply with the group conclusions in their actions, accepting at the same time the (meta) principle of precedence of the group knowledge over individual knowledge.
3. Pointing out non-dogmatic procedures of extracting collective knowledge from the individual knowledge.
4. Tolerance toward inconsistencies appearing on various levels in the knowledge of individual agents forming the group. This results in acceptance of paraconsistent knowledge bases.
5. The existence of information gaps, where some aspects of knowledge are not fully available in various applications. This results in acceptance of various forms of non-monotonic reasoning.

A switch from the perspective based on reasonings carried out in multi-modal systems of high complexity to reasonings based on queries directed to paraconsistent knowledge bases has been proposed by Dunin-Kęplicz and Szałas (2012). Their approach has led to a new formalization of complex belief structures. They introduced the notion of an *epistemic profile* which is thought of as a tool that fills a gap between the epistemic logic approach and various practical implementations aiming at transforming the initial data into the final ones. 4QL, a four-valued language of queries, built by Małuszyński and Szałas (2011a, b) lays foundations for creating workable paraconsistent knowledge bases. 4QL assures practical computability, being thus an underpinning enabling one to implement belief structures and epistemic profiles.

6.3.2 Frames for Non-monotonic Reasonings

When do non-monotonic reasonings come into sight then? The term 'non-monotonic logic' seems to be misleading. The term 'logic' is reserved for a consequence operation which is structural (that is, closed under endomorphisms of the propositional language). A better term is 'non-monotonic reasoning'. A reasoning is *non-monotonic* if some conclusions can be invalidated by adding more knowledge. Non-monotonic reasonings need not be structural in comparison with inference rules of logic. Non-monotonic forms of reasoning should be treated as pragmatic rules of acceptance of sentences (or propositions) in which the rules of inference of classical logic are fully utilized. Therefore, the formal constructs which are defined in the theory of non-monotonic reasonings may admit supraclassical modes of reasoning. (As is well known, due to the maximality of classical logic, these non-monotonic formal non-structural constructs extend the consequence operation of classical logic.) The clue is that acquired knowledge (in sharp contrast to mathematical theories) may (in part) be suspended in light of a new evidence; the latter may cancel the validity of some premises previously employed in the reasoning carried out by the agents. Therefore, reasoning based on the gathered and updated knowledge may be subject to the process of revision, which in turn may invalidate earlier conclusions.This process is not captured by the (monotone) consequence operation of classical logic (or any other logic). Reasoning based on definite clauses with negation as failure is non-monotonic. Non-monotonic reasoning is useful for representing defaults. A *default* is a rule that can be used unless it is overridden by an exception. The phenomenon of non-monotonicity explicitly appears especially in these situational contexts when one is faced with fast information flow and information processing. The problem of description of the dynamics of the logical structure of the future events knowledge is one of the most difficult and challenging problems that philosophical logic faces. As is known, sentences and utterances about future are often logically indetermined, that is, there is no way of assigning a logical value of truth or falsity to them at the moment they are uttered. Certain knowledge is the knowledge *ex post*. The knowledge about the future is entrenched in various modalities: *it is possible or it is probable that an event (a situation) will occur*. The grammar of many natural languages have a complex system of tenses and of future tenses in particular. Usually, sentences about the future express an intention or ability of the agent of doing (or refraining from doing) an action (e.g., *I will not go to school tomorrow*) or they express routine or repeatable events as e.g., *A new issue of the New York Times will appear tomorrow*. In the first case the situation is more complex because the future is not under the control of agents—just after declaring that *I will go to the cinema tomorrow* it may happen that I will be involved in a car accident, the event that will annihilate my plans for tomorrow. The description of the future evolution of knowledge may contain, apart from time parameter, epistemic operators as in the utterance *Next month I will know more about relativity theory*. The situation is more involved in case of a dialogue about the future conducted by many persons.

The theory of non-monotonic reasoning has its roots in a critical appraisal of the adequacy of 'traditional' logic in practical life and in AI. The systems of traditional logic are founded (explicitly or implicitly) on the notion of Tarskian consequence relation or on the formalisms of natural deduction. Both these formalisms are monotone in the well-known sense. Minsky (1974) maintains that monotonicity is a source of problems. He writes:

> In any logistic system, all the axioms are necessarily 'permissive'—they all help to permit new inferences to be drawn. Each added axiom means more theorems, none can disappear. There simply is no direct way to add information to tell such the system about kinds of conclusions that should be drawn! To put it simply: if we adopt enough axioms to deduce what we need, we deduce far too many things.

Minsky is also critical of the requirement of consistency, the property being at the core of mathematical logic.

As Alexander Bochmann writes in *Logic in Nonmonotonic Reasoning* (available in the Internet):

> On the most general level, nonmonotonic reasoning is ultimately related to the traditional philosophical problems of natural laws and *ceteris paribus* laws. A more salient feature of these notions is that they resist precise logical definitions, but involve a description of 'normal' or 'typical' cases. As a consequence, reasoning with such concepts is inevitably *defeasible*, that is they may fail to 'preserve truth' under all circumstances, which has always been considered a standard of logical reasoning.

On a more concrete level, the original motivation that resulted in the rise of the theory of non-monotonic reasoning was the inadequacy of first-order logic in solving the problems besetting Artificial Intelligence, especially those concerning the representation of knowledge. We will mention three such problems (they are not strictly separated): (1) the frame problem—a fluent (time-dependent fact) does not change unless some action (it may be the action of nature) modifies it; (2) the decision-making problem—some actions are planned in the absence of complete information (some unforeseen obstacles may emerge while performing a string of actions). (3) the usability problem and usability engineering, the latter being a collection of techniques to support effective management of available resources, especially of databases while performing actions.

The theory of non-monotonic reasoning is intended to be close to commonsense, practical reasoning, the latter being based on what typically or normally holds. According to this theory, the agents make plans and act being devoid of a complete picture of the world in which they live, they do not know details of states of affairs, and they make assumptions and decisions relying on their own experience and habits. There are many general situations when one applies non-monotonic reasoning: it may be a communication convention in which one makes default assumptions in the absence of explicit information or a representation of action plans when, e.g., the teacher declares that the math exam will be given on Tuesday unless another decision is explicitly made, etc. The resulting formal structures conventionally called non-monotonic logics encompass *circumscriptions* (McCarthy 1980), *default logic* (Reiter 1980), *autoepistemic logic* (Moore 1985), *cumulative consequences*

(Makinson 1989) and many others. (It should be underlined that the theory of non-monotonic reasoning itself is an 'ordinary' multifaceted mathematical theory, developed by means of the tools of classical, but not necessarily first-order logic. For instance, there are strong links between circumscriptions and second-order logic.)

Dov Gabbay (1985) was the first who put forward investigations into non-monotonic reasoning from a general inferential perspective. The point of departure is the classical notion of a consequence operation (relation) satisfying the well-known conditions of reflexivity, cut, and cumulation and operating in an arbitrary sentential language. (Such a language need not contain only classical connectives.) By way of relaxation or modification of the above conditions one arrives at an appropriate inferential unit that may constitute a sound basis of non-monotonic reasonings.

Some researchers point out that intuitions associated with consequence relations may fail when one tries to transfer them onto non-monotonic inferences. This especially concerns the cut rule which is validated by every consequence relation. In some cases involving non-monotonic arguments this rule does not work. Consider the sentences:

P: It is raining,
Q: I take an umbrella,
R: I will not be wet.

I have the habit that I take an umbrella whenever it rains. Thus I accept $P \vdash Q$. If it is raining and I take an umbrella, then I will not be wet. Consequently $P, Q \vdash R$. But the cut rule then yields the paradoxical conclusion that $P \vdash R$, that is, if it is raining then I will not be wet. To explain this 'paradox', it suffices to observe that in the inference $P \vdash Q$ the premises P is certain—if it is raining, that it is difficult to deny this fact. But the conclusion Q is not so certain—it is a consequence of my habit. But it may happen I was absent-minded and I forgot to take an umbrella. In this example the paradox results from the fact that the conclusion is less certain than the premises.

The above example points out some difficulties arising when one attempts to uniformly model non-monotonic reasoning, especially in an axiomatic framework, by way of weakening the well-known conditions imposed on the Tarskian notion of a consequence relation. An axiomatic theory of non-monotonic consequence relations patterned upon some finitistic ideas going back to Gentzen was suggested by Gabbay (1985). An approach patterned upon Tarski's theory of consequence operation was examined by Makinson (1989).

There is a vast literature concerning non-monotonic consequence relations, their inferential character and semantics. We mention here the classical papers by Gabbay (1985), Kraus et al. (1990), Makinson (1989, 2005) and Shoham (1987). Before passing to more formal aspects of non-monotonic reasoning, let us recall here some locutions which will lead us to the proper topic:

(a) It is the case that P, therefore one usually accepts Q.
(b) It is the case that P, therefore Q in normal circumstances.
(c) We know that P, therefore we should accept Q.

Omitting probabilistic reasonings (making judgements probable), non-monotonic forms of reasoning will be treated as those that are based on certain collections of formal rules of conduct. These rules may be of epistemic character (they are then pragmatic rules of acceptance of sentences or utterances in the light of available knowledge) or praxeological (as rules of acceptance on the basis of mastered skills). Praxeological rules are not purely verbal and they may refer to some nonverbal actions performed by agents. The rules of conduct may tolerate the principles of logic, the latter being identified with an appropriate finitary and structural consequence relation; they are not, however, identical with rules of inference of, say, classical logic. In other words, the bond holding between the premises and the conclusion of a rule of conduct does not yield that the conclusion logically follows from the premises. But it may happen that this bond is stronger than logical consequence. In this case one obtains supraclassical non-monotonic inferences. Moreover, rules of conduct are not deontologically loaded—it is safely assumed that all the actions the rule involves are permitted in the circumstances indicated by the rule.

The main aim of this paragraph is to present a new approach to non-monotonic forms of reasoning which links them with the theory of action. We shall first outline a *semantic* scheme of defining non-monotonic reasonings in terms of frames. Each frame $\boldsymbol{F}$ is a set of states W endowed with a family $\boldsymbol{R}$ of relations of a certain kind. If S is a propositional language, say, the language of classical propositional logic equipped with some additional connectives, then one may define valuations of S in a frame. In turn, frames and valuations determine in a natural way consequence-like operations on S. The latter do not generally exhibit all properties of consequence operations such as monotonicity. These operations thus exemplify non-monotonic patterns of reasoning. The definition we give is general and the class of resulting structures encompasses preferential model structures investigated by Makinson. It also encompasses supraclassical models of reasoning. Then, in the second step, we pass to elementary action systems. The crucial notion we investigate is that of a *rule of conduct* associated with a given action system $\boldsymbol{M}$. Each rule of conduct (also called a *pragmatic rule*) is a rendition of a certain infallible procedures inherently linked with $\boldsymbol{M}$. It may be for example a complex system of foolproof action plans that guarantee manufacturing cars in a net of cooperating plants. In a simplified version, pragmatic rules are labeled trees defined in a certain way. In other words, pragmatic rules are conceived of as systems of actions or conduct the agents are obeyed to in the circumstances pertaining to the rule. These circumstances are determined by the propositions being the labels attached to vertices and the actions assigned to the edges of the tree.

Definition 6.3.1 A *frame* is any pair

$$\boldsymbol{F} = (W, \boldsymbol{R}),$$

where

(1) W is a nonempty set,

(2) $\boldsymbol{R}$ is a binary relation between families of subsets of W and subsets of W; that is,

$$\boldsymbol{R} \subseteq \wp\big(\wp(W)\big) \times \wp(W).$$

Thus, if $(\boldsymbol{X}, \Phi) \in \boldsymbol{R}$, then $\boldsymbol{X}$ is a family of subsets of W and Φ is a subset of W. We write $\boldsymbol{R}(\boldsymbol{X}, \Phi)$ whenever $(\boldsymbol{X}, \Phi) \in \boldsymbol{R}$. As is customary, the elements of W are called *states*, for which we will use the letters u, w, u', w', etc. Sets of states are called *propositions*.

A frame $\boldsymbol{F}$ is *finitary* if $\boldsymbol{R}$ obeys the condition: for any $\boldsymbol{X} \subseteq \wp(W)$ and any $\Phi \subseteq W$, if $\boldsymbol{R}(\boldsymbol{X}, \Phi)$, then there is a finite subfamily $\boldsymbol{X}_f \subseteq \boldsymbol{X}$ such that $\boldsymbol{R}(\boldsymbol{X}_f, \Phi)$. A frame $\boldsymbol{F}$ is *reflexive* if for any $\boldsymbol{X} \subseteq \wp(W)$, $\boldsymbol{R}(\boldsymbol{X}, \Phi)$ holds for all $\Phi \in \boldsymbol{X}$. □

Example Given a nonempty set W, let $\boldsymbol{R}$ be defined as follows: for $\boldsymbol{X} \subseteq \wp(W)$ and $\Phi \subseteq W$,

$$\boldsymbol{R}(\boldsymbol{X}, \Phi) \quad \Leftrightarrow_{df} \quad \bigcap \boldsymbol{X} \subseteq \Phi.$$

Then $(W, \boldsymbol{R})$ is a frame. □

Let S be a sentential language and $(W, \boldsymbol{R})$ a frame. Any map $V : S \to \wp(W)$ is called a *valuation*. If V is a valuation, then the triple $(W, \boldsymbol{R}, V)$ is called a *model* (for S). If S is the language of classical propositional logic, each valuation is extended in a one-to-one manner to a homomorphism of S into the Boolean algebra of subsets of W. If S contains some other connectives as epistemic or dynamic ones, each valuation is extended to a homomorphism of S using some additional resources of $\boldsymbol{F}$ as e.g., various accessibility relations on W. But these relations do not belong to $\boldsymbol{R}$.

Definition 6.3.2 Let $\boldsymbol{F} = (W, \boldsymbol{R})$ be a frame. $\boldsymbol{F}^{\models}$ is the operation from $\wp(S)$ to $\wp(S)$ defined as follows:

$$\alpha \in \boldsymbol{F}^{\models}(X) \quad \Leftrightarrow_{df} \quad \begin{array}{l}\text{for every valuation } V \text{ of S in } \boldsymbol{F} \\ \text{it is the case that } \boldsymbol{R}\big(\{V(\beta) : \beta \in X\}, V(\alpha)\big).\end{array} \tag{6.3.3}$$

□

We shall first give a bunch of simple observations concerning $\boldsymbol{F}^{\models}$. The proofs are easy and are omitted.

Lemma 6.3.3 *If $\boldsymbol{F} = (W, \boldsymbol{R})$ is reflexive, then the operation $\boldsymbol{F}^{\models}$ satisfies inclusion; that is, $X \subseteq \boldsymbol{F}^{\models}(X)$ for all $X \subseteq S$.* □

Lemma 6.3.4 *Suppose $\boldsymbol{F} = (W, \boldsymbol{R})$ satisfies the condition: for all $\boldsymbol{X}, \boldsymbol{Y} \subseteq \wp(W)$,*

$$\text{if } \boldsymbol{X} \subseteq \boldsymbol{Y} \text{ and } \boldsymbol{R}(\boldsymbol{X}, \Psi) \text{ for all } \Psi \in \boldsymbol{Y}, \quad \text{then for all } \Phi \subseteq W,\ \boldsymbol{R}(\boldsymbol{Y}, \Phi) \text{ implies } \boldsymbol{R}(\boldsymbol{X}, \Phi).$$

Then $\boldsymbol{F}^{\models}$ is cumulative transitive, i.e., $X \subseteq Y \subseteq \boldsymbol{F}^{\models}(X)$ implies $\boldsymbol{F}^{\models}(Y) \subseteq \boldsymbol{F}^{\models}(X)$, for all $X, Y \subseteq S$. □

Corollary 6.3.5 *If $\boldsymbol{F}$ satisfies the assumptions of Lemmas 6.3.3 and 6.3.4, then $\boldsymbol{F}^{\models}$ is idempotent, i.e., $\boldsymbol{F}^{\models}(\boldsymbol{F}^{\models}(X)) = \boldsymbol{F}^{\models}(X)$.* □

Corollary 6.3.6 *If $\boldsymbol{F}$ satisfies the condition: for all $\boldsymbol{X}, \boldsymbol{Y} \subseteq \wp(W)$ and any $\Phi \subseteq W$,*

$$\boldsymbol{X} \subseteq \boldsymbol{Y} \text{ and } \boldsymbol{R}(\boldsymbol{X}, \Phi) \text{ imply } \boldsymbol{R}(\boldsymbol{Y}, \Phi),$$

then $\boldsymbol{F}^{\models}$ is monotone, i.e., $\boldsymbol{F}^{\models}(X) \subseteq \boldsymbol{F}^{\models}(Y)$ whenever $X \subseteq Y$. □

The frame $\boldsymbol{F}$ from the above example satisfies the hypothesis of Corollary 6.3.6. $\boldsymbol{F}$ thus yields a *monotone* consequence operation.

Note The hypothesis of Lemma 6.3.4 can be simplified as follows:

Lemma 6.3.7 *Suppose $\boldsymbol{F} = (W, \boldsymbol{R})$ satisfies the condition: for all $\boldsymbol{X}, \boldsymbol{Y} \subseteq \wp(W)$ and $\Psi \subseteq W$,*

$$\text{if } \boldsymbol{R}(\boldsymbol{Y}, \Phi) \text{ for all } \Phi \in \boldsymbol{X}, \text{ and } \boldsymbol{R}(\boldsymbol{X}, \Psi) \text{ then } \boldsymbol{R}(\boldsymbol{Y}, \Psi).$$

Then $\boldsymbol{F}^{\models}$ satisfies: $X \subseteq \boldsymbol{F}^{\models}(Y)$ implies $\boldsymbol{F}^{\models}(X) \subseteq \boldsymbol{F}^{\models}(Y)$, for all $X, Y \subseteq S$. □

Let W a nonempty set. Let $<$ be a binary relation on W which is called a *preference relation*. We define the relation $R_< \subseteq \wp(\wp(W)) \times \wp(W)$ as follows: for $\boldsymbol{X} \subseteq \wp(W)$ and $\Phi \subseteq W$,

$$R_<(\boldsymbol{X}, \Phi) \Leftrightarrow_{df} (\forall w \in W)\big(w \in \bigcap \boldsymbol{X} \,\&\, \neg(\exists w')(w' < w \,\&\, w' \in \bigcap \boldsymbol{X}) \Rightarrow w \in \Phi\big).$$

$\boldsymbol{F}_< := (W, \boldsymbol{R}_<)$ is called a *preference frame*.

We then define the consequence operation $\boldsymbol{F}_<^{\models}$ according to formula (6.3.3), that is, for every valuation V of S in $\boldsymbol{F}_<$, $\alpha \in \boldsymbol{F}^{\models}(X)$ if and only if it is the case that $\boldsymbol{R}_<(\{V(\beta) : \beta \in X\}, V(a))$.

The frame $\boldsymbol{F}_<$ is linked with preferential model structures in the sense of Makinson (2005, Chap. 3) in the following way. Let $V : S \to \wp(W)$ be a valuation in $\boldsymbol{F}_< = (W, \boldsymbol{R}_<)$. We define the *satisfaction relation* $\models^V \subseteq W \times S$ (depending on V):

$$w \models^V \alpha \quad \text{if and only if} \quad w \in V(\alpha).$$

Following Makinson, the triple $(W, \models^V, <)$ is called a *preferential model structure*. For any $X \subseteq S$ we put:

$$\begin{array}{lll} w \models^V X & \text{if and only if} & (\forall \alpha \in X)(w \models^V \alpha) \\ & (\text{if and only if} & w \in \bigcap_{\alpha \in X} V(\alpha)). \end{array}$$

We also define:

$$w \models^V_< X \quad \text{if and only if} \quad w \models^V X \;\&\; \neg(\exists w')(w' < w \;\&\; w' \models^V X)$$

and say that w *preferentially* satisfies X with respect to V.

Proposition 6.3.8 *Let X be a set of formulas and α a formula of S. For a given preference frame $\boldsymbol{F}_< = (W, \boldsymbol{R}_<)$ the following conditions are equivalent:*

(i) $\alpha \in \boldsymbol{F}^{\models}_<(X)$;
(ii) For any valuation $V : S \to \wp(W)$ in $\boldsymbol{F}_<$, it is the case that $(\forall w \in W)(w \models^V_< X \Rightarrow w \models^V \alpha)$.

Proof (i) $\Rightarrow$ (ii). Assume $\alpha \in \boldsymbol{F}^{\models}_<(X)$. This means that $\boldsymbol{R}_<\big(\{V(\beta) : \beta \in X\}, V(\alpha)\big)$ for any valuation $V : S \to \wp(W)$ in $\boldsymbol{F}_<$. Accordingly, for any valuation V in $\boldsymbol{F}_<$, it is the case that

$$(\forall w \in W)\big(w \in \bigcap_{\beta \in X} V(\beta) \;\&\; \neg(\exists w')\big(w' < w \;\&\; w' \in \bigcap_{\beta \in X} V(\beta)\big) \Rightarrow w \in V(\alpha)\big).$$

But $w \in \bigcap_{\beta \in X} V(\beta) \;\&\; \neg(\exists w')\big(w' < w \;\&\; w' \in \bigcap_{\beta \in X} V(\beta)\big)$ states that $w \models^V_< X$. Thus for any valuation V in $\boldsymbol{F}_<$, $(\forall w \in W)(w \models^V_< X \Rightarrow w \models^V \alpha)$. So (ii) holds.

The reverse implication (ii) $\Rightarrow$ (i) is also immediate. □

How should the above semantics be combined with action theory? A realistic theory of action should not assume the omnipotence and omniscience of agents. Some natural limitations should be assumed, which take into account the lives and professional experience of agents as well as accompanying principles of conduct, also in difficult situations. We shall not mean merely algorithmizable procedures but also those actions that require consideration and best fit to a new situation. Agents may trip up and some tasks may outgrow their action abilities and intellectual capabilities. But, on the other hand, the agents have the ability to overcome or get around problems and obstacles. Any reasonable theory should not ignore these aspects of action; it should however point to sources of the problems and to repair work.

6.3.3 Tree-Like Rules of Conduct

In this section the notion of a rule of conduct is introduced. The terms *rule of conduct*, *procedure*, and *pragmatic rule* are treated interchangeably. Rules of conduct are usually historically and socially changeable; they are not set for ever. They constitute, however, the basis of action. When speaking of rules of conduct we shall abstract from deontological aspects of action. Each rule of conduct in the semantic formulation is strictly conjoined with an elementary action system and it involves some actions of the system that are organized in a certain way. It is assumed that these actions are totally performable and hence permitted in the sense of the system. (Rules of conduct may be treated as norms when one is interested in examining various situations in which the actions the rule involves are obligatory; but this (difficult) problem is omitted here.) We shall consider simple rules founded on finite trees. (Rules based on more elaborate set-theoretic constructs like finite (or infinite) directed graphs are also conceivable; such rules are not discussed in this book.)

The notion of a rule of conduct is the basic semantic tool that underlies a conception of non-monotonic reasonings. This conception strictly links the theory of actions with the semantics of non-monotonic reasonings. This conception is expounded in this paragraph.

Let us consider the following sentences taken from everyday English:

If it is raining, I usually take an umbrella;

If we are sleepy, we go to bed;

If you have a headache, you take a pill;

He usually drinks tea for breakfast;

If you are a patient with end-stage heart failure or severe coronary artery disease, usually a heart transplant is performed;

If one is in cardiac arrest, cardiopulmonary resuscitation (CPR) *is performed* (*in an effort to manually preserve intact brain function until further measures are taken.*)

The above examples show that under the given circumstances one performs an appropriate action so that after performing it the required states of affairs are *usually* reached (I will not be wet; we will be well-rested; you will lose your headache, etc.) The proviso *usually* is relevant: I may forget to take the umbrella (hence I will be wet); the patient in cardiac arrest may die directly after the event despite the fact that CPR has been immediately performed, etc. We want to model the above situations on the grounds of our theory of actions, that is, we want to find set-theoretic entities that will represent the above procedures. The problem is that set theory does not deal with formulas of type ‘an element a *usually* belongs to a set A’. (This is not the case of fuzzy sets or a probabilistic grounded set theory; building a theory of sets which would capture meanings of such formulas is a program for the future.) We face therefore with a phenomenon that is an instance of the ramification problem. The solution we propose is simple: the above procedures are treated as infallible. Possible failures of them are neglected; exceptions to the rule (when the rule possibly fails) do not disqualify it if the rule is still in use.

A remark is appropriate here. Pseudo-Geber was a famous medieval alchemist (but also a chemist!) He assumed that all metals are composed of unified sulfur and mercury. He gave detailed descriptions of metallic properties in those terms. Today we know that the procedures (i.e., the rules of transmuting sulfur and mercury into various metals) based on his descriptions are utterly wrong. However, it should be added that “his practical directions for laboratory procedures were so clear that it is obvious he was familiar with many chemical operations” (from Wikipedia.)

Let $\boldsymbol{M} = (W, R, \boldsymbol{\mathcal{A}})$ be an atomic action system. Since deontological issues are not discussed here, it is assumed for simplicity that $\boldsymbol{M}$ is normal. Consequently, each atomic action of $\boldsymbol{\mathcal{A}}$ is totally performable and the relation R may be discarded. Therefore, the definitions that follow do not involve the relation of direct transition.

In a more formal framework, one may then argue that the above examples are instantiations of the following scheme of action. Let Φ and Ψ be propositions, i.e., subsets of W, and let A be an atomic action of $\boldsymbol{M}$. We consider the rule:

> 'It is the case that Φ; therefore one usually performs an action A in order to achieve a set of states Ψ'

(Since the notion of an agent is not incorporated into elementary action systems, we shall henceforth use the impersonal form 'one'.) The above rule of conduct should be infallible in the following sense: if the action A is performed in *whatever* state of Φ then, after performing A, a state in Ψ is reached. (As we assume that $\boldsymbol{M}$ is normal, A is included in the transition relation of the system $\boldsymbol{M}$.) This postulate is encapsulated by the inclusion

$$A[\Phi] \subseteq \Psi \tag{6.3.4}$$

($A[\Phi]$ is the A-image of Φ, $A[\Phi] := \{w \in W : (\exists u \in \Phi)\ (u, w) \in A\}$.) This inclusion expresses the fact that for every (realizable) performance (u, w) of A, if $u \in \Phi$, then $w \in \Psi$. In other words, by performing A in any state of Φ the agent always reaches a state belonging to Ψ, as the rule says. From the epistemic perspective (see Lemma 6.1.1), if $A = A_a$ is the accessibility relation of an agent a then (6.3.4) says that the agent a knows Ψ at every state $u \in \Phi$. But we shall use rules that are infallible in a possibly strongest way—these are rules in which (6.3.4) is lifted to the equality

$$A[\Phi] = \Psi. \tag{6.3.5}$$

Thus, if the action A is performed in *whatever* state of Φ then, after performing A, a state in Ψ is reached *and* every state of Ψ is obtained from Φ by performing the action A.

In fact, we shall consider more complex rules of conduct. These are tree-like structures. More precisely, the rules we shall define are certain labeled finite trees. (We may also define rules built on finite directed graphs but we shall confine ouselves to finite trees here.) The rules in which only one action is involved and which satisfy (6.3.5) are certain limit cases of more general tree-like rules of conduct which we shall define below. In fact, the former are rules defined on two-element, linear trees.

Let us quote the following passage from Wikipedia:

> An *intermediate product* is a product that might require further processing before it is saleable to the ultimate consumer. This further processing might be done by the producer or by another processor. Thus, an intermediate product might be a final product for one company and an input for another company that will process it further.

Suppose $c_1, \ldots, c_k$ are agents (producers or suppliers). Each of them manufactures an intermediate product of a certain kind which is supplied to another producer b. The agent b processes the products provided by $c_1, \ldots, c_k$ and produces a (more advanced) product which in turn is an input for another company operated by the agent a. We therefore have a simple tree showing this situation:

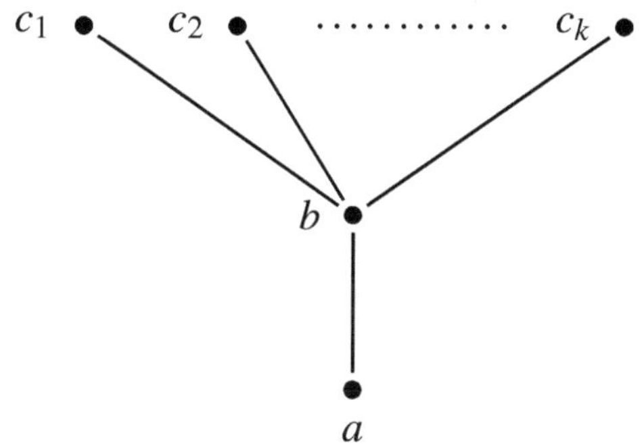

Fig. 6.1

The above tree may be a subtree of a larger tree representing the whole production process. The suppliers of raw materials are marked at the leaves of this larger tree. The final product (commodity) is manufactured by the agent at the root of the tree. The manufacturer c_i performs an action $A_{c_i b}$ in a definite set of states Φ_{c_i}, required by the production regime and supplies his intermediate product, being the result of $A_{c_i b}$, to b. In turn, the manufacturer b starts producing in a set of definite states Φ_b in which the intermediate products provided by the suppliers $c_1, \ldots, c_k$ are indispensable. As the output (result) of the action A_{ba} the agent b performs yet another product is manufactured he supplies to c (c may also have other suppliers.) The above fragment of the production scheme may be treated as a special infallible rule of action. (From the mathematical viewpoint, the above tree-like scheme is a simplification of the real manufacturing process because the intermediate product manufactured by a producer is often an input for more than one company that processes it further.)

From the more abstract perspective the above scheme is encapsulated by the following definitions.

We recall that a finite partially ordered set $(T, \leqslant, \mathbf{0})$ is a tree (with zero $\mathbf{0}$ as a designated constant) if one vertex has been designated (the root), in which case the edges have a natural orientation: *upwards* from the root. The root $\mathbf{0}$ is the least element of the tree. The *tree-order* is thus the partial ordering on the vertices of a tree with $x \leqslant y$ if and only if the unique path from the root to y passes through x.

In each finite tree, the *parent* of a vertex is the vertex connected to it on the path to the root; every vertex except the root has a unique parent. A *child* of a vertex x is a vertex of which x is the parent. *Leaves* are the vertices that do not have children.

A *labeled tree* is a tree $(T, \leqslant, \mathbf{0})$ in which each vertex is given a unique label. In Chap. 5 some preliminary remarks on tree-like positive norms were presented. We shall return to these norms. We shall define here the notion of a rule of conduct. A rule of conduct is a tree-like positive norm which is labeled in a certain canonical way.

Definition 6.3.9 Let $\boldsymbol{M} = (W, R, \boldsymbol{A})$ be a normal atomic action system. A *rule of conduct* for $\boldsymbol{M}$ is a finite labeled tree with root $\mathbf{0}$ in which to each vertex a a subset Φ_a of W is assigned (Φ_a is thus the label of a) and to each edge ab in which b is a child of a (and hence $a < b$) an atomic action A_{ba} of $\boldsymbol{M}$ (from child b to its parent a) is assigned so that the following condition is met:

(6.3.6) if $b_1, \ldots, b_k$ are the children of a, then $A_{b_1 a}[\Phi_{b_1}] \cap \cdots \cap A_{b_k a}[\Phi_{b_k}] = \Phi_a$.

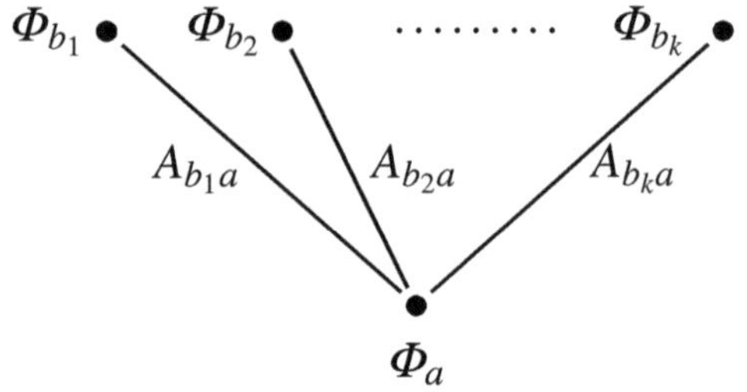

Fig. 6.2

As declared earlier, we shall interchangeably used the terms 'rule of conduct' and 'pragmatic rule'.

$\Phi_{\mathbf{0}}$ is the proposition assigned to the root $\mathbf{0}$. Intuitively, each rule assures that if the agents start working in states belonging to the propositions attached to the leaves of the tree and consecutively perform (nonlinearly) the actions A_{ba} for all a, b with b being a child of a, the system will eventually reach a state belonging to $\Phi_{\mathbf{0}}$. The proposition $\Phi_{\mathbf{0}}$ is thus regarded as the set of *final states* (or the *conclusion*) of the rule while the propositions being the labels of leaves are called initial *propositions* (or the *premises*) of the rule. The actions the rule involves thus guarantee a success; that is, reaching a final state if the agents start working in any states belonging to the initial propositions. Each rule is therefore *stable* in the sense that the actions performable by the children of a in the states belonging to the respective labels lead to the states attached to the label of the parent *a and* each state belonging to the label attached to a is obtained by performing the actions assigned to the children of a in some states belonging to the respective label.

If a rule r is built on a two-element tree, and therefore r is visualized as

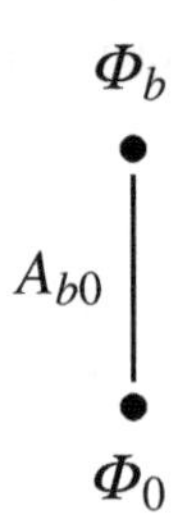

Fig. 6.3

we have that $\Phi_{\mathbf{0}} = A_{b\mathbf{0}}[\Phi_b]$.

If r is built on a one-element tree (consisting of the vertex $\mathbf{0}$ only), the set of children of $\mathbf{0}$ is empty and no atomic action is attached to the tree. Therefore the set of initial propositions (premises) of the rule is empty.

Let $\boldsymbol{P}$ be a collection of pragmatic rules associated with the system $\boldsymbol{M}$. We define a relation $\boldsymbol{R_P}$ between finite sets of propositions and individual propositions.

Definition 6.3.10 Let X be a finite set of proposition and Ψ a proposition. Then

$$\boldsymbol{R_P}(X, \Psi)$$

means that there is a rule of conduct r in $\boldsymbol{P}$ such that $\Psi = \Phi_0$ is the final proposition of r (= the label of the root) and X is the set of initial propositions of the rule (= the set of labels of the leaves of the tree).

The above definition is extended onto infinite sets X of propositions, using the well-known finitarization procedure, that is, $\boldsymbol{R_P}(X, \Psi)$ means that $\boldsymbol{R_P}(X_f, \Psi)$ holds (in the above sense) for some finite subset $X_f \subseteq X$. □

If $\boldsymbol{P}$ contains a rule built on a tree in which the label of each leaf is the empty set, then $\boldsymbol{R_P}(\emptyset, \Psi)$.

Finite, linearly ordered sets are trees. If $\boldsymbol{P}$ consists of labeled linear orders, $\boldsymbol{R_P}$ is a one-premises relation, that is, $\boldsymbol{R_P}(X, \Psi)$ implies that the family X consists of at most one element.

The above definition is correct and it implies that

$$\boldsymbol{F_P} := (W, \boldsymbol{R_P})$$

is a finitary frame assigned to the action system $\boldsymbol{M}$.

One of the operations performed on a tree is taking a subtree of the tree. In the context of pragmatic rules *special* subtrees are of some importance. (We confine ourselves to finite trees.) A special subtree of a finite tree $(P, \leqslant, \mathbf{0})$ with zero is a tree (with the same bottom element $\mathbf{0}$) being the result of consecutive applying the operation of cutting off the upper part of the tree $(P, \leqslant, \mathbf{0})$ (see the figure below) above a fixed vertex and leaving the rest of the tree.

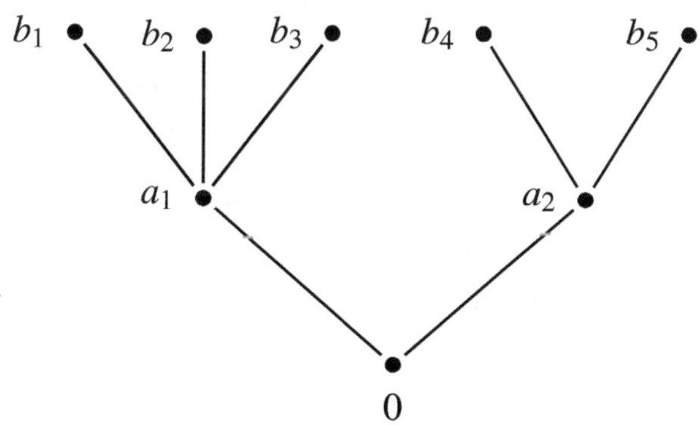

Fig. 6.4

By cutting off the three leaves over a_1 we get the tree

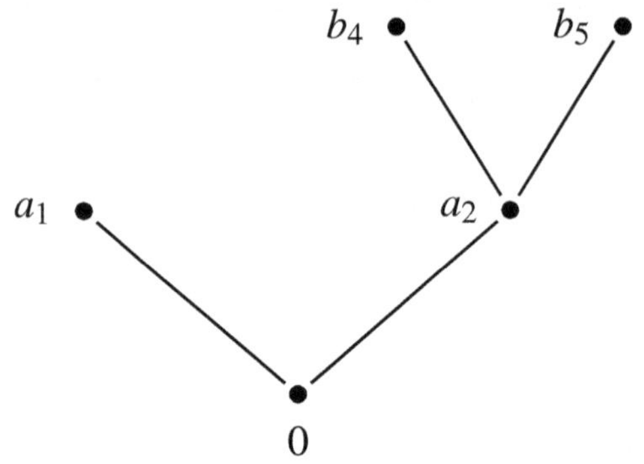

Fig. 6.5

It is easy to see that if r is a pragmatic rule, then any special subtree of the labeled tree of r is a labeled tree as well and it is also a rule, i.e., condition (6.3.6) is also met for the labels of the subtree. If the set $\boldsymbol{P}$ has the property that for every pragmatic rule r in $\boldsymbol{P}$, the rules based on the special subtrees of the labeled tree of r also belong to $\boldsymbol{P}$, we say that $\boldsymbol{P}$ is *closed with respect to special subtrees*. The closedness with respect to special subtrees is a quite natural condition that may be imposed on any set $\boldsymbol{P}$ of applicable pragmatic rules.

$\boldsymbol{R_P}$ is *monotone* if for any finite families $\boldsymbol{X}$, $\boldsymbol{Y}$ and any set $\Psi \subseteq W$, $\boldsymbol{X} \subseteq \boldsymbol{Y}$ and $\boldsymbol{R_P}(\boldsymbol{X}, \Psi)$ implies $\boldsymbol{R_P}(\boldsymbol{Y}, \Psi)$.

It is not difficult to produce action systems $\boldsymbol{M}$ and sets $\boldsymbol{P}$ of pragmatic rules such that $\boldsymbol{R_P}$ is *not* monotone. For example, if $\boldsymbol{P}$ is nonempty and consists of rules based on linearly ordered finite sets, then there are one-element families $\boldsymbol{X}$ and sets $\Psi \subseteq W$ such that $\boldsymbol{R_P}(\boldsymbol{X}, \Psi)$ holds. But $\boldsymbol{R_P}(\boldsymbol{Y}, \Psi)$ fails to hold if $\boldsymbol{Y}$ has at least two elements. Hence $\boldsymbol{R_P}$ is not monotone.

Suppose now that the atomic actions of the system $\boldsymbol{M}$ are reflexive. This implies that $\Psi \subseteq A[\Psi]$ for any set $\Psi \subseteq W$ and any atomic action A. Let $\boldsymbol{P}$ be any set of pragmatic rules associated with $\boldsymbol{M}$ and closed under special subtrees. Then

(6.3.7) $\boldsymbol{R_P}(\boldsymbol{X}, \Psi)$ *implies* $\bigcap \boldsymbol{X} \subseteq \Psi$ *for any finite family* $\boldsymbol{X}$ *and any* Ψ.

Indeed, let us assume that $\boldsymbol{R_P}(\boldsymbol{X}, \Psi)$, where $\boldsymbol{X} = \{\Psi_1, \ldots, \Psi_k\}$. There is a rule r in $\boldsymbol{P}$ such that $\boldsymbol{X}$ is the set of labels assigned to the leaves of the tree of r and Ψ is the label of the root of the tree. Induction on the height of the subtrees of the tree of r then shows that $\Psi_1 \cap \cdots \cap \Psi_k \subseteq \Psi$.

We are interested in reversing the implication (6.3.7) for an action system and a set $\boldsymbol{P}$ of rules, that is, we ask when

(6.3.8) $\bigcap \boldsymbol{X} \subseteq \Psi$ *implies* $\boldsymbol{R_P}(\boldsymbol{X}, \Psi)$ *for any finite family* $\boldsymbol{X}$ *and any* Ψ.

(6.3.8) is strictly linked with supraclassical non-monotonic forms of reasoning, the ones that accept the rules and axioms of classical logic.

Let $\boldsymbol{S}$ be the formula-algebra of classical propositional language. A *valuation* V is a mapping from the set of propositional variables of S into the power set $\wp(W)$. V is then extended in the standard (Boolean) way onto the entire set S.

Let $\boldsymbol{P}$ be a set of pragmatic rules and $\boldsymbol{F_P} = (W, \boldsymbol{R_P})$ the frame corresponding to $\boldsymbol{P}$. If V is a valuation, then the triple $(W, \boldsymbol{R_P}, V)$ is called a *model* (for S).

$\boldsymbol{F}_{\boldsymbol{P}}^{\models}$ denotes the operation from $\wp(S)$ to $\wp(S)$ defined according to formula (6.3.3) above; that is, for a finite set of formulas $X = \{b_1, \ldots, b_m\}$ and a formula α,

$$\alpha \in \boldsymbol{F}_{\boldsymbol{P}}^{\models}(\beta_1, \ldots, \beta_m) \quad \Leftrightarrow_{df} \quad \text{for every valuation } V \text{ of } S \text{ in } \boldsymbol{F_P}$$
$$\text{it is the case that } \boldsymbol{R_P}\big(\{V(\beta_1), \ldots, V(\beta_m)\}, V(\alpha)\big).$$

$\boldsymbol{F}_{\boldsymbol{P}}^{\models}$ is subsequently extended onto infinite subsets of S by means of the finitarization procedure.

In the light of the above remarks, $\boldsymbol{F}_{\boldsymbol{P}}^{\models}$ need not be a monotone operation, that is, $\alpha \in \boldsymbol{F}_{\boldsymbol{P}}^{\models}(X)$ and $X \subseteq Y$ do not yield $\alpha \in \boldsymbol{F}_{\boldsymbol{P}}^{\models}(Y)$.

The above constructions provide a bunch of systems of non-monotonic reasoning from the semantic perspective. This perspective is constituted by sets $\boldsymbol{P}$ of tree-like rules of conduct and the associated relation $\boldsymbol{R_P}$.

The rules of inference of classical logic can also be represented as tree-like rules of conduct on spaces of states. Let W be a nonempty set. States are *proof trees* whose edges are labeled by subsets of W. Each rule of inference of CPC is treated as an action on a proof tree which modifies this tree in an appropriate way. Rules of inference can be then represented as tree-like labeled rules of conduct acting on proof trees. Yet another option is to consider finite sequences of subsets of W which are (Hilbert-style) proofs. The details are omitted.

A thorough examination of the resulting systems of non-monotonic reasonings is beyond the reach of this book and requires a separate, detailed scrutiny. This is a project for a future investigation.

Bibliography

Ajdukiewicz, K. (1974). *Pragmatic logic*. Dordrecht: Reidel.

Apostel, L. (1978). The elementary theory of collective action. *Philosophica, 21*, 129–157.

Atienza, M., & Manero, J. R. (1998). *A theory of legal sentences*. Dordrecht: Kluwer.

Åqvist, L. (1974). A new approach to the logical theory of action and causality. Logical theory and semantic analysis (pp. 73–91). Dordrecht: D. Reidel.

Banachowski, L., Kreczmar, A., Mirkowska, G., Rasiowa, H., & Salwicki, A. (1977). An introduction to algorithmic logic. Metamathematical investigations in the theory of programs. In A. Mazurkiewicz & Z. Pawlak (Eds.), Mathematical foundations of computer science (Vol. 2, pp. 7–99). Warsaw: Banach Center Publications, PWN.

Barbiani, P., Herzig, A., & Troquard, N. (2007). Alternative axiomatics and complexity of deliberative STIT theories. *Journal of Philosophical Logic, 37*, 387–406.

Bartoszyński, R., Pawlak, Z., & Szaniawski, K. (1963). Matematyczne przyczynki do rozwoju prakseologii (Mathematical contributions to the development of praxiology) (in Polish). Warsaw.

Barwise, J., & Perry, J. (1983). *Situations and attitudes*. Cambridge: MIT.

Bergstöm, L. (1966). Alternatives and consequences of actions (dissertation). University of Stockholm. Stockholm Studies in Philosophy (Vol. 4). Stockholm: Almqvist and Wiksell.

Blikle, A. (1973). Wybrane zagadnienia lingwistyki matematycznej (Selected problems of mathematical linguistics) (in Polish). In A. Mazurkiewicz (Ed.), *Problemy przetwarzania informacji (Problems in information processing)* (Vol. 2, pp. 94–152). Warsaw: WNT.

Blikle, A. (1977). An analysis of programs by algebraic means. In A. Mazurkiewicz & Z. Pawlak (Eds.), *Mathematical foundations of computer science* (Vol. 2, pp. 167–213). Warsaw: Banach Center Publications, PWN.

Błaszczyk, A., & Turek, S. (2007). *Teoria mnogości (Set theory) (in Polish)*. Warsaw: PWN.

Boden, M. (1977). *Artificial intelligence and natural man*. Brighton: Harvester Press.

Bratman, M. E. (1999). *Intention, plans, and practical reason*. Stanford: CSLI Publications.

Brown, F. M. (Ed.) (1987). The frame problem in artificial intelligence. In Proceedings of the 1987 Workshop Held in Lawrence, Kan. Palo Alto: Morgan Kaufmann.

Cai, J., & Paige, R. (1992). Languages polynomial in the input plus output. In Second International (Ed.), *Conference on Algebraic Methodology and Software Technology (AMAST 91)* (pp. 287–300). Berlin: Springer.

Castilho, M., Herzig, A., & Varzinczak, I. (2002). It depends on the context! A decidable logic of action and plans based on a ternary dependence relation. Workshop on nonmonotonic reasoning (NMR).

J. Czelakowski, *Freedom and Enforcement in Action*, Trends in Logic 42,
DOI 10.1007/978-94-017-9855-6

Chang, C. C., & Keisler, H. J. (1973). *Model theory*. Amsterdam: North Holland and American Elsevier.
Cholvy, L. (1999). Checking regulations consistency by using SOL-resolution. In Proceedings of the 7th International Conference on AI and Law (pp. 73–79). Oslo.
Chomsky, N., & Miller, G. A. (1958). Finite state languages. Information and Control, 1(2), 91–112. (Cook, M.).
Chomsky, N., & Miller, G. A. (2004). Universality in elementary cellular automata. *Complex Systems*, *15*(1), 1–40.
Czelakowski, J. (1991). Prakseologia formalna i formalna teoria działania (Formal praxiology and formal action theory) (in Polish). *Prakseologia, Nr*, *1–2*(110–111), 221–255.
Czelakowski, J. (1993). Probabilistyczne teorie działania (Probabilistic theories of action) (in Polish). *Prakseologia, Nr*, *1–1*(118–119), 25–54.
Czelakowski, J. (1996). Elements of formal action theory. In A. Fuhrmann & H. Rott (Eds.), Logic, action and information. Essays on logic and artificial intelligence (pp. 3–62). Berlin: Walter de Gruyter.
Czelakowski, J. (1997). Action and deontology. In E. Ejerhed & S. Lindström (Eds.), *Logic, action and cognition* (pp. 47–88). Dordrecht: Kluwer.
Czelakowski, J. (2001). *Protoalgebraic logics*. Dordrecht: Kluwer.
Davey, B. A., & Priestley, H. (2002). *Introduction to lattices and order* (2nd ed.). Cambridge: Cambridge University Press.
Davidson, D. (1963). Actions, reasons, and causes. *Journal of Philosophy*, *60*, 685–700.
Davidson, D. (1967). The Logical form of action sentences. In N. Rescher (Ed.), *The logic of decision and action* (pp. 81–95). Pittsburgh: University of Pittsburgh Press.
Davidson, D. (2001). *Essays on actions and events* (2nd ed.). Oxford: Oxford University Press.
Desharnais, J., & Möller, B. (2005). Least reflexive points of relations. *Higher-Order and Symbolic Computation*, *18*, 51–77.
Dunin-Kęplicz, B. (2012). Epistemic profiles and belief structures. In G. Jezic et al. (Eds.), Agent and multi-agent systems. Technologies and applications. Proceedings of the 6th KES International Conference, KES-AMSTA 201. Dubrovnik, Croatia, 25–27 June 2012. Lecture Notes in Computer Science (Vol. 7327, pp. 360–369). Springer.
Eiter, T, Erdem, E., Fink, M, & Senko, J. (2005). Updating action domains descriptions. In L. Kaelbling & A. Saffioti (Eds.), Proceedings of the 19th International Joint Conference on Artificial Intelligence (IJCAI) (pp. 418–423). Morgan Kaufmann Publishers.
Finger, M., & Gabbay, D. (1992). Adding a temporal dimension to a logic system. *Journal of Logic, Language and Information*, *1*(3), 203–233.
Fischer, M. J., & Ladner, R. E. (1979). Propositional dynamic logic of regular programs. *Journal of Computer and System Sciences*, *18*, 194–211.
Gabbay, D. M. (1985). Theoretical foundations for non-monotonic reasoning in expert systems. In K. R. Apt (Ed.), Proceedings NATO Advanced Institute on Logics and Models of Concurrent Systems. (La Colle-sur-Loup, France) (pp. 439–457). Berlin: Springer.
Gabbay, D., Horty, J., Parent, X., van der Meyden, R., van der Torre, L. (Eds.), (2014–2015). Handbook of deontic logic and normative systems (Vol. 1). Available on Amazon UK/US (Vol. 2). Planned for publication 2014/2015. London: College Publications.
Gabbay, D., Pnueli, A., Shelah, S., & Stavi, J. (1980). The temporal analysis of fairness. In Seventh ACM Symposium on Principles of Programming Languages (pp. 163–173). Las Vegas, Nevada.
Genesereth, M. R., & Nilsson, N. J. (Eds.). (1987). *Logical foundations of artificial intelligence*. Palo Alto: Morgan Kaufmann.
Gettier, E. (1963). Is justified true belief knowledge? *Analysis*, *23*, 121–123.
Goldblatt, R. I. (1982a). Axiomatizing the logic of computer programming. Lecture Notes in Computer Science (Vol. 130). Berlin: Springer.
Goldblatt, R. I. (1982b). The semantics of Hoare's iteration rule. *Studia Logica*, *41*, 141–158.
Goldblatt, R. I. (1986). Review article. *The Journal of Symbolic Logic*, *51*, 225–227.
Goldblatt, R. I. (1987). Logics of time and computation. CSLI Lecture Notes (Vol. 7). Stanford.

Goldmann, A. I. (1970). *A theory of human action*. Englewood Cliffs: Prentice-Hall.

Goldman, E. (1971). *Theory of actions*. Ann Arbor: Michigan University Press.

Gonzalez, L. D. (2003). *The paradoxes of action: Human action, law and philosophy*. Dordrecht: Kluwer.

Greibach, S. A. (1965). A new normal form theorem for context-free phrase structure grammars. *Journal of the ACM, 12*, 42–52.

Grigoris, A. (1997). *Nonmonotonic reasoning*. Cambridge: The MIT Press.

Halmos, P. R. (1950). *Measure theory*. New York: Van Nostrand.

Halpern, J. Y. (Ed.) (1986). Theoretical aspects of reasoning about knowledge. In Proceedings of the 1986 Conference Held in Monterey, California 19–22 March 1986. Palo Alto: Morgan Kaufmann.

Hansson, S. O. (1986). Individuals and collective actions. *Theoria, 52*, 87–97.

Harel, D. (1979). First-order dynamic logic., Lecture Notes in Computer Science Berlin: Springer.

Harel, D. (1984). Dynamic logic. In D. Gabbay & F. Guenthner (Eds.), Handbook of philosophical logic. Extensions of classical logic (Vol. II, pp. 497–604). Dordrecht: D. Reidel.

Harel, D., Kozen, D., & Parikh, R. (1980). Process logic: Expresiveness, decidability, completeness. In Proceedings of the 21st Symposium on Foundations of Computer Science (pp. 129–142). Syracuse.

Harel, D., Pnueli, A., & Stavi, J. (1981). Propositional dynamic logic of context-free programs. Proceedings of the 22th Symposium on Foundations of Computer Science (pp. 310–321). Nashville: Tennessee.

Herzig, A., & Troquard, N. (2006). Knowing how to play: Uniform choices in logics of agency. In Proceedings of the Fifth International Joint Conference on Autonomous Agents and Multiagent Systems (pp. 209–216). ACM.

Hopcroft, J. E., & Ullman, J. D. (1979). Introduction to automata theory, languages, and computation. Reading: Addison-Wesley.

Hendricks, V. F. (2003). Active agents. In J. Van Benthem, & R. Van Rooy (Eds.), Journal of Logic, Language and Information, 12(4), 469–495.

Hintikka, J. (1962). *Knowledge and belief*. Ithaca: Cornell University Press.

Hornsby, J. (1980). *Actions*. London: Routledge.

Horthy, J. F. (1993). Deontic logic as founded on nonmonotonic logic. *Annals of Mathematics and Artificial Intelligence, 9*(1–2), 69–91.

Horthy, J. F. (2001). *Agency and deontic logic*. Oxford: Oxford University Press.

Hyman, J., & Steward, H. (Eds.). (2004). *Agency and action*. Cambridge: Cambridge University Press.

Kanger, S. (1957). New foundations for ethical theory, reprinted in R. Hilpinen (Ed.), Deontic logic: Introductory and systematic readings, 1970 (pp. 36–58).

Kleene, S. C. (1952). *Introduction to metamathematics*. New York: Van Nostrand.

Konieczny, J. (1970). Elementy prakseologii formalnej (Elements of formal praxiology) (in Polish). *Prakseologia, 35*, 65–131.

Konieczny, J., & Stonert, H. (1971). Prakseologia formalna (Formal praxiology) (in Polish). *Prakseologia, 37*, 39–52.

Kooi, B. P. (2011). Dynamic epistemic logic. In J. van Benthem & A. ter Meulen (Eds.), *Handbook of logic and language* (2nd ed.). Amsterdam: Elsevier.

Kotarbińsk, T. (1965). *Praxeology. An introduction to the science of efficient action*. New York: Pergamon Press.

Kotarbińsk, T. (1969). *Traktat o dobrej robocie (A treatise on good work) (in Polish)*. Wrocław: Ossolineum.

Kozen, D., & Parikh, R. (1981). An elementary proof of the completeness of PDL. *Theoretical Computer Science, 14*, 113–118.

Kraus, S., Lehmann, D., & Magidor, M. (1990). Nonmonotonic reasoning, preferential models and cumulative logics. Artificial Intelligence 44, 167–207. (The full bibliography of the subject till 1990 is available there.).

Lorenzen, P., & Lorenz, K. (1978). *Dialogische Logik*. Darmstadt: Wissenschaftliche Buchgesellschaft.
Makinson, D. (1989). General theory of cumulative inferences. In M. Reinfrank (Ed.), Nonmonotonic reasoning. Lecture Notes on Artificial Intelligence (Vol. 346). Proceedings of the Second International Workshop on Non-Monotonic Reasoning. Berlin: Springer.
Makinson, D. (2005). Bridges from classical to nonmonotonic logic. Text in Computing (Vol. 5). London: King's College.
Małuszyński, J., & Szałas, A. (2011a). Living with inconsistency and taming nonmonotonicity. In O. de Moor, G. Gottlob, T. Furche, & A. Sellers (Eds.), Datalog reloaded. LNCS (Vol. 6702, pp. 384–398). Berlin: Springer.
Małuszyński, J., & Szałas, A. (2011b). Logical foundations and complexity of 4QL, a query language with unrestricted negation. *Journal of Applied Non-Classical Logics*, *21*(2), 211–232.
Mazurkiewicz, A. (1972a). Iteratively computable relations. *Bulletin de l'Académie Polonaise des Sciences, Series, Sciences, Mathematics, Astronomy, Physics*, *20*, 793–798.
Mazurkiewicz, A. (1972b). Recursive algorithms and formal languages. Bulletin de l'Académie Polonaise des Sciences. *Series, Sciences, Mathematics, Astronomy, Physics*, *20*, 799–803.
Mazurkiewicz, A. (1974). Podstawy teorii programowania (Foundations of the theory of programming) (in Polish). In A. Mazurkiewicz (Ed.), *Problemy przetwarzania informacji (Problems in information processing)* (Vol. 2, pp. 39–93). Warsaw: WNT.
McCarthy, J. (1963). Situations, actions and causal laws. Technical report. Stanford University.
McCarthy, J. (1980). Circumscription: A form of non-monotonic reasoning. *Artificial Intelligence*, *13*(1–2), 23–79.
McCarthy, J. (1986). Applications of circumscription to common sense reasoning. *Artificial Intelligence*, *28*(1), 89–116.
Mele, A. R. (1997). *The philosophy of action*. Oxford: Oxford University Press.
Minsky, M. A Framework for representing knowledge. Technical report 306, Artificial Intelligence Laboratory, MIT, 1974. (Shorter versions. In: P. Winston (Ed.), The psychology of computer vision, McGraw-Hill 1975, J. Haugeland (Ed.), Mind design. MIT Press, 1981, and A. Collins and E. E. Smith (Eds.), Cognitive science. Morgan-Kaufmann 1992.).
Montague, R. (1970). Pragmatics and intensional logic. *Synthese*, *22*, 6–34.
Montague, R. (1974). In R. H. Thomason (Ed.), *Formal philosophy*. New Haven: Yale University Press.
Moore, R. C. (1985). Semantic considerations on nonmonotonic logic. *Artificial Intelligence*, *25*, 75–94.
Moschovakis, Y. N. (1994). *Notes on set theory*. New York: Springer.
Nowakowska, M. (1973a). *Language of motivation and language of actions*. The Hague: Mouton.
Nowakowska, M. (1973b). Formal theory of actions. *Behavioral Science*, *18*, 393–413.
Nowakowska, M. (1979). *Teoria działania (Theory of action) (in Polish)*. Warsaw: PWN.
O'Connor, T., & Sandis, C. (Eds.). (2010). *A companion to the philosophy of action*. Oxford: Wiley.
Parikh, R. (1978). The completeness of propositional dynamic logic. Mathematical foundations of computer science 1978. Lecture Notes in Computer Science (Vol. 64, pp. 403–415). Berlin: Springer.
Parikh, R. (1984). Logics of knowledge, games, and dynamic logic. In M. Joseph & R. Shyamasundar (Eds.), Foundations of software technology and theoretical computer science. Lecture Notes in Computer Science (Vol. 181, pp. 202–222). Berlin: Springer.
Pawlak, Z. (1969). *Matematyczne aspekty procesu produkcyjnego (Mathematical aspects of manufacturing processes) (in Polish)*. Warsaw: PWE.
Pnueli, A. (1977). The temporal logic of programs. In Proceedings of the 8th Symposium on Foundations of Computer Science (pp. 46–57). Providence.
Pratt, V. R. (1976). Semantical considerations on Floyd-Hoare logic. In Proceedings of the 17th Annual IEEE Symposium on Foundations of Computer Science (pp. 109–121).
Pratt, V. R. (1980). Applications of modal logics to programming. *Studia Logica*, *39*, 257–274.
Prior, A. N. (1967). *Past, present and future*. Oxford: Oxford University Press.

Reiter, R. (1980). A Logic for default reasoning. *Artificial Intelligence, 13*, 81–132.

Reiter, R. (1991). The frame problem in the situation calculus: A simple solution (sometimes) and a completeness result for goal regression. In V. Lifschitz (Ed.), Artificial intelligence and the mathematical theory of computation. Essays in honor of John McCarthy (pp. 359–380). New York: Academic Press.

Salomaa, A. (1969). *Theory of automata*. London: Pergamon Press.

Salwicki, A. (1970). Formalized algorithmic languages. Bulletin de l'Académie Polonaise des Sciences. *Series, Sciences, Mathematics, Astronomy, Physics, 18*, 227–232.

Samet, D. (2008). *The Sure-thing principle and independence of irrelevant knowledge*. The faculty of management: Tel Aviv University, available via the Internet.

Sandis, C. (2009). *New essays on the explanation of action*. Basingstoke: Palgrave Macmillan.

Scott, D. (1970). Advice on modal logic. In K. Lambert (Ed.), *Philosophical problems in logic* (pp. 143–173). Dordrecht: D. Reidel.

Segerberg, K. (1980). Applying modal logic. *Studia Logica, 39*, 275–295.

Segerberg, K. (1982a). The logic of deliberate action. *Journal of Philosophical Logic, 11*, 233–254.

Segerberg, K. (1982b). A completeness theorem in the modal logic of programs. In T. Traczyk (Ed.), *Universal algebra and applications* (Vol. 9, pp. 31–46). Warsaw: Banach Centre Publications, PWN.

Segerberg, K. (1982c). A deontic logic of action. *Studia Logica, 41*, 269–282.

Segerberg, K. (1985a). Models for action. In B. K. Matilal & J. L. Shaw (Eds.), *Analytical philosophy in comparative perspective* (pp. 161–171). Dordrecht: D. Reidel.

Segerberg, K. (1985b). *Routines. Synthese, 65*, 185–210.

Segerberg, K. (1988a). Actions in dynamic logic (abstract). *The Journal of Symbolic Logic, 53*, 1285–1286.

Segerberg, K. (1988b). Talking about actions. *Studia Logica, 47*, 347–352.

Segerberg, K. (1989). Bringing It about. *Journal of Philosophical Logic, 18*, 327–347.

Shoham, Y. A. (1987). Semantical approach to non-monotonic logics. In Proceedings Logics in Computer Sciences (pp. 275–279). Ithaca.

Suppes, P. (1984). *Probabilistic metaphysics*. Oxford: Blackwell.

Suszko, R. (1975). Abolition of the Fregean axiom. In P. Rohit (Ed.), Logic colloquium. Symposium on logic held at Boston, 1972–1973. Lecture Notes in Mathematics (Vol. 453, pp. 169–236). Berlin: Springer.

Talja, J. (1985). On the logic of omissions. *Synthese, 65*, 235–248.

Tarski, A. (1955). A lattice-theoretical fixpoint theorem and its applications. *Pacific Journal of Mathematics, 5*, 285–309.

Thielscher, M. (1997). Ramification and causality. *Artificial Intelligence, 89*(1–2), 317–364.

Thielscher, M. (2010). *A unifying action calculus. Artificial Intelligence, 175*(1), 120–141.

Thomason, R. H. (1970). Indeterminist time and truth-value gaps. *Theoria, 36*, 265–281.

Thomason, R. H. (1984). Combinations of tense and modality. In D. Gabbay & F. Guenther (Eds.), *The handbook of philosophical logic* (Vol. 2, pp. 135–165). Dordrecht: D. Reidel.

Trypuz, R. (Ed.). (2013). *Krister segeberg on logic of actions. Outstanding Contributions to Logic (Vol. 1)*. Berlin: Springer.

Trypuz, R. & Kulicki, P. (2014). A deontic logic of actions and states, available in the Internet.

van Benthem, J., & Liu, F. (2004). Diversity of agents in games. *Philosophia Scientiae, 8*(2), 163–178.

van Benthem, J., & Pacuit, E. (2006). The tree of knowledge in action: Towards a common perspective. In G. Governatori, I. Hodkinson, & Y. Venema (Eds.), *Advances in modal logic* (Vol. 6, pp. 87–106). London: College Publications.

van Benthem, J., Gerbrandy, J., Hoshi, T., & Pacuit, E. (2009). Merging framework for interaction. *Journal of Philosophical Logic, 38*, 491–526.

van Benthem, J. (2011). *Logic of dynamics of information and interaction*. Cambridge: Cambridge University Press.

van Benthem, J. (2011a). *Logic in games*. Texts in logic and games. FoLLI Series. Amsterdam: Springer.

van Benthem, J. (2013). Logic in games. MIT Press (to appear).

van Benthem, J., & Pacuit, E. (2013). Connecting logics of choice and change (to appear).

van Benthem, J., van Ditmarsch, H., van Eijck. J., & Jaspars, J. (2013). Logic in action. New edition. www.logicinaction.org/docs/lia.pdf.

Van der Hoek, W., & Wooldridge, M. (2003). Cooperation, knowledge and time: Alternating-time temporal epistemic logic and its applications. *Studia Logica*, *75*(1), 125–157.

van der Torre, L. (1997). Reasoning about obligations: Defeasibility in preference-based deontic logic. Thesis, available in the Internet.

van der Torre, L. (1999). Defeasible goals. In A. Hunter & S. Parsons (Eds.), *ESCQARU* (Vol. 1638). Berlin: Springer.

van der Torre, L., & Tan, Y. (1999). An update semantics for deontic reasonings. In P. McNamara & H. Prakken (Eds.), Norms, logics and information systems. New studies on deontic logic and computer science. Frontiers in Artificial Intelligence and Applications (Vol. 49). Amsterdam: IOS Press.

van Ditmarsch, H., van der Hook, W., & Kooi, B. (2013). Dynamic epistemic logic, available in the Internet.

van Leeuwen, J. (Ed.). (1990). *Handbook of theoretical computer science (Vol. A-B)*. Amsterdam/Cambridge: The MIT Press/Elsevier.

Varzinczak, L. (2006). What is a good domain descriptions? Evaluating and revising theories in dynamic logic. Ph.D. thesis. Université Paul Sabatier, Touluse.

Varzinczak, L. (2010). On action theory change. *Journal of Artificial Intelligence Research*, *37*, 189–246.

von Kutschera, F. (1980). Grunbegriffe der Handlungslogik. Handlungstheorien - interdisziplinär (Vol. 1, pp. 67–106). München, Band.

von Wright, G. H. (1963). *Norm and action*. London: Routledge and Kegan Paul.

von Wright, G. H. (1968). *An essay in deontic logic and the general theory of action*. Amsterdam: North Holland.

von Wright, G. H. (1971). *Explanation and understanding*. Ithaca: Cornell University Press.

von Wright, G. H. (1981). On the logic of norms and actions. In R. Hilpinen (Ed.), *New studies in deontic logic* (pp. 3–25). Dordrecht: D. Reidel.

Weinberger, O. (1985). Freedom, range for action and the ontology of norms. *Synthese*, *65*, 307–324.

Wolfram, S. (2002). *A new kind of science*. Champaign: Wolfram Media.

Wolper, P. (1983). Temporal logic can be more expressive. *Information and Control*, *56*, 72–99.

Wójcicki, R. (1984). Roman Suszko's situational semantics. *Studia Logica*, *43*, 326–327.

Wójcicki, R. (1984). Heuristic rules of inference in non-monotonic arguments. *Bulletin of the Section of Logic*, *17*(2), 56–61.

Wansing, H. (2006). Tableaux for multi-agent deliberative stit logic. In G. Governatori, I. Hodkinson, & Y. Venema (Eds.), *Advances in modal logic* (Vol. 6, pp. 503–520). London: College Publications.

Xu, M. (2012). Actions as events. *Journal of Philosophical Logic*, *41*, 765–809.

Zhang, D., Chopra, S., & Foo, N. (2002). Consistency of action descriptions. In M. Ishizuka, & A. Sattar (Eds.), Proceedings of the 7th Pacific Rim International Conference on Artificial Intelligence (PRICAI): Trends in Artificial Intelligence. LNCS (Vol. 2417, pp. 70–79). Berlin: Springer.

Zhang, D., & Foo, N. (2001). EDPL: A logic for causal reasoning. In B. Nebel (Ed.), *Proceedings of the 127th Joint Conference on Artificial Intelligence (IJCAI)* (pp. 131–138). San Francisco: Morgan Kaufmann Publishers.

Zhang, D., & Foo, N. (2002). Interpolation properties of action logic: Lazy-formalization to the frame problem. In S. Flesca, S. Greco, N. Leone, & G. G. Ianni (Eds.), Proceedings of the 8th European Conference on Logics in Artificial Intelligence (JELIA). LNCS (Vol. 2424, pp. 357–368). Berlin: Springer.

Symbol Index

J. Czelakowski, *Freedom and Enforcement in Action*, Trends in Logic 42,
DOI 10.1007/978-94-017-9855-6

Index of Definitions and Terms

J. Czelakowski, *Freedom and Enforcement in Action*, Trends in Logic 42,
DOI 10.1007/978-94-017-9855-6

Index of Names

J. Czelakowski, *Freedom and Enforcement in Action*, Trends in Logic 42,
DOI 10.1007/978-94-017-9855-6

MIX
Papier aus verantwortungsvollen Quellen
Paper from responsible sources
FSC® C105338

If you have any concerns about our products,
you can contact us on
ProductSafety@springernature.com

In case Publisher is established outside the EU,
the EU authorized representative is:
Springer Nature Customer Service Center GmbH
Europaplatz 3, 69115 Heidelberg, Germany

Printed by Libri Plureos GmbH
in Hamburg, Germany